ENTROPY MINIMAX SOURCEBOOK

VOLUME I: GENERAL DESCRIPTION

ENTROPY MINIMAX SOURCEBOOK

VOLUME I:
GENERAL DESCRIPTION

Ronald Christensen

FIRST EDITION

Entropy Limited
Lincoln, Massachusetts

FIRST EDITION
First Printing, April 1981

ISBN 0-938-87606-6
PRINTED IN THE UNITED STATES OF AMERICA

How can it be that that which is apparently
the last phase of this investigation is so
slow in materializing? It is best that I myself set out on a secret investigation to see
whether I cannot trace this murderer.

—Judge Dee (630-700 A.D.) in "The Case
of the Double Murder at Dawn",
authorship unknown, translated by
Robert Van Gulik

PREFACE

At various stages in life, there arises an irresistible urge to pull together, organize and wrap up things - a sort of entropy reduction urge. Perhaps it operates to free us of old ties, enabling an unincumbered start on a fresh track. This sourcebook resulted from a recent encounter I had with this urge.

The process has been simultaneously satisfying and frustrating. I started with the task of assembling and arranging materials which I had accumulated over the past twenty years. The more I assembled, the more I uncovered. The amount grew and grew. The more I organized and reorganized, the greater the need for further reorganization became apparent. The harder I tried to settle topics, the more loose ends materialized. At virtually every turn, my organizational efforts were challenged with new complexities and unexpected twists. A numerical analyst would shake his head and call the process sadly nonconvergent. That is probably what makes it interesting.

This sourcebook contains background materials on the nature and origins of entropy minimax.

Volume I: A general introduction to entropy minimax.

Volume II: Philosophical origins of entropy minimax.
            A collection of materials written during
            the period 1960-1965.

Volume III: Computerized implementation of entropy
            minimax and further theoretical develop-
            ments stimulated by results of applications
            to various practical problems. A collection
            of materials written during the period
            1967-1979.

Volume IV: Applications of entropy minimax. A collection
            of materials describing applications in the
            fields of biology, medicine, nuclear engineering,
            geophysics, social sciences, business and economics.

This volume (I) provides a general description of entropy minimax. It addresses the basic ideas, rather than technical details contained in the other volumes. The emphasis is placed on description of examples and discussion of underlying rationale. My original aim when I began writing last August was simply to present a brief explanatory summary. As a glance at the table of contents shows, the orientation has evolved in the writing.

There are many groups making valuable contributions related to entropy minimax. With each group can be associated a core of individuals, a way of looking at the problem, a set of ideas and techniques, a vocabulary of terms and meanings, and an enclave of conferences, journals and books through which they principally communicate. Although by no means total, the extent of group isolation is nevertheless remarkable for our 20th Century. I am thinking of the physicists, the statistical mechanics researchers, the mathematical statisticians, the multivariate data analysts, the communications engineers, the information theorists, the pattern recognition experts, the artificial intelligence designers, the philosophers of science and of language, the economists, the psychologists, the physiologists, the research neurologists, and the molecular biologists, just to name a few. This volume is intended not only to provide a general introduction to entropy minimax, but to present it from a sufficiently broad perspective to span intergroup cross-fertilization of ideas.

Terminology is one of the greatest difficulties in such an endeavor. Various disciplines often use different terms for the same concept (e.g., "feature" in pattern recognition and "independent variable" in multivariate analysis), or the same term for different concepts (e.g., "information" in communications theory and in statistical likelihood analysis). So a selection is necessary. I have attempted to adopt as standard a terminology as possible, although in some cases I have made slight revisions to correct misleading implications (e.g., the former "mutual information" has been amended to "mutual information exchange").

Finally, in justice, I should respond to the question sometimes asked: Do I *really* believe that we receive messages from the future? When we make predictions (categorical or probabilistic) about future events, we behave logically as though we receive messages across a span in time, just as when I drop a stone it behaves as though the earth acts upon it with a force of gravity across a distance in space. Physical action at a distance is mediated by particle (e.g., graviton) exchange. Conceptual action at a distance is mediated by entropy exchange.

R. Christensen
Lincoln, Massachusetts
April 1981

TABLE OF CONTENTS

PART III: Independent Variables: Minimum Entropy Partitions

11. Independent Variables and Feature Space

12. Minimum Entropy Partitions

13. Feature Selection

14. Potential Function Estimation

15. Uncertainty, Irrelevancy, Ambiguity and Non-Bayesian Aspects

PART IV: Entropy Minimax Predictive Models

16. Predictive Reliability and Definitiveness

17. Non-Information-Theoretic Clustering Methods

18. Non-Information-Theoretic Multivariate Analysis Methods

19. Chance Correlations and Contrivedness

20. Information-Theoretic Crossvalidation and Weight Normalization

PART V: Foundations of Entropy Minimax

21. Counterfactuals and the Meaning of Prediction

22. The Problem of Induction and Its Justification

23. Entropy Minimax Induction

24. The Evolution of Language

25. Entropy Minimax and Value Judgment

APPENDICES

1. Examples of Continuous Univariate Distributions

2. Examples of Discrete Univariate Distributions

3. Glossary and Symbols

LIST OF QUOTATIONS

PART I

OVERVIEW OF ENTROPY MINIMAX

CHAPTER 1

INTRODUCTION

CHAPTER 1

INTRODUCTION

A. The Prediction Game

Predicting the future is a favorite pastime of fortunetellers, mete-
orologists, economists, politicians, scientists, engineers, mothers,
fathers, doctors, lawyers, statisticians, gamblers, tipsters, underwriters, market analysts...just about everybody. We know that
it can rarely, if ever, be done with certainty. However, we believe that it should be possible to make justifiable probabilistic
forecasts.

We can explore the far corners of space and chart the course of history. But the future will always be the last frontier. We can communicate with it, but can it not communicate with us? If we are
insulated from any of its effects, as Einsteinian separability suggests, why are we so concerned about the future? Perhaps it is only
that we identify our*selves* with time distributed ensembles of mind-
bodies somehow discounted according to remoteness, even though this
means we do not know very well who we are. Perhaps the future communicates with our soul rather than our body.

The future is important to us because it has to do with life, death,
love, hate, health, disease, pain, pleasure, wealth, poverty,
success, failure, and the myriad other dimensions of our existence.

If we could predict the future a bit more reliably, we could improve
our expectations along these dimensions. In business we could base
our judgments on a more accurate picture of future trends. In every-
day life we could adapt our decisions to a more accurate set of
expectations. In gambling we could improve our betting odds.

In emergency preparedness we could base our planning on more reliable forecasts of probabilities of drought, flood, hurricane, and other phenomena. In agriculture, trucking, shipping, construction, maintenance, and manufacturing we could more rationally establish hiring policies, inventory levels, and production schedules.

If we could predict contingent futures more reliably, we could also choose more intelligently among alternatives in situations in which our choices affect the outcomes. We could improve our procedures in such diverse areas as medicine, teaching, marketing and materials science. We could make better career choices, employer-employee decisions, property assessments, and social/economic/military/political/legal judgments.

Entropy minimax is a method for predicting the future. It uses modern concepts drawn from information theory to make predictions based on observational data. In the extreme of plentiful data it is equivalent to conventional approaches. However, in most real-life situations data is limited, especially considering the number and complexity of variables influencing the outcomes. In such cases, entropy minimax provides an edge over the other techniques.

B.   Key Concepts of Entropy Minimax

The basic idea of entropy minimax is quite simple. It is that we should base our predictions about the future on all of, but no more than, the information we actually have.

Self-evident as this principle appears, it is not easy to follow. Sometimes nonexistent information is assumed, resulting in biased predictions. This can happen by such an innocent-appearing process as the use of words to describe observations. Additionally, important information may be unknowingly overlooked, yielding predictions based on an oversimplification of the actual circumstances. Sources of information include not only explicit data in observational samples, but also implicit data in the language used to describe past

experiences. In complex situations, there is frequently so much information that it is difficult to know what is important and what
can be safely ignored. Entropy minimax provides a means of efficient selection from and utilization of the available information.
The concept of *information* used by entropy minimax is taken from modern communications theory. Observation of an event is said to communicate more or less information to us depending upon the extent
to which we are surpised by the outcome of the event. The idea is
that if we are more surprised by the outcome, we learn more that we
did not know beforehand than if we are less surprised.

Surprise is measured by our pre-observation estimate of the probability of the particular outcome which did occur. The lower this
probability, the greater our surprise and, hence, the greater the
information gained.

Imagine that, prior to observing the event, we make an estimate of
the amount of information we expect to gain when we do observe the
event. We make this estimate the same way as a gambler estimates his
expected return when placing bets in roulette. He multiplies the
expected monetary return for each combination on which he bets by
the probability of that combination and sums the products. The total
is the expected future monetary value to him of the turn of the
wheel. Similarly, we multiply the probability of each outcome by
the information we will gain if it occurs, and sum the products.
The total is the expected future information value to us of the
event. This is called the *entropy* of the event.

Of course, in order to calculate the entropy of an event we need
to know its outcome probabilities. These are precisely the unknowns
we seek. Entropy minimax assists us in computing these probabilities. It does this in three ways.

First, it helps us determine which events in our past experience
are similar to the event for which we wish to make the prediction.
This is the role of entropy minimization. Entropy is a measure of

the amount of disorder in a classification system. Entropy minimization is an ordering principle by means of which we determine
which past events are most like our future events in ways that are
relevant to predicting the outcome.

Second, it helps us assign numerical values to the probabilities.
This is the role of entropy maximization. In addition to measuring
disorder, entropy also measures how much information we regard the
data as containing. By assigning the probabilities values which
maximize entropy, we assume the least about what we know, i.e., we
minimize the introduction of artificial information or bias.

Finally, it helps us determine the reality of our predictions, i.e.,
how much confidence we can place in them. This is accomplished by
comparing the entropy on our actual data to the expected value of
the entropy on random data.

With these three principles, entropy minimax has been used success-
fully to find predictive patterns in a number of difficult real-world
situations.   These include such diverse problems as long-range weather
forecasting, medical survival prognosis, and nuclear reactor fuel
failure. This volume (I) is written to provide a comprehensive
general introduction to entropy minimax. Volumes II, III and IV
provide documentation of its philosophical origins,* computer implementation, and applications, respectively.

C. Organization of This Volume

This volume is organized into five parts, each containing five chapters. Appendices 1 and 2 give examples of commonly used distribution functions. Frequently used terms are defined in the glossary.
The symbols list at the end of the glossary defines frequently used
mathematical symbols.

---

* A more complete treatment of a portion of the philosophical background is
contained in the *Foundations of Inductive Reasoning*.

A general introduction to entropy minimax is provided in Part I, focusing on historical development, epistemological perspective, computer implementation, and applications.

The principle of entropy maximization is covered in Part II, beginning with a discussion of samples and histograms. The concepts of probability, information and entropy are then defined and described. This is followed by a discussion of the principle of entropy maximization. The concluding chapters in this part provide a linkage to conventional statistical concepts of frequency distributions, goodness-of-fit tests, and parametric and nonparametric probability estimators.

The independent variables are taken up in Part III. The first two chapters include a discussion of feature space as it is constructed from the independent variables. This is followed by a discussion of the principle of entropy minimization. Also included in this part are chapters on potential function estimation, and on uncertainty, irrelevancy and ambiguity.

Topics related to the confidence one can have in predictions are covered in Part IV. Included is a discussion of means of assessing predictive reliability and definitiveness, a brief review of non-information-theoretic methods of generating predictive models, a discussion of chance correlations and contrivedness, and an explanation of information-theoretic crossvalidation and weight normalization.

Philosophical problems associated with entropy minimax are addressed in Part V. These include counterfactuals and the meaning of prediction, the problem of induction and its justification, entropy minimax induction, the evolution of language, and the problem of value judgment.

At the end of each chapter appear selected quotes from literature relevant to the topics covered, and citations to related references.

QUOTE

FISHER (1959)

I have stressed my conviction that the art of framing cogent
experiments, and that of their statistical analysis, can each
only establish its full significance as parts of a single process
of the improvement of natural knowledge...

> —Ronald A. Fisher (1890-1962)
> *Statistical Methods and Scientific Inference*, 2nd
> ed., Hafner Publ. Co., NY, 1959, pp. 6-7.

REFERENCES

Christensen, R.A., *Foundations of Inductive Reasoning*, Berkeley, CA, 1964.

_____, *Entropy Minimax Sourcebook, Vol. II: Philosophical Origins*, Entropy Limited, Lincoln, MA, 1980.

_____, *Entropy Minimax Sourcebook, Volume III: Computer Implementation*, Entropy Limited, Lincoln, MA, 1980.

_____, *Entropy Minimax Sourcebook, Volume IV: Applications*, Entropy Limited, Lincoln, MA, (in press).

Fisher, R.A., *Statistical Methods and Scientific Inference*, 2nd ed., Hafner Publ. Co., NY, 1959.

CHAPTER 2

BACKGROUND OF ENTROPY MINIMAX

CHAPTER 2

BACKGROUND OF ENTROPY MINIMAX

A. How It All Started

I returned to Harvard in the fall of 1960 after spending the summer at the Rand Corporation in Santa Monica, California. There,
in an atmosphere stimulated by such diverse persons as Kenneth Arrow,
Margaret Mead and Herman Kahn, I had attempted (rather fruitlessly)
to formulate a utility-type theory of "choice" to describe physical
processes of uncertain outcome, based on the idea that it may be as
enlightening to model matter in terms of will as it is to model mind
in terms of mechanics. Shortly after returning, I happened one lunch
hour to share a table at Harkness Commons with a fellow student who
seemed familiar. This encounter was to haunt me for years to come.

After a short "Haven't we met before?" introduction, we found that
we had both been at Rand that summer. I told him about my excitement
over mathematical decision theory and asked what he had worked on.

"My job was to teach a computer to tell the difference between a zero
and a one," he replied.

"Oh?" I responded politely, but somewhat disappointedly.

"Yes, but I didn't succeed."

My image of computers at the time was that of glorified adding machines. Noticing my expression, he continued, "It's not quite what
you think. You see the problem is that we have a square array of
light bulbs, ten on each side, 100 bulbs in all. Now each bulb can
be either on or off.

"What we do is display a sample, say 1,000, of random on-off patterns to a human judge. In each case the human decides whether the on-bulbs form a pattern which looks more like a '0' or more like a '1'. We give these judgments to the computer.

"Then we display a brand new on-off pattern and ask the computer whether it looks like a '0' or a '1'."

"So why didn't you just program all of the possibilities?" I asked.

"Too many," he replied, "two to the one-hundredth power. Even if we could handle that number, the objective is to develop a general method for any number of possibilities."

"Why not write some kind of explicit rule for distinguishing things like curved and straight lines?" I tried again.

"No good. First, such algorithms are virtually impossible to write for the general case considering all the odd patterns possible. Second, the idea is to use the human judgment sample. We want a procedure which will also work for 2's, 3's, A's, B's, and so forth."

The conversation concluded in this frustrating vein. Despite my initial conviction that the problem looked simple, it eluded every approach I suggested.

That evening, back in my room, I tackled it with pencil and paper. The next noon hour I posed my new solution. He quickly demolished it. I tried variations. One after another he demonstrated their inadequacies. This went on for several days. During this time I came to see that the innocent appearing "zero-one" problem mirrored many of the fundamental questions of the scientific method:

What is reality? What is causality?

How do humans perform inductive reasoning?

What are probabilities? How does one use observational samples to assign them numerical values?

What is the meaning of "similar" when we group similar events
for probability estimation purposes?

How can we justify any assignment of probability to a new
event?

The heart of the matter was Berkeley's denial of the possibility of
proving reality, and Hume's argument that attempting to justify any
generalization about matters of fact involves an insoluble dilemma.
After that week, our paths crossed on only a handful of occasions,
but the vexation of the unsolved problem lingered on.

During the next two years, I gathered materials on probability and
utility determination and their relation to decision-making. My
interests during this period leaned toward subjective probabilities
and the associated utilities. Ramsey, de Finetti, Savage, von
Neumann and Morgenstern were among my favorite authors. Objective
probabilities seemed well taken care of as a practical matter by the
frequency limit hypothesis, despite the theoretical enigma of Hume's
dilemma.

In the fall of 1962, I studied statistical mechanics at Berkeley and
learned of the relationship between entropy maximization and state
probabilities as worked out by Boltzmann, Gibbs, Darwin, Fowler and
others. The most probable microscopic state of an ensemble is a
state of uniformity described by maximizing its entropy subject to
constraints specifying its observed macroscopic condition.

This, together with the work of Shannon in 1948 extending the idea
of entropy to the communication of information as defined by Nyquist
in 1924 and Hartley in 1927, leads to the conclusion that entropy
maximization can be used as the basis for equiprobability assumptions generally. Information-theoretic entropy maximization subject to known constraints was explored by Jaynes in 1957 as a basis

for statistical mechanics, which in turn makes it a basis for thermostatics and thermodynamics.

The matter seemed settled.   There was a well-developed procedure for
assigning outcome probabilities to events.   This procedure allows one
to take auxiliary information into consideration in the form of "constraints" specifying expectation values for measurable consequences
of the outcome mix for sets of events.

But when, in the winter of 1962-63, I tried to use this as the basis
of a solution to the zero-one problem, I ran up against two unresolved
questions.   First, how do we take into consideration the information
in the empirical sample?   Second, how do we resolve the arbitrariness of our definitions of the events and their outcomes?

The empirical sample question arises as follows: Assume that there
are t ways in which an outcome can be true and f ways in which it
can be false, and that we have no auxiliary constraint information.
Then entropy maximization yields

$$P = \frac{t}{t+f} .$$

This is the equiprobability statement, used since Fermat and Pascal
in 1654, Huygens in 1657 and De Moivre in 1738.

Alternatively, assume that our auxiliary information comes in the
form of a sample of x observations of the outcome occurring among n
observations of similar events.   Then conventional practice uses
such information to justify the estimate

$$P = \frac{x}{n} .$$

This is the position implied, in various stages of development, by
maximum likelihood advocates since Gauss in 1809, and by frequency
limit hypothesis proponents since Poisson in 1837, Ellis in 1842 and
Cournot in 1843.   How do we reconcile the two alternative formulae?

The second unresolved question, concerning outcome classification arbitrariness, arises as follows: The principle of insufficient reason, recognized by Bernoulli in 1713, makes the probabilities equal regardless of how the states are defined. Entropy maximization does not resolve the difficulty. If one makes a nonlinear transformation of phase space, entropy maximization in a uniform discretization of the new space leads to a result incompatible with that of entropy maximization in a uniform discretization of the original space.

The event definition arbitrariness is equally perplexing. The frequency limit hypothesis, accepted with various interpretations by scientists such as Gibbs, statisticians such as Fisher, and philosophers such as Reichenbach, says that the ratio x/n is a good estimate of P at large n. But the values of x and n that we use in making a prediction for a particular future event depend upon how we classify the conditions of that event. What we count as "similar" events in our historical experience can significantly affect the ratio x/n. Yet to try to base judgments of what *is* similar on experience alone is to become hopelessly enmeshed in circular reasoning.

The reconciliation of entropy maximization and data samples came first. I had been fascinated by an oddity of probability theory called the Rule of Succession. Derived by Laplace in 1774, it says that if we make n observations of an event and a particular outcome occurs in x instances, then the expected value of the outcome probability is

$$P = \frac{x+1}{n+2}.$$

This seemed to connect the equiprobabilities result for a two-way outcome classification $P = 1/2$ to the frequency posit result $P = x/n$. So I asked myself: How is it connected with entropy maximization? Can it be generalized to an arbitrary classification?

Looking into the derivation of the Rule of Succession, I learned that it is based upon a postulate made by Bayes in 1763 that equal intervals of probability are themselves equally probable. This is its connection to entropy maximization, namely, maximizing the entropy of a distribution over probability itself.

A bit of formal manipulation with probabilistic logic then yielded the desired generalization:

$$P = \frac{x+t}{n+t+f} \, .$$

This result is satisfying for a number of reasons. It combines the purely theoretical $P = t/(t+f)$ and the purely empirical $P = x/n$ in a very natural way. It has a solid basis in entropy maximization. It approaches the frequency limit as the amount of data becomes arbitrarily large. Its functional form goes back at least as far as 1889 when Hardy suggested a beta distribution for initial class densities.

On the other hand, it still suffers from arbitrariness. Without a principle to guide outcome classification, the values of t and f are arbitrary. Without a principle to guide condition classification, the values of x and n are also arbitrary.

From quantum mechanics I had learned the importance of discretizing mathematically continuous variables into intervals of observational uncertainty. An observationally justifiable model can be no more fine-structured than the observational reality being modeled.

Applying this idea to Bayes' postulate, I asked myself whether our intervals of unresolved uncertainty for probability are of equal length. If our data contain n observations, we have a grid of 1/n length intervals in which the frequency can be resolved. This grid is uniform over the entire range from 0 to 1. Thus, treating frequency as reality, and probability as merely a model, for any finite sample size we must represent the situation with equal interval discretization.

This left the condition classification problem. I decided to study a specific example. A ten-by-ten array of 100 lights was too large to cope with manually. So I set up a miniature version of the zero-one problem with only six lights. I constructed an arbitrary rule for sorting events into the categories zero=false and one=true. Then I flipped a coin 90 times and obtained a random sample of 15 events. These I sorted according to the rule.

The game was to devise a procedure for guessing (probabilistically) how the rule would sort another randomly selected event. The only information allowed was the 15-event sample, the rule being treated as unknown.

Armed with $P = (x+t)/(n+t+f)$ and Ockham's razor, I proceeded in March and April 1963, to construct elaborate schemes for generating and assessing candidate rules.

Later in April, I finally saw what should have been obvious. I was trying to classify the data into a more ordered state. I was trying to *lower* the entropy.

Thus was entropy minimax born. Entropy maximization disordering, used as early as 1957 by Jaynes in providing statistical mechanics with an information-theoretic interpretation, gives the values of the outcome probabilities. The converse, entropy minimization ordering, used as early as 1958 (I was to learn later) by Maccacaro and by Tanimoto in constructing biological hierarchies, defines the event classes for which these probabilities are computed.

The remaining loose end was the lack of a justification for basing beliefs upon entropy minimax probabilities. In the final analysis, this amounts to justifying the use of one categorization of events over another, and hence of thinking or analyzing in one language rather than another. Language is a conceptual system. The second law of thermodynamics, anchored in statistical mechanics by Boltzmann and Gibbs, explains how closed physical systems evolve. They tend toward greater entropy. Is there a similar law that explains the evolution of the conceptual systems through which we view the physical world? Do they tend toward lesser entropy as our vision clears?

Comparing this to Darwinian analysis, it became evident that this is what is meant by *adapting* to the environment for a conceptual system. Its structure is adjusted by thinking beings to organize the things and events in the world with respect to those conditional probabilities which are used in decision-making. The behavior guided by the conceptual system stimulates feedback driving its evolution.

During the next three months, I wrote a description of the generic induction problem and the principles involved in its solution: 1) entropy maximization, 2) probability models of frequency, 3) entropy minimization, and 4) the evolution of language. By the end of the following year, 1964, this had grown to a ten-chapter treatise on the foundations of inductive reasoning.

These four principles seemed to nail down the fundamentals. After that it was only a matter of working out details and writing computer software to implement automated inductive reasoning.

Wrong! Six years later, in the winter of 1970-71, it became obvious that contrivedness was a serious threat to the foundations of entropy minimax. The computerized implementation was becoming increasingly sophisticated. Eventually it would acquire the ability to find pattern matches for any past data simply by making the pattern boundaries complex enough. My inductive reasoning machine would become a numerologist.

At first I responded by denying that contrivedness posed a fundamental problem. It appeared to be simply a matter of meta-analysis.

By performing an entropy minimax analysis of the shapes of successful patterns found by past entropy minimax analyses, one should be
able to classify types of classifications and thereby overcome contrivedness.

This stratagem failed both on the theoretical ground of entailing
an infinite regress, and on the practical ground of being unworkable. A year later, in 1972, I tried to formulate explicit measures of contrivedness based upon a combination of background information and information in the sample for the pattern. This involved
generalizing the induced probability formula to read

$$P_k = \frac{x_k + w_k}{n + w} \quad ,$$

where the total weight $w = \sum w_k$ incorporates virtual events in the
background information. This estimator has "regression to a multi-
pattern mean" aspects similar to Stein-James-Lindley estimators suggested by Lindley in 1962 based on Stein's 1955 proof that the conventional pattern average estimator does not necessarily have minimum expected error squared when there are several patterns in
the feature space partition. However, both contrivedness measures
that I succeeded in devising with this estimator entailed the use
of unknown parameters. With this failure, the means of contrivedness avoidance used in the computer implementation of entropy minimax continued to be heuristic restrictions on pattern dimensionality
and shape.

In 1973 I was working with Tom Reichert at Carnegie-Mellon, analyzing
a data set he had obtained from Linus Pauling's orthomolecular research group at Stanford. They were searching for indicators of
certain forms of mental disease using urine chromatography data for
a population of mentally retarded (unknown biochemical defect), phenyl-
ketonuric, and normal subjects.

When Reichert had shown our preliminary results to Arthur Robinson
at Stanford, Robinson had raised the question: "What confidence is

there that the pattern discoverer has not just unearthed a chance correlation?"

It was time to quit burying the problem of contrivedness under the rug of heuristic restrictions. The risk of chance correlation is too serious a matter. Contrivedness and predictive reliability are the opposing characteristics of systematized anecdotalism and science. Contrivedness lurks in the background as the potential spoiler of predictive reliability in all attempts at scientific inference. It is not an easy foe to bring under control.

The conventional approach of post-pattern discovery hypothesis testing by computing critical values of chosen test-statistics for assumed distributions proved inadequate. If the patterns pass such a test, it does not tell us whether better patterns exist. If they fail, it does not recommend a remedy. Further--which test-statistic and why? What was needed was a mechanism for making the criteria guiding the numerous decisions underlying hypothesis testing an integral part of the pattern discovery process.

My first formal approach for dealing with contrivedness was to define an entropy of chance correlation as its measure. The idea is to penalize patterns for contrivedness by discounting their classification entropies according to this measure.

Numerical experimentation led to the interesting observation that for many situations the entropy of chance correlation is a minimum for a specific number of variables used to make up the patterns. The number depends upon the data and the intercorrelations. Contrivedness is greater if we use either more or fewer variables. This is what underlies the observation by Hughes in 1968 that the squared error of a pattern recognizer may have a minimum when treated as a function of the number of variables used.

Unfortunately, analytic computation of the entropy of chance correlation proved intractable in practice. This is because it involves

dealing with the entire equation set for the pattern search algorithm. So I resorted to the development of a Monte-Carlo routine.
On analyses conducted during 1973-74, this introduced an explicit
tendency toward patterns of somewhat lower dimensionality. Heuristic restrictions could be relaxed. The routine required, however,
an undesirable amount of computer time.

Finally, in 1975, I turned back to the link between the a priori weights
and contrivedness. This time I used contrivedness to fix the weights
rather than vice versa. Treating the weight normalization as a
parameter, I used crossvalidation to estimate its value. This approach turned out to have a number of major advantages. First, it
serves the purpose. Chance correlation patterns have a low probability of crossvalidating. Second, it is computationally economical. Crossvalidation adds only incrementally to computer run time.
Third, it provides a test of the predictability of the patterns on
data not used in their generation (subject to using up one degree of
freedom to fix the weight normalization).

Crossvalidation has an additional important implication. It sets
the scale for sample size. This scale, which determines the degree
of reality of our general concepts, is thereby made an explicit function both of the data available to us and of the purposes for which
we wish to make predictions. Thus entropy minimax tells us not only
how to categorize events and compute probabilities for our predictive purposes, but also how many events we need to observe to achieve
a desired level of reliability.

B. Where Is It All Going?

It is now seven years since those initial discussions with Reichert
concerning contrivedness in independent variable space. Recently
we began analyzing data on coronary artery disease, searching for
survival prognosis patterns. In the initial analyses, I defined a
set of dependent variables as "survival to at least X percent of average remaining lifetime for persons of the same age". I let X range

over the values 10%, 20%, etc. Reichert pointed out that medical
statistics for such patients are generally given in terms of a fixed
two-year survival, necessitating yet another analysis with this definition of the dependent variable.

Then Frank Harrell, who had originally supplied us with the data
from the Duke University Medical Center cardiology computer files,
made an interesting observation: There can be contrivedness in a dependent as well as an independent variable.

If we sequence through a sufficiently great number of questions that
might conceivably be answered by the data, there is a likelihood of
hitting upon a question that the data seem to answer very well by
chance correlation alone. So when using a particular set of data
to support the significance of an hypothesis with respect to a particular dependent variable, we should be careful to take into consideration the number of hypotheses formulations studied using the
data. The validity of any result must be discounted for the equivalent number of independent hypotheses tested. Failure to discount
would be analogous to asking a witness a number of random questions,
then introducing into evidence a response that happened to be to one's
liking, eliminating the remainder.

Harrell's observation, Reichert points out, leads to the following
paradox: Suppose a researcher tests a few hypotheses on his data.
He is satisfied that he has verified one hypothesis in particular
with an adequate level of statistical confidence. He locks the data
in his desk drawer and then goes home. During the night a burglar
breaks in, pries open the drawer, and tests another hypothesis. Has
the burglar stolen statistical confidence from the researcher?

If confidence has not been stolen, then the validity of the researcher's
results is independent of the extent of contrivedness of the question asked. (We extend this to contrivedness in the complexity of
the individual hypothesis as well as the number tested.) This,

however, ignores the problem of chance correlation, and gives researchers unbridled license to contort questions to fit what happens to be the data. As Tukey warned: "Bending the data to fit the
analysis can be vital...But bending the question to fit the analysis is to be shunned at all costs."

If confidence has been stolen, then the reality of the subject matter, about which the data supplies information, is brought into question. How can something be real when its properties depend upon
the knowledge of the observer?

It is not difficult to accept the idea that observation is essential
to knowledge of physical systems, and even the idea that observation must almost invariably impact, to some extent at least, upon
the observed system. However, should we also accept the proposition
that the nature of the physical system depends upon how we analyze
the data obtained by observation?

It appears to me that the answer is yes. The conundrum of quantum
theory is not restricted to microscopic processes. It lurks behind
the most common of conceptual interpretations of experience, rais-
ing its head whenever we attempt to understand a complex system with
relatively little data. We see it every day manifest in differences
of opinion concerning the implications of past events in engineering, economics, medicine, law, business, finance, government, sports,
and other activities. What must yield is the law of the excluded
middle. It is an incomplete analysis to say that an event either
occurs or does not occur. Reality has degrees. It is the extent
of independence of what we apparently can know about events from
what we do happen to know, measured by the difference between the
chance correlation entropy and the entropy on our data.

All this is merely an extension of a philosophy of quantum theory
to describe fundamental limits to small sample knowledge about complex systems. But the story is not over.

Originally my intention was to keep this volume "pure", in the sense of focusing on probabilities as distinct from utilities. However, Terry Oldberg at the Electric Power Research Institute, in discussions during 1979-80 concerning the implications of patterns in materials failure, urged that the relationship of entropy minimax probabilities to expected disutility minimization also needed elucidation. Else how do we understand their relation to decision-making? This prompted me to outline a discussion of this relationship for inclusion in the last chapter. For the most part, it turns out to be simply a standard decision-theoretic optimization problem. But on one point a new puzzle has emerged. This is a puzzle peculiar to small sample knowledge. It is seen when we ask: How small is small?

Whenever we use crossvalidation to determine the weight normalization, which fixes the scale for sample size, we employ a measure of predictive performance. This measure inherently depends upon the relative utilities of different types of correct predictions. Thus, the sample size scale factor depends upon value judgments. This means that the probabilities we assign to events are, to some extent at least, inherently dependent upon these judgments. As sample size grows, these effects may be washed out, but when we have relatively little data they can be quite pronounced.

Small sample knowledge of complex systems has an aspect which depends intrinsically upon value judgment. Perplexing as this conclusion is, there appears to be no way around it. Suppose two people with differing values come to different conclusions about the outcome probabilities for a complex process based on a small sample of data. Are there no objective means by which we can compare them? It is irrelevant to consider making a large number of test observations to determine which of the two sets of probabilities is closer to the "long-run" frequencies. Our task is one of before-the-fact comparison, not after-the-fact assessment. True, with a large sample we might wash out the difficulty. The problem, however, is to decide what probabilities to believe with only the available data on the particular process in question.

Perhaps we can examine a large ensemble of "other" small sample experiments, and determine which set of values generally resulted in more accurate predictions. This puts us in a double bind. If the small sample experiments involved processes different from the one of ultimate concern, then the outcome of the ensemble is irrelevant to the question we are trying to answer. If, on the other hand, processes similar to the one in question are included in sufficient quantity to justify extension of the results to it, then we have merely ignored the real issue by moving from an observed small sample to an assumed large sample.

We have come full circle. The viewpoint of Bernoulli that expected value is the truly primitive notion, which saw its last exponent in Bayes (whose adoption of this perspective was to be noted two centuries later by Jeffreys as an "elaborate argument"), is indeed correct, physically as well as psychologically. We were led astray by Leibniz, De Moivre, Laplace and their equiprobability followers. We were led even farther astray by Poisson, Ellis, Cournot, Gibbs, Fisher, von Mises, Reichenbach and others of the various frequency limit schools. Venn and Keynes illuminated pointers back to the road, but we did not take heed. We listened momentarily, then went on as before computing probabilities for categories however we felt like defining them. I must concede that this has been the natural way for language to evolve. Only the theoretically greater freedom of the computer presents residual difficulty to understanding knowledge. Or does it also simply see another reality?

QUOTES

FEYNMAN (1965)

We have to find a new view of the world that has to agree with everything that is known, but disagree in its predictions somewhere, otherwise it is not interesting. And in that disagreement it must agree with nature. If you can find any other view of the world which agrees over the entire range where things have already been observed, but disagrees somewhere else, you have made a great discovery. It is very nearly impossible, but not quite, to find any theory which agrees with experiments over the entire range in which all theories have been checked, and yet gives different consequences in some other range, even a theory whose different consequences do not turn out to agree with nature. A new idea is extremely difficult to think of.

—Richard Feynman (1918-   )
The Character of Physical Law, The MIT Press, Cambridge, MA, 1965, pp. 171-172.

REFERENCES

Bayes, Thomas, "An Essay Towards Solving a Problem in the Doctrine of Chances" (posthumous, 1763), *Philosophical Transactions of the Royal Society of London*, *53*, London, 1763, pp. 370-418.

Berkeley, George, *Three Dialogues Between Hylas and Philonous* (1713), *Berkeley, Essay, Principles, Dialogues with Selection from Other Writings*, ed. by M. Calkins, Charles Scribner's Sons, NY, 1929, pp. 99-343.

Bernoulli, Jacob (James), *Ars Conjectandi* (posthumous, Basel, 1713), tr. by H.E. Wedeck, *Classics in Logic*, D.D. Runes, ed., Philosophical Library Inc., NY, 1962.

Boltzmann, Ludwig, *Lectures on Gas Theory* (1896-98), tr. by S. Brush, Univ. of Calif. Press, Berkeley, CA, 1964.

Christensen, R.A., correspondence June 23, 1962 through January 3, 1964. [*Chapter 5 of Volume II.*]

_____, "Induction and the Evolution of Language," Physics Dept., Univ. of Calif., Berkeley, CA, July 19, 1963. [*Chapter 7 of Volume II.*]

_____, *Foundations of Inductive Reasoning*, Berkeley, CA, 1964.

_____, "A General Approach to Pattern Discovery," Tech. Report No. 20, Computer Center, Univ. of Calif., Berkeley, CA, June 29, 1967 (revised Nov. 15, 1967). [*Chapter 1 of Volume III.*]

_____, "Entropy Minimax Method of Pattern Discovery and Probability Determination," Arthur D. Little, Inc., Acorn Park, Cambridge, MA, March 7, 1972. [*Chapter 4 of Volume III.*]

_____, "Entropy Minimax, A Non-Bayesian Approach to Probability Estimation from Empirical Data," *Proc. of the 1973 International Conference on Cybernetics and Society*, IEEE Systems, Man and Cybernetics Society, Nov. 5-7, 1973, Boston, MA, 73 CHO 799-7-SMC, pp. 321-325. [*Chapter 5 of Volume III.*]

_____, "Contrived Patterns, Trying to Avoid Them Without Trying Too Hard," Tech. Report No. 40.12.75, Carnegie-Mellon Univ., Pittsburgh, PA, June 19, 1975. [*Chapter 9 of Volume III.*]

_____, "Crossvalidation: Minimizing the Entropy of the Future," *Information Processing Letters*, *4*, Dec. 1975, pp. 73-76. [*Chapter 10 of Volume III.*]

Cournot, Augustin, *Exposition de la theorie des chances et des probabilités*, Paris, 1843.

Darwin, Charles R., *The Origin of Species by Means of Natural Selection* (1859), The Modern Library, NY, 1958.

Darwin, C.G., "Free Motion in Wave Mechanics," *Proc. Roy. Soc. A*, *117*, 1927, p. 258.

Darwin, C.G. and R.H. Fowler, "Fluctuations in an Assembly in Statistical Equilibrium," *Proc. Camb. Phil. Soc.*, *21*, 1922, p. 391.

De Moivre, A., *The Doctrine of Chances* (London, 1738), reprinted by Chelsea Publ. Co., NY, 1967.

Ellis, Leslie, "On the Foundations of the Theory of Probabilities" (paper presented Feb. 1842), *Trans. Camb. Phil. Soc., 8*, 1844, pp. 1-6.

Feynman, Richard, *The Character of Physical Law*, The MIT Press, Camb., MA, 1965.

de Finetti, B., "Fondamenti Logici del Regionamento Probabilistico," *Boll. Un. Mat. Ital., 9* (Ser. A), 158-161 (1930); "La Prevision: Ses Lois Logiques, Ses Sources Subjectives," *Ann. Inst. Henri Poincaré, 7*, 1937, pp. 1-68.

Fisher, R.A., "Theory of Statistical Estimation," *Proc. Camb. Phil. Soc., 22*, 1925, pp. 700-725.

Fowler, R.H., *Statistical Mechanics* (1929), 2nd ed., Cambridge Univ. Press, Cambridge, 1936.

Gauss, Karl Friedrich, *Theory of the Motion of the Heavenly Bodies Moving About the Sun in Conic Sections* (1809), Dover Pubs., Inc., NY, 1963.

_____, "Theory of the Combination of Observations Which Leads to the Smallest Errors," *Gauss' Work (1803-1826) on the Theory of Least Squares*, tr. by H.F. Trotter, Tech. Rept. No. 5, Dept. of Army Project No. 5B99-01-004, Ordinance R and D Project No. PB 2-0001, OOR Project No. 1715, Contract No. DA 36-034-ORD 2297, Statistical Techniques Research Group, Sec. on Math. Stats., Dept. of Math., Princeton Univ., Princeton, NJ, Aug. 5, 1957.

Gibbs, J. Willard, *Elementary Principles in Statistical Mechanics* (1902), Dover Pubs., Inc., NY, 1960.

Hardy, G.F., in correspondence in Insurance Record, 1889, reprinted in *Trans. Fac. Actuaries, 8*, 1920.

Hartley, R.V.L., "Transmission of Information," *Bell System Technical Journal, 7*, 1928, pp. 535-563.

Hughes, G.F., "On the Mean Accuracy of Statistical Pattern Recognizers," *IEEE Trans. Info. Theory, IT-4*, Jan. 1968, pp. 55-63.

Hume, David, *A Treatise of Human Nature* (1736), Oxford Univ. Press, London, 1888.

_____, *Enquiries Concerning the Human Understanding and Concerning the Principles of Morals* (1777), Oxford Univ. Press, London, 1961.

Huygens, Christian, *Calculating in Games of Chance* (1657), *Oeuvres Complètes, XIV*, The Hague, 1888-1950.

Jaynes, E.T., "Information Theory and Statistical Mechanics," *The Physical Review, 106*, 1957, pp. 620-630 and *108*, 1957, pp. 171-190.

_____, "Information Theory and Statistical Mechanics," *Statistical Physics* (1962 Brandeis Lectures), ed. by K. Ford, W.A. Benjamin, Inc., NY, pp. 181-218.

Jeffreys, Harold, *Theory of Probability*, Oxford Univ. Press, London, 1961.

Keynes, John Maynard, *A Treatise on Probability* (1921), Macmillan & Co., Ltd., London, 1957.

Laplace, Pierre Simon de, *Mem. de l'Acad. R. d. Sci.*, *6*, Paris 1774, p. 621.

Leibniz, Gottfried Wilhelm, *De incerti aestimatione* (Sept. 1678), tr. by K. Biermann and M. Faak, *Forschungen und Fortschritte*, *31*, 1957, pp. 45-50.

Lindley, Dennis V., "Discussion on Professor Stein's Paper," *J. of the Royal Stat. Soc.*, Ser. A, *125*, 1962, pp. 265-296.

Maccacaro, G.A., "La misura della informazione contenuta nei criteri di classificazione," *Ann. Microbiol. Enzimol.*, *8*, May 1958, pp. 231-239.

von Mises, Richard, *Probability, Statistics and Truth* (1928), tr. by J. Neyman, D. Scholl and E. Rabinowitsch, George Allen & Unwin Ltd., London, 1957.

von Neumann, John and Oskar Morgenstern, *Theory of Games and Economic Behavior* (1944), Princeton Univ. Press, Princeton, NJ, 2nd ed., 1947.

Ockham, William of, *Sentences; Summa totius logices; Quodlibeta septem; Centiloquium theologicum*, Oxford, c. 1327.

_____, *Tractatus de Successivis* (c. 1327), *Philosophical Review*, *54*, 1945, pp. 519-520.

_____, *Tractatus de Praedestinatione* (c. 1327), *Philosophical Review*, *55*, 1946, pp. 446-448.

Poisson, S.D., *Recherches sur la Probabilité des Jugements en Matière Criminelle et en Matière Civile*, Paris, Bachelier, Imprimeur-Libraire, 1837.

Ramsey, Frank P., "Truth and Probability" (1926), *The Foundations of Mathematics and Other Logical Essays*, Littlefield Adams & Co., Paterson, NJ, 1960.

_____, "Variable Hypotheticals," *The Foundations of Mathematics*, Kegan Paul, Trench, Trubner & Co., Ltd., London, 1931.

Reichenbach, Hans, *The Theory of Probability*, Univ. of Calif. Press, Berkeley, CA, 1949.

Savage, Leonard J., *The Foundations of Statistics*, J. Wiley & Sons, NY, 1954.

Shannon, Claude E. and Warren Weaver, *The Mathematical Theory of Communication*, The Univ. of Illinois Press, Urbana, IL, 1949.

Stein, Charles M., "Inadmissibility of the Usual Estimator for the Mean of a Multivariate Normal Distribution," *Proc. of the Third Berkeley Symposium on Mathematical Statistics and Probability*, Dec. 1954 and July-Aug. 1955, Vol. 1, Univ. of Calif. Press, Berkeley, CA, 1956, pp. 197-206.

Tanimoto, T.T., "An Elementary Mathematical Theory of Classification and Prediction" (1958), *The IBM Taxonomy Application*, M & A-6, ed. by T.T. Tanimoto and R.G. Loomis, Mathematics and Applications Dept., IBM, New York, 1960, pp. 30-39.

Todhunter, Isaac, *A History of the Mathematical Theory of Probability from the Time of Pascal to that of Laplace* (1865), Chelsea Publ. Co., NY, 1965.

Tribus, Myron, "Information Theory as the Basis for Thermostatics and Thermodynamics," *Jour. Appl. Mech.*, *28*, March 1961, pp. 1-8.

_____, *Thermostatics and Thermodynamics*, D. Van Nostrand Co., Inc., NY, 1961.

Tukey, J.W., "Analyzing Data: Sanctification or Detective Work?" *American Psychologist*, *24*, 1969, pp. 83-91.

CHAPTER 3

EPISTEMOLOGICAL PERSPECTIVE

CHAPTER 3

EPISTEMOLOGICAL PERSPECTIVE

A. The Strange World of Complex Systems in the Dim Light of
Small Samples

There are many instances in which principles which apply to situations in the range of our common experience must be modified when
we try to apply them to situations beyond this range.

o *Velocity*

Within the range of "ordinary" velocities we have the
laws of Newtonian physics. When we go beyond these to
relativistic velocities (those approaching the speed
of light in vacuum), material objects appear to become
"distorted" in shape, size, mass, etc. The laws must
be modified in accord with the principles of special
relativity to properly describe their states and behavior.
Masses increase. Rates slow down. Lengths shrink in
the direction of motion. Shapes warp.

o *Mass*

Within the range of "ordinary" masses, Newtonian physics
is adequate to describe the behavior of objects. However,
for very small and for very large masses the behavior is
distorted from what one would anticipate using this physics. Objects with very small masses assume the peculiar
indeterministic behavior of quantum mechanics. Objects
with very large masses may gravitationally collapse and
become black holes. What is an "ordinary" mass depends
not only upon the magnitude of the mass itself but also
upon another variable, the velocity. At very high velocities, even otherwise ordinary masses become distorted.

o *Size*

Measurement of "ordinary" sizes involves Euclidean
geometry, once thought to be a "necessary" mathematical
description of the universe, like two-plus-two-equals-
four. However, measurement of very large sizes, especially in the presence of gravitational fields, involves
new principles such as general relativity. Straight
lines become curved and measurements do not agree with
Pythagorus' theorem. When the size being measured is of
the order of the limits of current telescopes, then the
matter of the curvature of the universe must be considered. At the other extreme, when we attempt to measure
very small sizes, we run into quantum mechanics. It

becomes unclear whether we are measuring a well-defined
object or an amorphous wave. When we push this to extremely small sizes, we run into particle production.
In addition, as with the case of mass, what is "ordinary"
size may depend upon other variables. Even otherwise
ordinary sizes become distorted at high velocities.

o *Time*

Ordinary time durations, from fractions of a second to
many generations, are handled under ordinary circumstances
by the idea of a periodic "clock" without conceptual difficulty. However, like mass and size, the idea of time
involves new principles for magnitudes outside this usual
experience. These include time intervals of very short
duration, of very long duration, and under conditions
of very high velocity. Periods become dilated at high
velocities or near massive bodies. Time takes on new
meanings for durations as long as the age of the universe
or as short as the interval it takes light to travel the
radius of an elementary particle.

o *Energy*

Under "ordinary" circumstances, energy can be described
as kinetic or potential (gravitational, electro-chemical,
thermal, etc.). The behavior of objects can be described
by classical mechanics, thermodynamics, electrodynamics,
etc., involving the conservation of energy. However,
under "exceptional" circumstances energy may disappear
and matter appear in its place. A verbal distinction
between matter and energy then becomes an inaccurate
means of description. At sufficiently high energies
the strengths of the fundamental interactions alter.

o *Entropy*

Our ordinary ideas of cause and effect dictate that the
hypothesis with the greatest likelihood of accurately
predicting future states of a system is one which most
accurately describes the observed past states of similar
systems. But this is true only when the number of independent descriptors is small compared to the number of
possible states and the number of past observations. In
complex situations, the best predictive hypothesis would
appear to be distorted were it used to describe the past.
Flat distributions become curved. Probabilities do not
agree with Bayes' theorem. The rule of succession warps.
The boundaries specifying similarity become altered.
Probabilities for particularized events shrink toward
values for more general ensembles. What is a "small"
sample for a given complexity becomes dependent upon another variable: utility. Even otherwise large samples
become small for extreme utility differences.

B.   Mini and Max

The most frequently asked question about entropy minimax is, "What is minimized and what is maximized?"   In both cases the answer is entropy, in one case "global," in the other "local."   The key difference is what is being varied as one extremum or the other is sought.   See Fig. 1.

Being maximized is the unconditional entropy defined for any subset of events.   This is a "local" entropy.   The variables, with respect to which this maximization is performed, are the probabilities for these events.   The maximization principle selects values for the probabilities which maximize this entropy, subject to any given constraints.

Being minimized is the conditional outcome entropy defined for events, given their classification in terms of characteristics other than the outcome.   This is a "global" entropy.   The variables, with respect to which this minimization is performed, are class boundaries used to partition events into different subsets.   The minimization principle selects boundaries for which this entropy is a minimum.

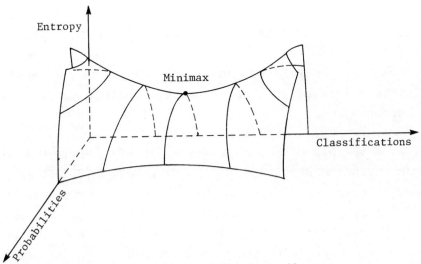

Fig. 1.   Entropy Mini and Max (actually two
          different entropies are involved)

C. Max

Entropy maximization is based on the following principles:

o Frequencies are "reality". On a specific 100 tosses
   of a specific coin, we observed heads 54% of the time.
   That is reality.

o Probabilities are our models of reality, but they are
   not real themselves. They do not exist outside our
   minds. Hence, it makes no sense to speak of the "true
   probability" for an event, or to speak of frequencies
   as "approaching" probabilities.

o Probabilities are functions of our current and past
   (background) observational data about reality.

o Probabilities should only reflect as much data as we
   actually have about reality. If they are made to reflect more, then we are modeling an imaginary world,
   not the one in which we live.

Two people with different data may, in general, assign different
probabilities to the same events. Yet they both may be entirely
correct. Probabilities represent their knowledge, not a property
of the events independent of their knowledge. See Fig. 2.

It is misleading to say, for example, that a particular coin toss
has a probability of one-half of coming up heads. It is more accurate to say we *assign* this probability to the coin toss.

Entropy maximization tells us what numerical values we assign to
specific probabilities when we accurately reflect our data (current
and background).

Entropy is a measure of the expected value of information we would
gain from further observation. Both extremely high and extremely
low probabilities give low entropies. They reflect little uncertainty. Either we are very sure the outcome will occur, or we are
very sure it will not. Intermediate probabilities give high entropies. They indicate that we are more uncertain and hence can
expect to gain more information from further observation.

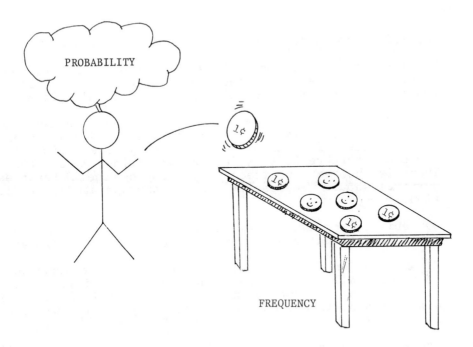

Fig. 2.   Probabilities Are Mental Images of Frequencies

To assign values to probabilities which maximize entropy subject
to available data is to give them values which represent no more information than is contained in the available data.   The reason we
use the idea of entropy to accomplish this is that entropy is a convenient concept to express situations mathematically, enabling us
to compute numerical values for probabilities based on information
in real-world data files.

D. Mini

Entropy minimization is based on the following principles:

o Reality consists of particulars: particular objects, particular events, etc. We see a particular chair. That is
reality. We observe a particular coin toss. That is reality.

o Classes are our conceptual groupings of real particulars,
but they do not exist outside of our minds. Such groupings
are convenient for purposes of recording observations for
our own future recall, or for communicating them to others.
There is, however, no reason why groupings useful for these
purposes should necessarily also be appropriate for prediction.

o Our ability to classify real objects and events depends upon
our observational capabilities and our observational data.

o Classification for prediction purposes should be made so as
to reflect all of the data we have about reality. To base
decisions on forecast estimates other than this is to opti-
mize for a different world from the one in which we live.

Two points should be especially emphasized: i) outcome probabilities for a particular event depend strongly upon the class to which
we assign the event, and ii) there are no a priori "correct" ways
of classifying events. See Fig. 3.

The probability of heads we assign to a particular coin toss, for
example, depends upon what *other* events we decide to put into the
same class as the particular coin toss.

Suppose I make the following classification: "Flat circular metal
objects, presently on this table, used in commercial exchange tossed
at least two feet into the air." The following objects qualify:

o Fifty pennies with a "head" on one side, and

o Fifty bus tokens with a "head" on neither side.

I pick up a penny and make a qualifying toss. While it is still in the
air I claim: "The probability of heads for *this coin toss* is one-
fourth."

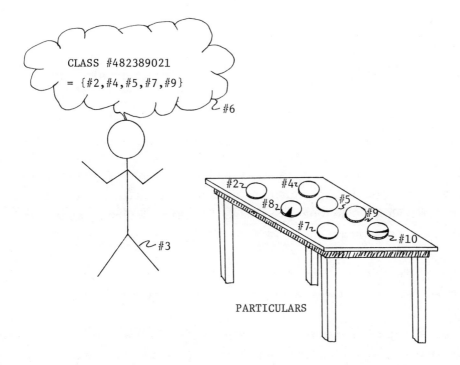

Fig. 3.   Classes Are Mental Groupings of Particulars

This example illustrates the fact that probability is as much a statement of what is going on in my mind as it is a statement about the particular event to which I assign it.

One might suppose that it would be easy to design a maximum information classification. We simply classify the events as finely as our observational capabilities and data will allow.

This approach, however, clearly fails for finite data samples (hence for any real-world situation). We need some principle for aggregating events *despite* our ability to distinguish among them, some principle for sacrificing event classification information in order to enhance outcome classification information.

The principle which does this job is entropy minimization.

We have two types of information:

o   Outcome classification information: This pertains to
    the *dependent* variable--what we want to predict.

o   Event classification information: This pertains to the
    independent variables--the given characteristics of the
    event which may help us in making our prediction.

Mutual entropy exchange is a measure of the amount of information
one message contains about another. We are interested in the amount
of information about the outcome classification which is contained
in the event classification.

The key point is that we can control entropy exchange, since it
depends upon how we *define* the event classification.

What we do is adjust this definition so that the event classification contains a maximum amount of information about the outcome classification.   To do otherwise would be not to utilize fully the available data to help us make our predictions.

It can be shown (as we will see in Chapter 7) that entropy exchange
is the difference between the entropy of the outcome classification
and the conditional entropy of the outcome classification given the
event classification.   Thus, if we adjust the event classification
so as to minimize this conditional entropy, we accomplish our objective.

Minimizing conditional entropy can be interpreted directly. This
entropy is the *currently* expected value of the information that we
would gain were we to learn the outcome for a future event, given
its event classification.   This expected information value of future observation depends upon how we classify our present and past
data.   We maximize the amount of information we extract from this
data by classifying it in such a way as to minimize the amount we
would learn by further observation.

QUOTES

SEXTUS EMPIRICUS (c. 200 A.D.)

...When they propose to establish the universal from the particulars by means of induction, they will effect this by a review either of all or of some of the particular instances. But if they review some, the induction will be insecure, since some of the particulars omitted in the induction may contravene the universal; while if they are to review all, they will be toiling at the impossible, since the particulars are infinite and indefinite. Thus on both grounds, as I think, the consequence is that induction is invalidated.

> —Sextus Empiricus (160-210)
> *Outlines of Pyrrhonism*, Chapter II, (c. 200 A.D.),
> tr. by R.G. Bury, *Sourcebook in Ancient Philosophy*,
> rev. ed., C. Bakewell, ed., Charles Scribner's Sons,
> NY, 1939, pp. 363-364.

JAYARĀŚI BHAṬṬA (7th Century A.D.)

A relation subsisting between one pair of terms cannot serve as the ground of inference for another.

*   *   *   *

From this it follows also that there is no possibility of understanding the relation of cause and effect.

> —Jayarāśi Bhaṭṭa (7th century A.D.)
> *Refutation of Inference*, Ch. VII (India, 7th Century
> A.D.), *A Sourcebook in Indian Philosophy*, ed. by
> S. Radhakrishnan and C. Moore, Princeton Univ. Press,
> Princeton, NJ, 1957, pp. 238, 242.

HUME (1739)

...there can be no *demonstrative* argument to prove, *that those instances, of which we have had no experience, resemble those, of which we have had experience.*

> —David Hume (1711-1776)
> *A Treatise on Human Nature* (1739), L.A. Selby-Bigge,
> ed., Oxford at the Clarendon Press, London, 1888,
> p. 89.

FREUD (1899)

Delboeuf (1885, 84) arrives at the same conclusion after somewhat different psychological arguments. We believe in the reality of dream-images, he says, because in our sleep we have no other impressions with which to compare them, because we are detached from the external world. But the reason why we believe in the truth of these hallucinations is not because it is impossible to put them to the test *within* the dream. A dream can seem to offer us such tests: it can let us touch the rose that we see - and yet we are dreaming. In Delboeuf's opinion there is only one valid criterion of whether we are dreaming or awake, and that is the purely empirical one of the fact of waking up.

> —Sigmund Freud (1856-1939)
> *The Interpretation of Dreams* (1899), ed. and tr. by J. Strachey, Avon Books, Discus Edition, NY, 1965, p. 84.

FREUD (1911)

...what is meant when some of the content of a dream is described in the dream itself as 'dreamt' - the engima of the 'dream within a dream'...

> —Sigmund Freud (1856-1939)
> *The Interpretation of Dreams* (1899), excerpt from paragraph added in 1911, tr. and ed. by J. Strachey, Avon Books, Discus Edition, NY, 1965, p. 373.

RUSSELL (1914)

It is...the principle of induction, rather than the law of causality, which is at the bottom of all inferences as to the existence of things not immediately given. With the principle of induction, all that is wanted for such inferences can be proved; without it, all such inferences are invalid. The principle has not received the attention which its great importance deserves.

> —Bertrand Russell (1872-1970)
> *Our Knowledge of the External World,* The Open Court Pub. Co., London, 1914, p. 226.

WITTGENSTEIN (1921)

The procedure of induction consists in accepting as true the *simplest* law that can be reconciled with our experiences.

This procedure, however, has no logical justification but only a psychological one.

WITTGENSTEIN (1921, cont.)

It is clear that there are no grounds for believing that the simplest eventuality will in fact be realized.

> —Ludwig Wittgenstein (1889-1951)
> *Tractatus Logico-Philosophicus* (1921), tr. by
> D.F. Pears and B.F. McGuinness, Routledge and
> Kegan Paul, London, 1961, p. 143.

POPPER (1934)

...the principle of induction must be a universal statement in its turn. Thus if we try to regard its truth as known from experience, then the very same problems which occasioned its introduction will arise all over again. To justify it we should have to employ inductive inferences; and to justify these we should have to assume an inductive principle of a higher order; and so on. Thus the attempt to base the principle of induction on experience breaks down, since it must lead to an infinite regress.

> —Karl R. Popper (1902-   )
> *The Logic of Scientific Discovery* (1934), Science
> Editions, Inc., NY, 1961, p. 29.

AYER (1936)

The problem of induction is, roughly speaking, the problem of finding a way to prove that certain empirical generalizations which are derived from past experience will hold good also in the future. There are only two ways of approaching this problem on the assumption that it is a genuine problem, and it is easy to see that neither of them can lead to its solution. One may attempt to deduce the proposition which one is required to prove either from a purely formal principle or from an empirical principle. In the former case one commits the error of supposing that from a tautology it is possible to deduce a proposition about a matter of fact; in the latter case one simply assumes what one is setting out to prove.

                    *   *   *   *

Thus it appears that there is no possible way of solving the problem of induction, as it is ordinarily conceived. And this means that it is a fictitious problem, since all genuine problems are at least theoretically capable of being solved...

> —Alfred J. Ayer (1910-   )
> *Language, Truth and Logic* (1936), Dover Pubs., Inc.,
> NY, 1946, pp. 49-50.

WILLIAMS (1947)

To the scandal of philosophy, however, after a billion years of
animal learning, twenty-five hundred years of science and philosophy, and some three hundred years of deliberate modern empiricism
and logical analysis, no one knows quite what is the rationale of
the inductive process, why it has worked in the past, and how its
working in the past can supply a reason for trusting it now and in
the time to come.

> —Donald Cary Williams (1899-   )
> *The Ground of Induction*, Harvard Univ. Press, Cam-
> bridge, MA, 1947, p. 4.

REICHENBACH (1949)

Ever since his [Hume's] famous criticism, philosophers have regarded the problem of induction as an unsolved riddle precluding
the completion of an empiricist theory of knowledge.

> —Hans Reichenbach (1891-1953)
> *The Theory of Probability*, Univ. of Calif. Press,
> Berkeley, CA, 1949, p. 470.

BARKER (1957)

Inevitably it will be asked why should we prefer simpler systems.
...This question perhaps is too fundamental to admit of any sharp
answer.

> —Stephen F. Barker (1927-   )
> *Induction and Hypothesis*, Cornell Univ. Press,
> Ithaca, NY, 1957, p. 181.

SEBESTYEN (1962)

Generally speaking, no unifying concept, general theory, or common systematic approach is at yet evident in work in the field of
pattern recognition.

> —George S. Sebestyen (1930-   )
> *Decision-Making Processes in Pattern Recognition*,
> The Macmillan Co., NY, 1962, p. 2.

ALLAIS (1966)

...the performance of a recognition system sometimes deteriorates
when additional measurements are included in the pattern analysis.
This strange phenomenon...

> —David Charles Allais (1933-   )
> "The Problem of Too Many Measurements in Pattern
> Recognition and Prediction," *IEEE Intl. Convention
> Report*, Part 7 (Discrimination, Measurements),
> March 21-25, 1966, p. 124.

BLACK (1967)

> ...no wholly satisfactory philosophy of induction is yet available.
>
> —Max Black (1909-   )
> "Induction," *The Encyclopedia of Philosophy*, Vol. 4,
> ed. by P. Edwards, Macmillan Pub. Co., NY, 1967,
> p. 179.

DEMPSTER (1971)

> The conceptual tools for approaching many parameters are not yet
> well formed, but two basic themes can be identified. The first of these
> is the need for measures of the confusion affecting questions of inter-
> est which is due to the uncertainty about many parameters.
>
> *   *   *   *
>
> The second theme is the need for guidelines in the use of confus-
> ion indices when they are available. If confusion is large, then
> one should consider simplifying a model but this can be done only
> at the cost of making the model less realistic. At present, the
> tradeoff between realism and clarity of message can only be made
> intuitively.
>
> —Arthur P. Dempster (1929-   )
> "An Overview of Multivariate Data Analysis," *Journal
> of Multivariate Analysis, 1,* 1971, pp. 336, 337.

CORMACK (1971)

> The growing tendency to regard numerical taxonomy as a satisfac-
> tory alternative to clear thinking is condemned.
>
> —Richard Melville Cormack (1935-   )
> "A Review of Classification," *J. of the Roy. Stat.
> Soc., 34,* Part 3, 1971, p. 321.

KANAL and CHANDRASEKARAN (1971)

> In statistical classification, estimation, and prediction, it has
> often been noted that, with finite samples, performance does not
> always improve as the number of variables is arbitrarily increased.
> Sometimes it may even deterioriate.
>
> —Laveen Kanal (1931-   ) and
> B. Chandrasekaran (1942-   )
> "On Dimensionality and Sample Size in Statistical
> Pattern Classification," *Pattern Recognition, 3,*
> Pergamon Press, London, 1971, p. 225.

COOLEY and LOHNES (1971)

We simply must plan for large samples (and good ones) if we are
going to use any of the function-fitting methods and expect the
functions to be taken seriously as generalizations to populations.

> —William W. Cooley (1930-   ) and
> Paul R. Lohnes (1928-   )
> *Multivariate Data Analysis*, J. Wiley & Sons, Inc.,
> NY, 1971, p. 56.

TOPLISS and COSTELLO (1972)

...the greater the number of variables tested, the greater role
chance will play in the observed correlation.

> —John G. Topliss (1930-   ) and
> Robert J. Costello (1939-   )
> "Chance Correlations in Structure-Activity Studies
> Using Multiple Regression Analysis," *J. of Medi-
> cinal Chemistry*, *15*, no. 10, 1972, p. 1006.

BISHOP, FIENBERG and HOLLAND (1975)

*...the "overall variability" of the estimates from the simpler
model about the "true" values for the cells is smaller than the
"overall variability" for the model with more parameters requiring estimation.* We have no general proof of this...

> —Yvonne M.M. Bishop (1925-   ),
> Stephen E. Fienberg (1942-   ) and
> Paul W. Holland (1940-   )
> *Discrete Multivariate Analysis: Theory and Prac-
> tice*, The MIT Press, Cambridge, MA, 1975, p. 313.

DAVIS (1977)

Complex systems requiring much specification information have a
fundamentally limited predictability when specified by real observations. The best model describing the evolution of the pre-
dictable part of such systems may differ qualitatively from the
detailed dynamics of the complete system: in general progressive
loss of predictability imparts irreversible or dissipative dynamics to the observable system.

> —Russ E. Davis (1941-   )
> Techniques for Statistical Analysis and Prediction
> of Geophysical Fluid Systems," *Geophys. Astrophys.
> Fluid Dynamics*, *8*, 1977, pp. 249-250.

REFERENCES

Allais, D.C., "The Selection of Measurements for Predictions," Rept. SEL-64-115 (TR No. 6103-9), Standard Electronics Laboratory, Stanford, CA, Nov. 1964.

_____, "The Problem of Too Many Measurements in Pattern Recognition," *IEEE Intl. Convention Report*, Part 7 (Discrimination, Measurements), March 21-25, 1966, pp. 124-130.

Ayer, Alfred J., *Language, Truth and Logic* (1936), Dover Pubs., Inc., NY, 1946.

Barker, Stephen F., *Induction and Hypothesis*, Cornell Univ. Press, NY, 1957.

Bishop, Yvonne M.M., Stephen E. Fienberg and Paul W. Holland, *Discrete Multivariate Analysis: Theory and Practice*, The MIT Press, Camb., MA, 1975.

Black, Max, "Induction," *The Encyclopedia of Philosophy*, Vol. 4, ed. by P. Edwards, Macmillan Pub. Co., NY, 1967, pp. 168-181.

Christensen, R.A., "Induction and the Evolution of Language," Physics Dept., Univ. of Calif., Berkeley, CA, July 19, 1963. [*Chapter 7 of Volume II.*]

_____, *Foundations of Inductive Reasoning*, Berkeley, CA, 1964.

_____, "Inductive Reasoning and the Evolution of Language," Physics Dept., Univ. of Calif., Berkeley, CA, Dec. 1964. [*Chapter 8 of Volume II.*]

_____, "A General Approach to Pattern Discovery," Tech. Rept. No. 20, Computer Center, Univ. of Calif., Berkeley, CA, June 29, 1967 (revised Nov. 15, 1967). [*Chapter 1 of Volume III.*]

_____, "A Pattern Discovery Program for Analyzing Qualitative and Quantitative Data," *Behavioral Science*, *13*, Sept. 1968, pp. 423-424. [*Chapter 2 of Volume II.*]

_____, "Seminar on Entropy Minimax Method of Pattern Discovery and Probability Determination," presented at Carnegie-Mellon Univ. and MIT, Feb. and March 1971, Tech. Rept. No. 40.3.75, Carnegie-Mellon Univ., Pittsburgh, PA, April 7, 1975. [*Chapter 3 of Volume III.*]

_____, "Entropy Minimax Method of Pattern Discovery and Probability Determination," Arthur D. Little, Inc., Acorn Park, Cambridge, MA, March 7, 1972. [*Chapter 4 of Volume III.*]

_____, "Entropy Minimax: A Non-Bayesian Approach to Probability Estimation from Empirical Data," *Proc. of the 1973 International Conference on Cybernetics and Society*, IEEE Systems, Man and Cybernetics Society, Nov. 5-7, 1973, Boston, MA, 73 CHO 799-7-SMC, pp. 321-325. [*Chapter 5 of Volume III.*]

_____, "Contrived Patterns, Trying to Avoid Them Without Trying Too Hard," Tech. Rept. No. 40.12.75, Carnegie-Mellon Univ., Pittsburgh, PA, June 19, 1975. [*Chapter 9 of Volume III.*]

_____, "Crossvalidation: Minimizing the Entropy of the Future," *Information Processing Letters*, *4*, Dec. 1975, pp. 73-76. [*Chapter 10 of Volume III.*]

Cooley, William W. and Paul R. Lohnes, *Multivariate Data Analysis*, J. Wiley & Sons, NY, 1971.

Cormack, R.M., "A Review of Classification," *Journal of the Royal Statistical Society, 34,* Part 3, 1971, pp. 321-367.

Davis, Russ E., "Techniques for Statistical Analysis and Prediction of Geophysical Fluid Systems," *Geophys. Astrophys. Fluid Dynamics, 8,* 1977, pp. 245-277.

Dempster, A.P., "An Overview of Multivariate Data Analysis," *Journal of Multivariate Analysis, 1,* 1971, pp. 316-346.

Eilbert, R.F. and R. Christensen, "Contrivedness: The Boundary between Pattern Recognition and Numerology," *Pattern Recognition* (to be published).

Freud, Sigmund, *The Interpretation of Dreams* (1899), tr. and ed. by J. Strachey, Avon Books, Discus Edition, NY, 1965.

Hume, David, *A Treatise of Human Nature* (1739), L.A. Selby-Bigge, ed., Oxford at the Clarendon Press, London, 1888.

Jayarāśi Bhaṭṭa, *Refutation of Inference* (India, 7th Century A.D.), *A Sourcebook in Indian Philosophy,* ed. by S. Radhakrishnan and C. Moore, Princeton Univ. Press, Princeton, NJ, 1957.

Kanal, Laveen and B. Chandrasekaran, "On Dimensionality and Sample Size in Statistical Pattern Classification," *Pattern Recognition, 3,* Pergamon Press, London, 1971, pp. 225-234.

Popper, Karl R., *The Logic of Scientific Discovery* (1934), Science Editions, Inc., NY, 1961.

Reichenbach, Hans, *The Theory of Probability,* Univ. of Calif. Press, Berkeley, CA, 1949.

Russell, Bertrand, *Our Knowledge of the External World,* The Open Court Pub. Co., London, 1914.

Sebestyen, George S., *Decision-Making Processes in Pattern Recognition,* The Macmillan Co., NY, 1962.

Sextus Empiricus, *Outlines of Pyrrhonism* (c. 200 A.D.), tr. by R.G. Bury, *Sourcebook in Ancient Philosophy,* C. Bakewell, ed., Charles Scribner's Sons, NY, 1935.

Topliss, John G. and Robert J. Costello, "Chance Correlations in Structure-Activity Studies Using Multiple Regression Analysis," *J. of Medicinal Chemistry, 15,* no. 10, 1972, pp. 1066-1068.

Williams, Donald, *The Ground of Induction,* Harvard Univ. Press, Camb., MA, 1947.

Wittgenstein, Ludwig, *Tractatus Logico-Philosophicus* (1921), tr. by D.F. Pears and B.F. McGuinness, Routledge and Kegan Paul, London, 1961.

CHAPTER 4

COMPUTER IMPLEMENTATION

CHAPTER 4

COMPUTER IMPLEMENTATION

A. Inception

Al Blumstein should have been from Missouri. It was the summer of 1966 at the Institute for Defense Analyses in Arlington, Virginia. I showed him my *Foundations*. "You say it is possible to program a computer to do inductive reasoning based on entropy minimax?" he asked.

"Yes," I answered.

"Well, have you ever done it?"

"No. I'm satisfied with a solution in principle," I replied.

"Then how do you know it really can be done?" he pressed on.

"It's a completely defined concept. It's only a matter of writing a sufficiently comprehensive pattern search algorithm," I insisted.

"Then how do you know there isn't some hitch you haven't anticipated?"

His question was unanswerable.

I decided it was time to learn about computers. So I got a BASIC manual and started programming SWAPDP, a Step-Wise Approximation Pattern Discovery Program implementation of entropy minimax.

The first FORTRAN version of SWAPDP came later during stops on the trip from Arlington back to the university. Driving three thousand five hundred miles alone, one needs something to think about anyway.

So I would work on a routine for an hour or two and then stop for coffee for fifteen minutes or so and write it down. When I final-ly reached Berkeley, my front seat was full of scraps of paper con-taining a draft of the program.

It is now fourteen years and many generations of SWAPDP later. Com-puters produce a prodigious amount of paper. I'm not sure I could even guess the total weight of all the versions generated.

The main program, SWAPDP, consists of approximately 7,000 lines of code, organized into over 100 subroutines. It operates on seven input files, and generates one output file.

The input files are prepared by another program, PDPREP. It is with PDPREP that the user generally interfaces on the input end.

A third program, PDPLZR, is used to condense the descriptive output from SWAPDP into a more computationally manageable form. This con-densed version is used by a pattern recognition program which ac-tually makes the predictions for future events.

Volume III of this Entropy Minimax Sourcebook series contains flow-charts and mathematical descriptions of the SWAPDP implementation of entropy minimax. The next section contains a brief overview of the organization of SWAPDP and supporting programs.

B. Organization

From an end-user viewpoint, SWAPDP can be regarded as a generator
of pattern files for use by prediction systems.

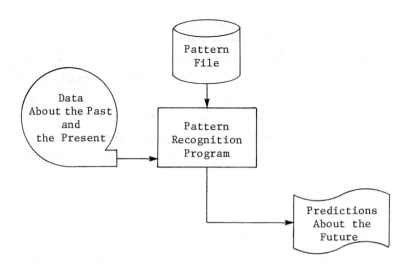

Fig. 4. Prediction System Schematic

Fig. 4 shows a prediction system schematic. It receives, from memory
access and sensory input, data about the past and the present. These
are fed into a pattern recognition program. This program compares
the data to a file of patterns. When it finds a match, it outputs
a prediction about the future.

SWAPDP's role is to generate the pattern file. Each pattern in this file consists of the following:

o Specifications which must be met by the past and present in order for the pattern to be matched.

o Future outcome probabilities for events matching the pattern.

o Uncertainties in these probabilities.

Fig. 5 illustrates the pattern file generation procedure. First the historical data are randomly split into two parts. One part is labeled "training". It is used to build the entropy minimax patterns. The other part is labeled "test". It is further subdivided by separating the dependent variable data from the independent variable data. This is done so that the test sample can be used for cross-validation purposes to fix an important parameter--the weight normalization.

An initial value for the normalization is selected using an algorithm based on such factors as sample size and number of independent variables. The entropy minimax patterns are then found for the training data using this value.

These initial patterns are used to predict the test data dependent variable values, given the independent variables. The predictions are then compared to the actual values, the informational errors computed, and the results stored.

This process is repeated for variations in the value of the normalization. Finally, the patterns with least error are selected. These are entered into the pattern file to be used by the pattern recognition program in the prediction system.

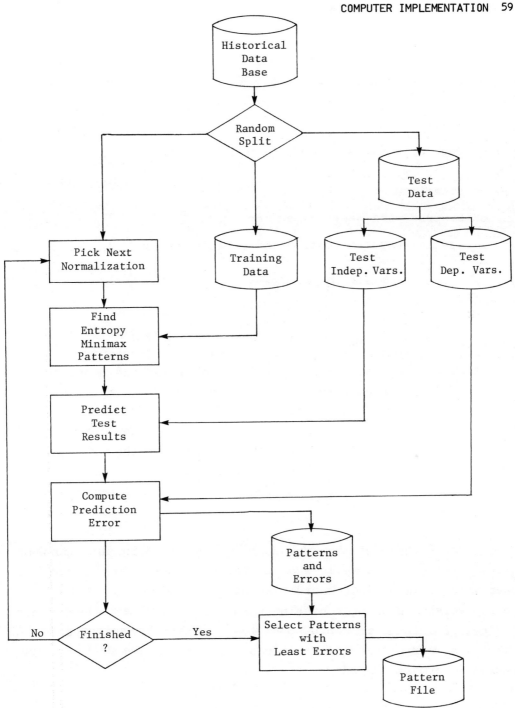

Fig. 5. Pattern Discovery System Schematic

C.   The Role of Human Judgment

Human judgment plays important roles in the pattern discovery process as implemented in the SWAPDP software. These roles can be grouped
as follows:

o Dependent variable selection and classification

o Independent variables selection and specification of combinations

o Program parameters selection

o Special pattern restrictions specification

Fig. 6 shows how they relate to the overall flow of the pattern
discovery process.

D.   Dependent Variable

Selection of the dependent variable is the first and most basic
step in defining a pattern discovery problem. This specifies the
question to which an answer is being sought.

The exact definition of the dependent variable should be formulated in such a way that there is an adequate amount of data on it
for the past, and so that knowledge of it for the future would be
useful.

The classification of the dependent variable is also an important
user decision. This tells what it is about the variable that the
user wants to know. It would take an infinite amount of data to
learn everything about a continuous variable. So with limited historical data, the user should subdivide the dependent variable into as
small a number of categories as possible, consistent with decision-
making information needs.

Human selection of dependent variables might someday be replaceable
by a programmed prime directive such as survival, altruism, or some
other ultimate criterion, together with an algorithm for generating

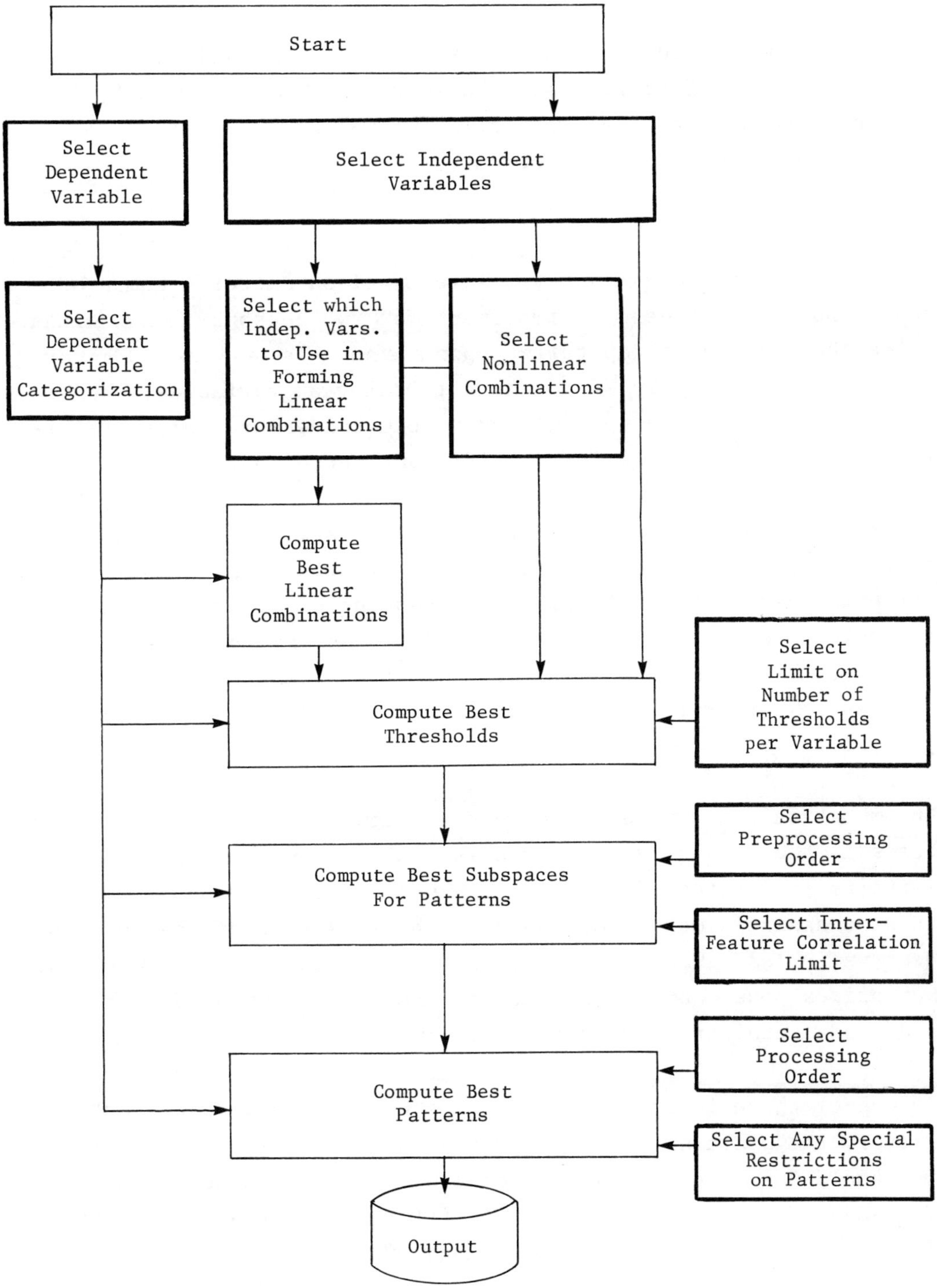

Fig. 6. Key Human Inputs (heavy boxes) to
SWAPDP Pattern Discovery Process

and assessing candidate dependent variables. But self-directedness
is another stage of artificial intelligence. For the present, we are
dealing with an automated mechanism for making generalizations addressed to human selected problems.

E.   Independent Variables

Selection of the independent variables is also a matter of human judgment.   Human input here is important for two reasons. First, it enables the machine to capitalize upon the human's sensory and memory
links to a vast supply of potentially relevant variables. Only in
highly structured and specialized circumstances are computers directly wired to sensors for a sufficient set of variables.

Second, the current level of programming is still sufficiently crude
that the computer generally needs a considerable amount of feature
selection assistance.   Automatic processing of linear combinations
is implemented, once the component variables are specified.   Nonlinear combinations are left for user selection.   Some subclasses
such as low order polynomials can be handled automatically, but the
general class is too broad.

An important part of the selection of independent variables in many
cases is the use of existing models.   For example, when seeking
patterns for weather forecasting, one can use the outputs from various mechanistic models of the atmosphere as some of the independent variables.   When seeking patterns to predict future stock market prices, one can use the outputs from a selection of technical
and fundamental trading rules as some of the independent variables.

If one of these independent variables happens to be a perfect predictor, the pattern discovery process will identify it. Thereafter,
one would use it directly, at least until circumstances changed to
degrade its performance, or until different information is needed.
For most complex problems, however, a perfect predictor is not known.

One can, nevertheless, facilitate the pattern discovery process by augmenting raw data independent variables with informationally enriched outputs from existing theories and hypotheses, both mechanistic and statistical.

F.   Program Parameters

A principal use of program parameters is as controls on computer run time. They enable tradeoffs to be made between run time and extensiveness of pattern search.

The key parameters are:

o   Number of thresholds per quantitative variable

   This tells the computer how finely it can partition the independent variables. (The positions of the thresholds are, however, calculated variables rather than input parameters.)

o   Pre-processing order

   This specifies the interrelationship level that is searched when sequencing through the input variables to select subspaces in which to form the patterns.

o   Inter-feature correlation limit

   This says that, if two independent variables are highly correlated, computer time can be saved by processing only one since the other probably has very little additional information.

o   Processing order

   This specifies the maximum dimensionality of any subspace in which a pattern can be formed. (Different subspaces can, however, be used for different patterns.)

G.   Pattern Restrictions

In order to cover situations in which there are special user requirements or special auxiliary information, the ability to impose restrictions on pattern formation is provided. Examples include:

o   Pattern population

   The user is given the ability to specify that a pattern will be rejected if it does not contain a minimum number of training sample events. Typical minima are in the range of 5-30 events.

o Suppression of multi-modal patterns

The user is given the ability to specify that the distribution for training sample events in any pattern along the
dependent variable will not be strongly multi-peaked. (A
measure of contour is defined and the user is permitted to
specify a lower limit to it for pattern acceptance.)

o Pattern dimensionality

The user is permitted to limit the number of features which
can be used to specify any single pattern.

Some of these restriction options are historical artifacts (left
in following Archie Bunker's astute admonition "If it works, don't
fix it"), dating back to the use of heuristically imposed limitations
to reduce the risk of chance correlation. Others are options enabling
the user to input auxiliary information which may be available about
the system under investigation. Additionally, there is a specific
routine in which the user may enter coded instructions to define
other forms of restrictions such as feature polarity with respect
to outcome probabilities.

H. Further Automation of Pattern Discovery

There are several evident benefits to be derived from further automation of the human role in pattern discovery.

o More economical processing of very large numbers of commonly encountered problems (e.g., curve-fitting, factor analysis).

o More economical processing of problems involving large numbers of variables and/or large amounts of data (e.g., meteorological, economic, psychological, medical problems).

o More uniformity of results of analyses conducted by different persons, thus facilitating independent validation.

One approach is to design specialized algorithms for sets of well-
defined situations. Examples include curve-fitting, informational
assessment of individual variables, and time-series and generalized
waveform analyses. Work by McConnell on piecewise-polynomial curve-
fitting, for example, is described in Volume IV of this series.

A significant portion of the human role is too difficult or impractical to automate. It is thus important that computer programs en-
gaged in pattern discovery be able to interface usefully with this
important source of experience and intuition. One approach, suggested by Eilbert, is based on the ancient Solonic code of law. Instead of seeking one or two "best" partitions of the space of independent variables, he suggests use of a multitude of partitions.
These can be generated, for example, by multiple SWAPDP pattern
searches with a variety of manually selected program parameters
and pattern restrictions. The results of these runs would then
be amalgamated to produce a grand compromise. The technique of
amalgamation would be similar to that described in Chapter 12 of
this volume, presently used to combine probabilities based on conditional entropy minimizing partitions with those based on entropy
exchange maximizing partitions.

This approach is carried even further in a suggestion by Reichert
to put user feedback in the SWAPDP partition-formation loop. With
users able to directly interact at the crucial points of hypothe-
sizing patterns and of accepting or rejecting them, their experience and judgment is effectively utilized. The trend to date has
been on increasing automation of the discovery process. In the
future we may see a shift toward more extensive and flexible per-
son-machine interaction.

QUOTES

LUCRETIUS (c. 60 B.C.)

...the nature of mind and soul is bodily. For when it is seen
to push on the limbs, to pluck the body from sleep, to change
the countenance, and to guide and turn the whole man--none
of which things we see can come to pass without touch, nor
touch in its turn without body--must we not allow that mind and
soul are formed of bodily nature?

> —Titus Lucretius (96-55 B.C.)
> *On the Nature of Things* (c. 60 B.C.), tr. by
> C. Bailey, Oxford at the Clarendon Press, London,
> 1921, p. 111.

DESCARTES (1637)

...although machines can perform certain things as well as or
perhaps better than any of us can do, they infallibly fall short
in others, by the which means we may discover that they did not
act from knowledge, but only from the disposition of their
organs. For while reason is a universal instrument which can
serve for all contingencies, these organs have need of some
special adaptation for every particular action. From this it
follows that it is morally impossible that there should be sufficient diversity in any machine to allow it to act in all the
events of life in the same way as our reason causes us to act.

> —Rene Descartes (1595-1650)
> "Discourse on the Method of Rightly Conducting the
> Reason and Seeking for Truth in the Sciences" (1637),
> *Philosophical Works of Descartes*, Vol. I, tr. by
> E. Haldane and G. Ross, Dover Pubs., Inc., NY,
> 1955, p. 116.

JEVONS (1874)

It is true that Swift satirically described the Professors of
Laputa as in possession of a thinking machine, and in 1851
Mr. Alfred Smee actually proposed the construction of a Relational machine and a Differential machine, the first of which
would be a mechanical dictionary and the second a mode of comparing ideas; but with these exceptions I have not yet met with
so much as a suggestion of a reasoning machine. It may be
added that Mr. Smee's designs, though highly ingenious, appear
to be impracticable, and in any case they do not attempt the
performance of logical inference.

> —W. Stanley Jevons (1835-1882)
> *The Principles of Science, A Treatise on Logic and
> the Scientific Method* (1874), Dover Pubs., Inc.,
> NY, 1958, p. 107.

FREUD (1899)

...our schematic picture of the psychical apparatus.

...at first the apparatus's efforts were directed towards keep-
ing itself so far as possible free from stimuli; consequently
its first structure followed the plan of a reflex apparatus, so
that any sensory excitation impinging on it could be promptly
discharged along a motor path.

> —Sigmund Freud (1856-1939)
> *The Interpretation of Dreams* (1899), ed. and tr.
> by J. Strachey, Avon Books, Discus Edition, NY,
> 1965, p. 604.

McCULLOCH (1961)

There is now underway a whole tribe of men working on artificial
intelligence--machines that induce, or learn--machines that
abduce, or make hypotheses. In England alone, there are Ross
Ashby, MacKay, Gabor, Andrews, Uttley, Graham Russel, Beurle,
and several others--of whom I could not fail to mention Gordon
Pask and Stafford Beer. In France, the work centers around
Schützenberger. The Americans are too numerous to mention.

I may say that there is a whole computing machinery group, followers of Turing, who build the great deductive machines. There
is Angyan, the cyberneticist of Hungary, now of California, who
had reduced Pavlovian conditioning to a four-bit problem, em-
bodied in his artificial tortoise. Selfridge, of the Lincoln
Laboratory, M.I.T.--with his Pandemonium and his Sloppy--is
building abductive machinery. Each is but one example of many.
We know how to instrument these solutions and to build them
in hardware when we will.

But the problem of insight, or intuition, or invention--call it
what you will--we do not understand, although many of us are
having a go at it. I shall not here name names, for most of us
will fail miserably and be happily forgotten. Tarski thinks
that what we lack is a fertile calculus of relations of more
than two relata. I am inclined to agree with him...

> —Warren S. McCulloch (1868-1969)
> "What Is a Number, that a Man May Know It, and a
> Man, that He May Know a Number?" (the Ninth
> Alfred Korzbybski Memorial Lecture), *General Se-
> mantics Bulletin*, Nos. 26 and 27, (Lakeville, CT:
> Inst. of General Semantics, 1961), pp. 7-11, in
> *Embodiments of Mind*, MIT Press, Camb., MA, 1965,
> pp. 13-14.

DANTZIG (1963)

Building a mathematical model often provides so much insight
into a system and the organization of knowledge about a system
that it is considered by many to be more important than the
task of mathematical programming which it precedes. The model
is often difficult to construct because of the richness,
variety, and ambiguity of the real world.

> —George B. Dantzig (1914-   )
> *Linear Programming and Extensions*, Princeton
> Univ. Press, Princeton, NJ, 1963, p. 34.

FOGEL, OWENS and WALSH (1966)

*Artificial intelligence is realized only if an inanimate machine
can solve problems that have, thus far, resisted solution by
man; not because of the machine's sheer speed and accuracy,
but because it can discover for itself new techniques for
solving the problem at hand.*

> —Lawrence J. Fogel (1928-   ),
> Alvin J. Owens (1918-   ) and
> Michael J. Walsh (1918-   )
> *Artificial Intelligence Through Simulated
> Evolution*, J. Wiley & Sons, Inc., NY, p. 8.

MINSKY and PAPERT (1969)

Rosenblatt's* schemes quickly took root, and soon there were
perhaps as many as a hundred groups, large and small, experi-
menting with the model either as a "learning machine" or in the
guise of "adaptive" or "self-organizing" networks or "automatic
control" systems.

The results of these hundreds of projects and experiments were
generally disappointing, and the explanations inconclusive.
The machines usually work quite well on very simple problems
but deteriorate very rapidly as the tasks assigned to them get
harder. The situation isn't usually improved much by increasing
the size and running time of the system. It was our suspicion
that even in those instances where some success was apparent, it
was usually due more to some relatively small part of the network, and not really to a global, distributed activity.

> —Marvin Minsky (1927-   ) and
> Seymour Papert (1928-   )
> *Perceptions*, The MIT Press, Cambridge, MA, 1969,
> p. 19.

* Rosenblatt, Frank (1959), "Two theorems of statistical separa-
  bility in the perceptron," *Proceedings of Symposium on the
  Mechanization of Thought Processes*, Her Majesty's Stationary
  Office, London, pp. 421-456.

SIMON (1969)

...there exist today a considerable number of examples of actual design processes, of many different kinds, that have been defined fully and cast in the metal, so to speak, in the form of running computer programs: optimizing algorithms, search procedures, and special-purpose programs for designing motors, balancing assembly lines, selecting investment portfolios, locating warehouses, designing highways, and so forth.

Because these computer programs describe complex design processes in complete, painstaking detail, they are open to full inspection and analysis, or to trial by simulation. They constitute a body of empirical phenomena to which the student of design can address himself and which he can seek to understand. There is no question, since these programs exist, of the design process hiding behind the cloak of "judgment" or "experience". Whatever judgment or experience was used in creating the programs must now be incorporated in them and hence be observable. The programs are the tangible record of the variety of schemes that man has devised to explore his complex outer environment, and to discover in that environment the paths to his goals.

> --Herbert A. Simon (1916-   )
> *The Sciences of the Artificial*, The MIT Press,
> Cambridge, MA, 1969, p. 80.

PATRICK (1972)

The first operation, invariably called preprocessing or feature extraction, is highly problem dependent and usually has not been conveniently described mathematically; it has been an art.

> —Edward A. Patrick (1937-   )
> *Fundamentals of Pattern Recognition*, Prentice-Hall,
> Inc., Englewood Cliffs, NJ, 1972, p. 2.

FUKUNAGA (1972)

In practical applications, effective features have primarily been found by the designer's intuition. The difficulty of solving a nonlinear problem is common in all engineering areas. However, this is particularly true in pattern recognition because the number of variables is large.

> —Keinosuke Fukunaga (1930-   )
> *Introduction to Statistical Pattern Recognition*,
> Academic Press, NY, 1972, p. 288.

ANDREWS (1972)

Human "pre-processing" or feature selection is a highly complex
and apparently nonlinear operation which has not, as yet, been
duplicated through mathematical formalism. In fact when a
human is used to replace the feature selector in a pattern
recognition machine, he evinces a heuristic feature selection
based upon his past experience, and the selection process is
difficult if not impossible to describe as he himself often
cannot explain upon what specific attributes he based his decision.

> —Harry C. Andrews (1943-   )
> *Introduction to Mathematical Techniques in Pattern
> Recognition,* J. Wiley & Sons, Inc., NY, 1972, p. 16.

YOUNG and CALVERT (1974)

The selection of measurements is very important in designing a
pattern recognition machine, and is a major factor in its success
or failure. There is no general theory on measurement selection
because the selection usually depends on the particular pattern
recognition problem under consideration.

> —Tzay Y. Young (1933-   ) and
> Thomas W. Calvert (1936-   )
> *Classification, Estimation and Pattern Recognition,*
> Am. Elsevier Pub. Co., NY, 1974, p. 3.

KANAL (1974)

What is a pattern that a machine may know it, and a machine that
it may know a pattern?

> —Laveen Kanal (1931-   )
> "Patterns in Pattern Recognition: 1968-1974,"
> *IEEE Trans. on Info. Theory, IT-20,* Nov. 1974,
> p. 697.

REFERENCES

Andrews, Harry C., *Introduction to Mathematical Techniques in Pattern Recognition*, J. Wiley & Sons, Inc., NY, 1972.

Christensen, R.A., "A General Approach to Pattern Discovery," Tech. Rept. No. 20, Computer Center, Univ. of Calif., Berkeley, CA, June 29, 1967 (revised Nov. 15, 1967). [*Chapter 1 of Volume III.*]

_____, "A Pattern Discovery Program for Analyzing Qualitative and Quantitative Data," *Behavioral Science, 13,* Sept. 1968, pp. 423-424. [*Chapter 2 of Volume III.*]

_____, "Seminar on Entropy Minimax Method of Pattern Discovery and Probability Determination," presented at Carnegie-Mellon Univ. and MIT, Feb. and March 1971, Tech. Rept. No. 40.3.75, Carnegie-Mellon Univ., Pittsburgh, PA, April 7, 1971. [*Chapter 3 of Volume III.*]

_____, "Entropy Minimax Method of Pattern Discovery and Probability Determination," Arthur D. Little, Inc., Acorn Park, Cambridge, MA, March 7, 1972. [*Chapter 4 of Volume III.*]

_____, "Entropy Minimax, A Non-Bayesian Approach to Probability Estimation From Empirical Data," *Proc. of the 1973 International Conference of Cybernetics and Society,* IEEE Systems, Man and Cybernetics Society, Nov. 5-7, 1973, Boston, MA, 73 CHO-799-7-SMC, pp. 321-325. [*Chapter 5 of Volume III.*]

_____, "Entropy Minimax Processing of Incomplete Data Sets," *Proc. of the 1973 International Conference on Cybernetics and Society,* IEEE Systems, Man and Cybernetics Society, Nov. 5-7, 1973, Boston, MA, 73 CHO 799-7-SMC. [*Chapter 5 of Volume III.*]

_____, "Entropy Minimax Determination of Threshold Values for Quantitative Variables," Tech. Rept. No. 40.8.75, Carnegie-Mellon Univ., Pittsburgh, PA, April 12, 1975. [*Chapter 7 of Volume III.*]

_____, "Entropy Minimax Rotations: The Extraction of Features from Sets of Quantitative Variables," Tech. Rept. No. 40.9.75, Carnegie-Mellon Univ., Pittsburgh, PA, April 28, 1975. [*Chapter 8 of Volume III.*]

_____, "Contrived Patterns, Trying to Avoid Them Without Trying Too Hard," Tech. Rept. No. 40.12.75, Carnegie-Mellon Univ., Pittsburgh, PA, June 19, 1975. [*Chapter 9 of Volume III.*]

_____, "Crossvalidation: Minimizing the Entropy of the Future *Information Processing Letters, 4,* Dec. 1975, pp. 73-76. [*Chapter 10 of Volume III.*]

Christensen, R.A. and T. Reichert, "Unit Measure Violations in Pattern Recognition: Ambiguity and Irrelevancy," *Pattern Recognition, 8,* 1976, pp. 239-245.

Dantzig, George B., *Linear Programming and Extensions,* Princeton Univ. Press, Princeton, NJ, 1963.

Descartes, Rene, "Discourse on the Method of Rightly Conducting the Reason and Seeking for Truth in the Sciences" (1637), *Philosophical Works of Descartes,* Vol. I, tr. by E. Haldane and G. Ross, Dover Pubs., Inc., NY, 1955.

Eilbert, Richard, "Multi-Partitioning of Feature Space: The Solonic Approach to Probability Estimation," Technical Memorandum, Entropy Limited, Lincoln, MA, March 2, 1981.

Fogel, Lawrence J., Alvin J. Owens and Michael J. Walsh, *Artificial Intelligence Through Simulated Evolution*, J. Wiley & Sons, Inc., NY, 1966.

Freud, Sigmund, *The Interpretation of Dreams* (1899), tr. and ed. by J. Strachey, Avon Books, Discus Edition, NY, 1965.

Fukunaga, Keinosuke, *Introduction to Statistical Pattern Recognition*, Academic Press, NY, 1972.

Jevons, W. Stanley, "On the Mechanical Performance of Logical Inference," *Phil. Trans. of the Royal Soc. of London, 160,* 1870, pp. 497-518.

_____, *The Principles of Science, A Treatise on Logic and the Scientific Method* (1874), Dover Pubs., Inc., NY, 1958.

Kanal, Laveen, "Patterns in Pattern Recognition: 1968-1974," *IEEE Trans. on Info. Theory, IT-20,* Nov. 1974, pp. 697-722.

Lucretius, Titus, *On the Nature of Things* (c. 60 B.C.), tr. by C. Bailey, Oxford at the Clarendon Press, London, 1921.

McConnell, Robert K., "Minimum Description Analysis of Faulted Data Sets," paper presented to the Canadian Exploration Geophysical Society - American Geophysical Union, Mining Geophysics Symposium, Toronto, Canada, May 22-23, 1980.

McCulloch, Warren S., "What Is a Number, that a Man May Know It, and a Man, that He May Know a Number?" *General Semantics Bulletin,* nos. 26 and 27 (Lakeville, CT: Inst. of General Semantics, 1961), pp. 7-11, in *Embodiment of Mind,* The MIT Press, Cambridge, MA, 1965.

Minsky, Marvin and Seymour Papert, *Perceptrons,* The MIT Press, Cambridge, MA, 1969.

Oldberg, S. and R. Christensen, "Dealing With Uncertainty in Fuel Rod Modeling," *Nuclear Technology, 37,* 1978, pp. 40-47.

Patrick, Edward A., *Fundamentals of Pattern Recognition,* Prentice-Hall, Inc., Englewood Cliffs, NJ, 1972.

Simon, Herbert A., *The Sciences of the Artificial,* The MIT Press, Cambridge, MA, 1969.

Smee, Alfred, *The Process of Thought Adapted to Words and Language, Together with a Description of the Relational and Differential Machines,* Longmans and Co., London, 1851.

Young, Tzay Y. and Thomas W. Calvert, *Classification, Estimation and Pattern Recognition,* Am. Elsevier Pub. Co., NY, 1974.

CHAPTER 5

APPLICATIONS

CHAPTER 5

APPLICATIONS

A. How Small Is Small?

Entropy minimax probabilities differ from sample average estimates
in two important respects:

- o   The maximum entropy probabilities are not simply fre-
      quencies for past events in a class most similar to
      the future event in question. Rather they are shrunk
      toward the frequencies for a more general class of
      events. This shrinkage is a function of our data about
      past events.

- o   What past events are classed as "similar" to the future
      event in question is determined by entropy minimization.
      Event classification is not an absolute, but rather is
      relative both to our data about past events, and to what
      it is that we wish to predict about future events.

For cases with large samples, however, we postulate a frequency-correspondence principle:

   Maximum entropy probabilities for a defined event class
   approach data frequencies in the limit of large sample
   sizes.*

The question of practical application is: What is the correspondence
threshold? What is the scale for sample size? How large would the
sample have to be before we could reliably use sample averages as
our probability estimates? What is the range of sample sizes over
which entropy minimax probabilities differ significantly? What are
typical situations in which our available data fall below the correspondence threshold?

---

*   R.A. Christensen, "Induction and the Evolution of Language," Physics Dept.,
Univ. of Calif., Berkeley, CA, 1963, *(Volume II*, pp. 134, 138); *Foundations of
Inductive Reasoning*, Berkeley, CA, 1964, Sec. 7.B; "A General Approach to Pat-
tern Discovery," Tech. Rept. No. 20, Computer Center, Univ. of Calif., Berkeley,
CA, 1967, *(Volume III*, p. 17).

There is no fixed number, such as 1,000 events, below which entropy minimax makes a significant difference and above which one is secure simply using sample averages. The threshold depends upon a weight normalization which is a function of several factors, including:

- o The number of and intercorrelations among known independent variables to which the outcomes may be related.

- o The distribution on the dependent (outcome) variable for fixed values of these known independent variables.

- o The amount of uncertainty regarded as "significant" by the user for the particular application intended.

- o The number of predictions to be made in the intended application.

In economic and business applications, where one's benefits derive from a marginal advantage over a large number of decisions, a small difference in accuracy of probabilistic forecasts can make a significant difference in total result. In such cases the correspondence threshold can be quite large.

In medical applications, where a decision may have serious health implications, one wants the uncertainty in probabilistic prognoses to be low. Here also the threshold can be large.

In engineering applications with a large number of variables upon which the result may depend, where the interrelations among these variables may be insufficiently understood or too complex for deterministic modeling, the threshold can be large.

In biological, behavioral and social science applications, where many unknown variables may strongly affect the outcome, the threshold can be large.

How complex can a problem be and still be amenable to a more "conventional" approach? A general proviso often heard is that linear methods, for example, should be used "with caution" for any but very large sample-to-variate number ratios. However, even if we were to accept such a criterion with some minimum ratio, it would be difficult to apply considering all the independent variates which the human investigator assesses and discards prior to formal mathematical analysis.

The statistical mechanics researcher working with only a handful of variates and sample sizes on the order of Avogadro's number has a comparatively easy job. This enabled Gibbs, for example, to state that he was concerned only with proportions and not with the absolute numbers of systems. Most researchers must cope with far more disadvantageous sample-to-variate number ratios. In these cases, absolute numbers are important.

Even if we have a sample size which is large in the context of a particular application requiring a certain number of predictions, it may be small in the context of an application requiring a much larger number of predictions (unless, of course, there are no utility differences among outcomes).

Ultimately, one can determine the frequency-correspondence threshold for a particular application only by actually conducting an information-theoretic analysis of the data available for that application.

What are our alternatives? Consider, for example, the implications of exhaustive use of sample averages.

Suppose that we have M independent variables, each with R significant ranges of values, and that events tend to be evenly distributed over these ranges. The number of different range combinations is R^M. If we assume Poisson statistics, the uncertainty in an outcome probability for events of any given range combination will be roughly

$$U = \frac{1}{\sqrt{N}},$$

where N is the number of sample events with this range combination. Thus, to hold the uncertainty down to U for all possible range combinations by exhaustive use of sample average estimates, one needs a total sample size of

$$N = R^M U^{-2}.$$

For example, if there are M = 20 independent variables and R = 5 ranges per variable, then to hold this uncertainty down to 5% we need a total sample size of N = $5^{20} (0.05)^{-2}$ = 3.8×10^{16} events!

If this typified the usual real-world situation, there would virtually be no case in which we could depend solely upon use of sample averages. Our sample size would almost always be well below the correspondence threshold. So there must be something critical missing from this reasoning.

Not so! The fact is that this *does* typify the usual real-world situation. Generally there are far fewer events in the sample than would be required to hold uncertainties down to acceptable levels by exhaustive application of sample averages estimation to each particularized event class. This is so even when the classification is restricted to use of variables known to affect the outcome significantly.

How then do we explain our ability to cope with probability estimation at all? The explanation is that we use heuristic entropy minimization. We aggregate the data into categories in terms of features which appear to differentiate the outcome of interest. By so organizing the data, we lower the entropy. But we take care not to refine the categorization too much, lest we lose sample size and get lost in the noise of the data.

We are already playing the entropy minimax game, although in a largely intuitive way. The applications issue is not: "On what problems is entropy minimax potentially useful?" Rather, it is: "For what problems do we not already have adequate heuristic entropy minimax procedures, so that we can significantly improve probability estimates with a computerized information-theoretic approach?"

B. Experience

Applications of computerized entropy minimax have included the following:

- o Calculating probabilities of fuel failure during operation of light water nuclear power reactors. (Project for the Electric Power Research Institute.)

- o Estimating fission gas release from LWR fuel as a function of burnup and operating temperature history in experimental test assemblies. (Analysis for the Halden Reactor Project, Halden, Norway.)

- o Conducting statistical analysis of coolant channel closure due to rod bow in nuclear reactor fuel assemblies. (Study by Science Applications, Inc.)

- o Finding patterns in diametral expansion, axial elongation and cracking of UO_2 pellets under conditions of direct electrical heating. (Joint project by Argonne National Laboratories and Entropy Limited.)

- o Finding patterns related to strain measured in Loss-Of-Coolant-Accident burst simulation experiments. (Analysis for the U.S. Atomic Energy Commission--now the Nuclear Regulatory Commission--using data specifically prepared by Battelle Northwest Laboratory.)

- o Assessing risks in consumer credit decision-making. (Research supported by National Science Foundation and M.I.T.)

- o Assessing criteria for pre-screening drugs for probable anti-cancer activity on the basis of physical/chemical properties and molecular structure features. (Study for the National Cancer Institute.)

- o Classifying Fisher's iris data. (Analysis by Carnegie-Mellon University Biotechnology Program on a National Institutes of Health grant.)

o Providing information to use in differential diagnosis
   of three diseases of the cervical spine: cervical spon-
   dylosis, multiple sclerosis, and amyotrophic lateral
   sclerosis. (Joint project by Pittsburgh University
   School of Medicine and Carnegie-Mellon University
   Biotechnology Program.)

o Determining ranges of presenting signs and symptoms over
   which differential disease diagnosis is essentially unchanged. (Research at Carnegie-Mellon University Biotechnology Program.)

o Finding patterns distinguishing normal users from overusers of health care clinic facilities. (Study at the
   Carnegie-Mellon University Biotechnology Program.)

o Identifying signs and symptoms associated with estimated
   severity of injury for patients with possible skull
   fracture. (Study at M.I.T.'s Sloan School using data
   collected at Harborview Medical Center and University
   of Washington's University Hospital, Seattle.)

o Separating patients with low probability of cortical defect from those with higher probabilities, as one component of information available to consider in assessing
   advisability of a radionuclide brain scan. (Study by
   Dept. of Radiology, Univ. of Mich., sponsored by National
   Cancer Institute and U.S. Dept. of Energy.)

o Identifying patient groups with very low probability of
   carcinoma/adenoma, as one component of information available to consider in assessing advisability of a radionuclide thyroid scan. (Study by Dept. of Radiology, Univ.
   of Mich., sponsored by U.S. Dept. of Energy.)

o Assessing urine chromatogram indicators of specific types
   of mental illness. (Analysis at Carnegie-Mellon University
   Biomedical Engineering Program using data from Stanford
   University.)

o Finding patterns in vector cardiogram data which tend to
   separate normal from abnormal heart conditions. (Analysis
   at Carnegie-Mellon University Biomedical Engineering
   Program.)

o Finding patterns in displacement cardiogram data which
   tend to separate normal from abnormal heart conditions.
   (Analysis at Carnegie-Mellon University Biomedical
   Engineering Program.)

o Estimating survival probabilities for patients who have
   been treated for coronary artery disease. (Analysis by
   Entropy Limited in cooperation with the Duke University
   School of Medicine.)

o Predicting probable direction of price movement one day
  in advance for a number of commodities futures. (Analysis
  of trading exchange data.)

o Computing conditional probabilities associated with the
  self-information of the cytochrome-c molecule for various species in an evolutionary hierarchy. (Research
  supported by National Science Foundation.)

o Identifying optimal thresholds for 58 symptom variables
  relative to 12 diseases in a set of 1430 pelvic surgery
  cases, and assessing the information content of each
  symptom. (Research at M.I.T. supported by the National
  Science Foundation and the National Institutes of Health.)

o Estimating α-helix classification probabilities for
  adenyl cyclase residues. (Research supported by National
  Institutes of Health.)

o Finding patterns in historical time-series data related
  to weather and climate, permitting more definitive prediction of wet-year vs. dry-year probabilities for a
  water resource region in California, with lead times of
  one, two or three years. (Project sponsored by the Office
  of Water Research and Technology, U.S. Department of the
  Interior.)

o Finding multivariate patterns in drug abuse among persons
  in the armed services. (Analysis for an Arthur D. Little,
  Inc., study sponsored by the U.S. Dept. of Defense.)

o Finding patterns related to crime rates in different
  cities. (Analysis at Institute for Defense Analyses in
  connection with activities of the Science and Technology
  Task Force, National Commission on Law Enforcement and
  the Administration of Justice.)

o Finding multivariate relationships in the behavior of
  women offenders in a state penal institution. (Analysis
  of data supplied by Iowa State Reformatory for Women.)

o Performing multivariate analysis of employment following
  higher education, for former students who had received
  financial aid from a specific federal program. (Analysis
  for a Law Enforcement Assistance Administration project
  conducted by Arthur D. Little, Inc.)

o Finding patterns in musical tastes evidenced by data from
  a sociological survey. (Research project at Univ. of
  Calif., Berkeley, CA.)

o Finding judicial decision patterns in a line of Supreme Court
  cases. (Analyses of data supplied by F. Kort and R. Lawlor.)

The data bases for these applications involved sample sizes ranging from less than 100 to nearly 20,000, generally in the range of
500-1,500. The number of independent variables ranged from 3 to
over 100. Typically, 10-20 variables contributed significantly to
the patterns.

When compared with other approaches applied to the same or similar
data, entropy minimax has demonstrated improved prediction performance, ranging from slight improvement to major improvement.   (In
a few cases, "conventional" estimators were included among the independent variables for the entropy minimax analysis. In such cases,
entropy minimax would necessarily do no worse than the best of these
estimators. In general it was able to find patterns with improved
performance.)

In wet/dry California climate prediction with lead times of one or
more years, the overall accuracy of entropy minimax patterns was
63% on independent test data. (The accuracy on extreme-wet/extreme-
dry years was 78%.) This is compared to an overall accuracy of 50%
for a random predictor and 54% for a persistence hypothesis predictor. For price-swing direction prediction, the accuracy aver-
aged about 70%, compared to a random 50%.

The prediction error rate on fuel failure in nuclear reactors by
entropy minimax was half that of the best alternative (a heuristically designed parameter tuning model). The error rate of entropy
minimax in classifying Fisher's iris data was the lowest of the approaches tried, including principle components analysis, 3-nearest
neighbor analysis, and linear discriminant analysis. The α-helix
misclassification rate in adenyl cyclase residues was lower for entropy minimax than for any of the heuristic schemes applied.

The entropy minimax analysis of health care facilities overuse followed a discriminant analysis which had been interpreted as identifying characteristics of a stereotypic overuser. The entropy minimax

pattern definitions enables this interpretation to be corrected by showing that only classes of predominantly nonoverusers could be typified by the data. The remainder of the data contained insufficient information to typify overusers.

The entropy minimax pattern in radionuclide thyroid scanning data for low probability of carcinoma/adenoma was found after both a multivariate regression analysis and a Bayesian conditional probability analysis had failed to produce any useful information from the data.

Entropy minimax analysis of detailed medical data on a sample of 1213 patients who underwent cardiac catheterization during the period 1969-1978 produced a total of 40 patterns and associated short-term (2-year) and long-term (20% of average remaining lifetime for persons of same age) survival probability estimates. Tested against another sample, reserved for crossvalidation, of 1223 patients who also underwent catheterization during the same period, the patterns were found to have greater validated accuracy than previous techniques for predicting survival for coronary artery disease patients.

In several cases, entropy minimax pattern discovery uncovered relationships not previously anticipated (even though explanations could be offered once they were pointed out). Examples include the following:

o   A strong relationship between failure likelihood of zircaloy clad UO_2 nuclear fuel rods and axial convexity of circumferential cladding creep in commercial power reactors.

o   A high sensitivity of unclad UO_2 fuel pellet cracking in a Direct-Electrical-Heating experimental setup to trace amounts of moisture in the helium atmosphere.

o   A pattern, matched on one out of four years, involving histories of precipitation at Nevada City and Colfax, tree ring widths at Truckee, and Pacific Ocean sea surface temperatures in belts along the equator indicating a high probability (0.78 ± 0.15) for the next year being drier than average for the Sacramento/Tahoe region of California.

Volume IV of this Entropy Minimax Sourcebook series contains descriptive details on applications.

QUOTE

MINSKY and PAPERT (1969)

Good theories rarely develop outside the context of a background of well-understood real problems and special cases.

—Marvin Minsky (1927-   ) and
Seymour Papert (1928-   )
Perceptrons, The MIT Press, Cambridge, MA, 1969,
p. 3.

REFERENCES

Boyle, Brian E., "Symptom Partitioning By Information Maximization," Dept. of Elec. Engr., MIT, Cambridge, MA, 1972.

_____, "The Decision to Grant Credit," Ph.D. Dissertation, Operations Research, Dept. of Elec. Engr., MIT, Cambridge, MA, 1974.

Christensen, R.A., "Induction and the Evolution of Language," Physics Dept., Univ. of Calif., Berkeley, CA, July 19, 1963. *[Chapter 7 of Volume II.]*

_____, *Foundations of Inductive Reasoning*, Berkeley, CA, 1964.

_____, "A General Approach to Pattern Discovery," Tech. Rept. No. 20, Computer Center, Univ. of Calif., Berkeley, CA, June 29, 1967 (revised Nov. 15, 1967). *[Chapter 1 of Volume III.]*

_____, "Entropy Minimax Analysis of Simulated LOCA Burst Data, Report Prepared for Core Performance Branch, Directorate of Licensing, U.S. Atomic Energy Commission," Contract No. AT(49-24)-0083, Entropy Limited, Belmont, MA, Dec. 31, 1974.

_____, "Statistical Model of UO_2 Nuclear Fuel Under Direct Electrical Heating," *Thermal Mechanical Behavior of UO_2 Nuclear Fuel, Vol. I - Statistical Analyses of Acoustic Emission, Diametral Expansion, Axial Elongation, and Crack Characteristics*, a report to EPRI under Contract No. RP508-2, Entropy Limited, Lincoln, MA, June 1978.

_____, "Nuclear Fuel Rod Failure Hazard Axes," *Fuel Rod Mechanical Performance Modeling, Task 3: Fuel Rod Modeling and Decision Analysis*, Entropy Limited, Lincoln, MA, FRMPM33-2, Aug. 1979 - Oct. 1979, pp. 7.1-7.11 and FRMPM34-1, Nov. 1979 - Jan. 1980, pp. 40-48.

_____, "Assessment of Statistical Models," *SPEAR Fuel Performance Reliability Code System: General Description*, ed. by R. Christensen, EPRI NP-1378, Electric Power Research Institute, Palo Alto, CA, March 1980, pp. 2.1-2.17.

Christensen, R.A. and R. Ballinger, "In-Service Predictions," paper presented at the Joint EPRI/DOE Fuel Performance Contractors' Overview Meeting, Atlanta, GA, April 8, 1980.

Christensen, R.A. and E. Duchane, "Element-Specific Failure Time Estimation from Ensemble Statistics," *Fuel Rode Mechanical Performance Modeling, Task 3: Fuel Rod Modeling and Decision Analysis, FRMPM32-2*, Entropy Limited, Lincoln, MA, May 1979 - July 1979, pp. 6.34-6.50.

Christensen, R.A. and R. Eilbert, "Estimating Chance Correlation Likelihood for Hazard Axes Analysis," *Fuel Rod Mechanical Performance Modeling, Task 3: Fuel Rod Modeling and Decision Analysis*, Entropy Limited, Lincoln, MA, FRMPM33-2, Aug. 1979 - Oct. 1979, pp. 7.12-7.47, and FRMPM34-1, Nov. 1979 - Jan. 1980, pp. 32-39.

Christensen, R., R. Eilbert, O. Lindgren and L. Rans, "An Exploratory Application of Entropy Minimax to Weather Prediction: Estimating the Likelihood of Multi-Year Droughts in California," report to the Office of Water Research and Technology, U.S. Dept. of the Interior, under Contract No. 14-34-001-8409, Entropy Limited, Lincoln, MA, Sept. 1980.

Christensen, R., R. Eilbert, O. Lindgren and L. Rans, "Successful Hydrologic Forecasting for California Using an Information Theoretic Model," *Journal of Applied Meteorology* (to be published).

Christensen, R.A., R. Eilbert and S.T. Oldberg, "Entropy Minimax Hazard Axes for Failure Analysis," *Fuel Rod Mechanical Performance Modeling, Task 3: Fuel Rod Modeling and Decision Analysis*, FRMPM 32-2, Entropy Limited, Lincoln, MA, May 1979 - July 1979, pp. 6.20-6.33.

Christensen, R.A. and T. Reichert, "Multivariate Analysis Methodology," *A Study of Department of Defense Drug Abuse Prevention and Control Programs, Vol. III, Methodology and Instrumentation*, a report to the Office of the Asst. Sec. of Defense (Health and Environment) under contract No. DAHC 15-73-C-0304, Arthur D. Little, Inc., Cambridge, MA, Jan. 19, 1975.

_____, "A Preliminary Entropy Minimax Search for Patterns in Structural, Physio-Chemical and Biological Features of Selected Drugs That May Be Related to Activity in Retarding Lymphoid Leukemia, Lymphocytic Leukemia and Melanocarcinoma in Mice," report to the National Cancer Institute under subcontract No. A10373 of Contract No. N01-CM-23711, Entropy Limited, Pittsburgh, PA, June 30, 1975.

England, S., T.A. Reichert, and R. Christensen, "Entropy Minimax Classification of Adenyl Cyclase Residues," Bio-Medical Engineering Program, Carnegie-Mellon Univ., Pittsburgh, PA, June 1976.

Franklin, D., "Analysis of Fission Gas Release Data by a Pattern Recognition Technique," IDG Note 1433, OECD Halden Reactor Project, Halden, Norway, Jan. 11, 1978.

Gift, D., J. Gard and W. Schonbein, "Thyroid Scanning--Pursuing the Relationships of Signs and Symptoms to Nucleide Uptake and Scan Interpretation," report prepared under DOE Grant No. EX-76-5-01-1777.A003, Dept. of Radiology, Mich. State Univ., E. Lansing, MI, 1980.

Gift, D.A. and W. Schonbein, "Diagnostic Yield Analysis of Indicators for Radionuclide Brain Scanning," report prepared under DOE Contract No. E(11-1)2777, Dept. of Radiology, Michigan State Univ., E. Lansing, MI, 1980.

Gift, D.A., W. Schonbein and E.J. Potchen, "An Introduction to Entropy Minimax Pattern Detection and Its Use in the Determination of Diagnostic Test Efficacy," report prepared under NCI Grant No. CA 18871-02ODHEW and DOE Grant No. EX-76-S-02-2777.A0012, Dept. of Radiology, Michigan State Univ., E. Lansing, MI, 1980.

Hirschman, Alan D., "An Application of Entropy Minimax Pattern Discovery in a Multiple Class Electrocardiographic Problem," Bio-Medical Engineering Program, Carnegie-Mellon Univ., Pittsburgh, PA, Dec. 18, 1975.

_____, "Methods for Efficient Compression, Reconstruction, and Evaluation of Digitized Electrocardiograms," Ph.D. Dissertation, BioMedical Engr. Program, E.E. Dept., Carnegie-Mellon Univ., Pittsburgh, PA, Oct. 24, 1977.

Leontiades, K., "Computationally Practical Entropy Minimax Rotations, Applications to the Iris Data and Comparison to Other Methods," Bio-Medical Engineering Program, Carnegie-Mellon Univ., Pittsburgh, PA, 1976.

Lim, E., *Coolant Channel Closure Modeling Using Pattern Recognition*, FRMPM 21-1, SAI-175-79-PA, Science Applications, Inc., Palo Alto, CA, June 1979.

McConnell, R.K., "Minimum Description Analysis of Faulted Data," paper presented to the Canadian Exploration Geophysical Society-American Geophysical Union, Mining Geophysics Symposium, Toronto, Canada, May 22-23, 1980.

Minsky, Marvin and Seymour Papert, *Perceptrons*, The MIT Press, Cambridge, MA, 1969.

Oldberg, S.T., "Probabilistic Code Development," *Planning Support Document for the EPRI Light Water Reactor Fuel Performance Program*, prepared by J.T.A. Roberts, R.E. Gehhaus, H. Ockan, N. Hoppe, S.T. Oldberg, G.R. Thomas and D. Franklin, EPRI NP-737-SR, Electric Power Research Institute, Palo Alto, CA, Jan. 1978, pp. 2.52-2.60.

_____, "SPEAR Methodology," *SPEAR Fuel Performance Reliability Code System: General Description*, ed. by R. Christensen, EPRI NP-1378, Electric Power Research Institute, Palo Alto, CA, March 1980, pp. 2.1-2.17.

Oldberg, S.T. and R. Christensen, "Dealing With Uncertainty in Fuel Rod Modeling," *Nuclear Technology, 37*, Jan. 1978, pp. 40-47.

Potchen, E.J., "Study on the Use of Diagnostic Radiology," *Current Concepts in Radiology, 2*, ed. by E. James Potchen, The C.V. Mosby Co., St. Louis, MO, 1975, pp. 18-30.

Potchen, E.J. and W.R. Schonbein, "A Strategy to Study the Use of Radiology as an Information System in Patient Management," M.S. Thesis, Sloan School for Management, MIT, Cambridge, MA, 1973.

Rans, L.L., "Employment Patterns for Criminal Justice Personnel Following LEAA Supported Higher Education," *Criminal Justice Higher Education Programs in Maryland*, report to LEAA under Contract No. C-76186, Arthur D. Little, Inc., Cambridge, MA, Oct. 31, 1973, pp. 7-61.

_____, "Population Profiles: Iowa Woman's Reformatory 1918-1975," a report based on data acquired under U.S. Dept. of Labor, Manpower Administration, Grant No. 21-25-75-11, August 1976.

Reichert, T.A., "Patterns of Overuse of Health Care Facilities--A Comparison of Methods," *Proc. of the 1973 Intl. Conf. on Cybernetics and Society*, Boston, MA, IEEE Systems, Man and Cybernetics Society, 73 CHO 799-7-SMC, Nov. 5-7, 1973, pp. 328-329.

_____, "The Security Hyperannulus--A Decision Assist Device for Medical Diagnosis," *Proc. of the Twenty-Seventh Annual Conf. on Engineering in Medicine and Biology*, Philadelphia, PA, Oct. 6-10, 1974, p. 331.

Reichert, T.A. and R. Christensen, "Validated Predictions of Survival in Coronary Artery Disease," Duke Medical Center, Durham, NC, and Entropy Limited, Lincoln, MA, 1981.

Reichert, T.A. and A.J. Krieger, "Quantitative Certainty in Differential Diagnosis," *Proc. of the Second Intl. Joint Conf. on Pattern Recognition*, Copenhagen, Denmark, 74-CHO-885-46, Aug. 13-15, 1974, pp. 434-437.

Reichert, T.A., J.M.C. Yu and R. Christensen, "Molecular Evolution as a Process of Message Refinement," *Journal of Molecular Evolution, 8*, 1976, pp. 41-54.

Schonbein, William R., "Identification of Patterns in Diagnostic Attributes in Skull Trauma Cases Using an Entropy Minimax Approach," AEC Document No. 00-2427-2, presented at the 1973 International Conference on Cybernetics and Society, Boston, MA, Nov. 5-7, 1973. [*Paper not included in conference proceedings.*]

_____, "Analysis of Decisions and Information in Patient Management," *Current Concepts in Radiology*, *2*, ed. by E. James Potchen, The C.V. Mosby Co., St. Louis, MO, 1975, pp. 31-58.

Vas, R. and R. Christensen, "Entropy Minimax Analysis of Displacement Cardiograph Data," Technical Memorandum, BioMedical Engineering Program Carnegie-Mellon Univ., Pittsburgh, PA, July 1975.

Verbeek, P. and J. Van Vliet, "Entropie Minimax," Rept. No. 980.00/100/n/056, Div. Etudes Combustible, Belgonucleaire S.A., Bruxelles, June 9, 1977.

PART II

THE DEPENDENT VARIABLE: MAXIMUM ENTROPY PROBABILITIES

CHAPTER 6

SAMPLES AND HISTOGRAMS

CHAPTER 6

SAMPLES AND HISTOGRAMS

A.   Samples

Suppose you want to predict something: for example, the total precipitation (rain, snow, etc.) in a region of California next year;
the price change for gold bullion tomorrow; the fraction of failed
fuel assemblies in a nuclear power reactor; or the cause of your
death.

As a basis for making such predictions, you acquire information
about what happened in similar circumstances.   For example, you go
to the weather bureau and obtain statistics on precipitation for
previous years.   Table 1 illustrates this for the period 1852-1977.

Table 1.   Total Precipitation vs. Year

Year	Inches of Precipitation
1977	17.44
1976	16.79
1975	42.97
1974	52.88
⋮	⋮
1852	41.65

You go to the library, look up past prices of gold bullion and tabulate the changes.   Table 2 illustrates this for the year 1980.

Table 2.   Change in Gold Bullion Price vs. Date

Date	Price Change from Previous Date ($ / oz)
Dec. 31, 1980	− 0.25
Dec. 30, 1980	− 4.00
Dec. 29, 1980	−10.75
Dec. 24, 1980	+ 2.00
⋮	⋮
Jan. 3, 1980	+74.50

You check records on nuclear fuel assemblies operated in commercial and experimental reactors. See Table 3.

Table 3. Fraction Nuclear Fuel Assemblies Failed

Location in Reactor	Failure Fraction
1019-1	0.000
1022-1	0.125
1023-3	0.125
1030-1	0.000
⋮	⋮
Y47133	0.333

You check death statistics for persons of your age, sex, race and geographical locale. See Table 4.

Table 4. Cause of Death for Various Persons

Name	Cause of Death
John Doe	Heart
Mary Jones	Cancer
Rich Taylor	Accident
Tom Black	Cardiovascular Disease
⋮	⋮
Jane Smith	Pneumonia

In each of these four examples, you have accumulated a *sample* of events more or less similar to the event you wish to predict. There are, of course, many important questions about how you de- fine the sample, i.e., how you decide what is important and what is not important in determining whether a past event should be in- cluded as *similar* to the one you are trying to predict. The ques- tion of similarity will be discussed in Part III of this volume. For the present, we shall assume that after due consideration you have settled upon your definition of the sample.

Each data table in our four illustrative problems consists of two columns and a number of rows. Each row corresponds to an *event* in

the sample. The entry in the left column is the event *name*. It
identifies a single unique event. Thus the ordering of the rows
could be scrambled without affecting the tabulated information
(except to make it more cumbersome to cross-check). The entry in
the right column is the value of the *dependent variable*. The value
for a particular event is referred to as the *outcome* of the event.

We ignore, for the time being, the possibility that the name may
contain *independent* variable information. (We take up the subject
of independent variables in Part III.)

B. Histograms

Since no particular ordering of the events is implied in our tabulations, we are free to rearrange them. For example, we may
arrange Tables 1 through 3 in order of decreasing magnitude of the
dependent variable. In the case of Table 4, where the dependent
variable is qualitative rather than quantitative, we may use an
ordering which puts events with similar outcomes together but is
otherwise arbitrary. These rearrangements yield Tables 5 through 8.

Table 5		Table 6		Table 7		Table 8	
1868	86.50	1/18/80	+ 85.00	S2:LR2	1.000	M.J.	Heart
1853	84.20	1/16/80	+ 75.40	S3:LR3	1.000	J.S.	Heart
1862	83.62	1/ 3/80	+ 74.50	S4:LR4	1.000	Q.W.	Heart
1890	77.21	2/21/80	+ 59.00	S5:LR5	1.000	F.S.	Heart
:	:	:	:	:	:	:	:
1976	16.79	1/22/80	-112.50	Y47134	0.000	R.T.	Other

Next, assume that the event names are of no significance to our
analysis. Recall that we are ignoring independent variable information at this point in the analysis, so the names are nothing more
than identifiers. This allows us to drop the first column alto-
gether and transform Tables 5 through 8 into Tables 9 through 12.
Granting the fact that we are provisionally pretending that the
names do not contain independent variable information, this is not

Table 9	Table 10	Table 11	Table 12
86.50	+ 85.00	1.000	Heart
84.20	+ 75.40	1.000	Heart
83.62	+ 74.50	1.000	Heart
77.21	+ 59.00	1.000	Heart
⋮	⋮	⋮	⋮
16.79	-112.50	1.000	Other

without danger. Whether or not states are separately identifiable can significantly affect behavior of quantum mechanical processes, statistical mechanical systems, etc.* However, we can back up and incorporate these complications later. At this point we are just trying to get a rough picture of some prediction problems without getting bogged down with attempting a complete description of the class of all possible situations.

Having dropped the event names from our data, we next introduce names for different *classes* of dependent variable values. We shall call such a system of names a classification system. In the case of Table 12 there is an obvious classification system, i.e., the individual dependent variable values (heart, cancer, etc.). Other classifications can be defined for this sample by specifying aggregations of these values (for example, by lumping all types of accidental deaths under a common heading).

In the case of continuous variables given in Tables 9 through 11, there is an infinite number of possible classification systems. Particular classification systems can be defined for these samples in a variety of ways. For example, we may select a number K and decompose the range from the minimum possible value to the maximum possible value into K equal segments. Alternatively, we may segment some monotonic function of the variable into equal length segments.

---

* States which are permanently distinguishable from one another obey Maxwell-Boltzmann statistics. States which are not obey Bose-Einstein statistics (if their eigensolutions are symmetric) or Fermi-Dirac statistics (if their eigensolutions are antisymmetric). See Tolman for a discussion.

Sample-dependent classification schemes may also be defined. One example is to define a sequence of segments such that there is exactly one segment-to-segment boundary between each adjacent pair of events in the sample. Another is to chop the interval spanned by the data into K equally-populated segments. (This reduces to the previous case when K equals the number of different dependent variable values occurring in the sample.)

Selection of a classification system for the dependent variable is an important part of the data analysis process. It requires careful consideration of the purposes for which the prediction is wanted as well as of statistical aspects of the data.

Consider, for example, the problem of predicting next year's rainfall. A pertinent classification system to a farmer whose crops need at least 40 inches of rainfall, but would be flooded out at 80 inches, is shown in Fig. 7.

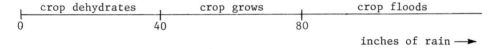

Fig. 7. Classification System for a Particular Farmer

The hydrologist responsible for the performance of a network of dams may classify the variable quite differently, as shown in Fig. 8.

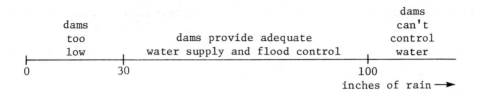

Fig. 8. Classification System for a Particular Hydrologist

Similar differences arise in other cases such as the gold price
prediction problem. Consider a trader holding gold through a
price rise who wants to sell just before it turns downward. A
relevant classification system is that shown in Fig. 9.

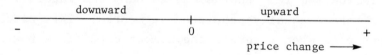

Fig. 9.   Classification System for
a Particular Gold Seller

A potential buyer, on the other hand, may not wish to buy unless
the expected price change is positive enough to meet a certain
rate of return, as shown in Fig. 10.

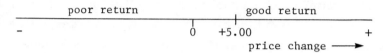

Fig. 10.   Classification System for
a Particular Gold Buyer

The obvious question is: Why not just divide the variable infinitely finely in such cases? Then surely we would have a classification system with which any potential user would be satisfied.

Given an infinite amount of data, this would indeed be the best
way to do it. However, real life sample sizes are frequently quite
small, especially when the event definition is particularized adequately to provide a reasonable level of repetition reliability.*

With a finite sample size, we must be parsimonious in partitioning
the dependent variable. It is best to use the minimum number of
classes adequate to make the distinctions essential for the user's
decision-making purposes. In regions where there is a large amount

---

* We take up in Part III the problem of determining the extent to which event
definitions are particularized to accomplish this.

of data, it may be desirable to subdivide the variable further to gain additional refinement. However, this should be carefully limited to avoid excessive reduction in class sizes.

With that digression, let us return again to the problem of defining classes for Tables 9 through 12. We adopt a sample-independent classification scheme with K = 5 equal segments between arbitrary limits. Now we may re-tabulate the data, listing the number of events in each class. This process yields Tables 13 through 16.

Table 13*			Table 14*			Table 15*			Table 16	
class	#		class	#		class	#		class	#
80-90	3		above +30	13		0.80-1.00	17		Cardiovascular	441
60-70	7		+10 to +30	58		.60- .79	4		Cancer	182
40-59	58		-10 to +10	117		.40- .59	13		Other disease	186
20-39	54		-30 to -10	50		.20- .39	31		Accident	49
0-19	4		below -30	15		0- .19	119		Other	24
	126			253			184			882

We have lost some information by aggregating the data this way. We will see later that it is important to "lose" some, but not too much, information by aggregation.

Histograms provide an alternative representation of sample frequency information such as that in Tables 13-16. See Fig. 11.

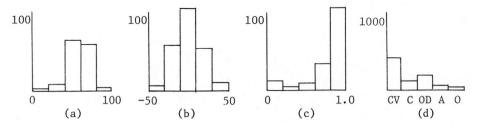

Fig. 11. Histograms of Counts

Equivalently, we may relabel the plots to give the fraction of the
sample in each class. See Fig. 12. The sample size is noted on
each of these figures to ensure that there is no information loss.

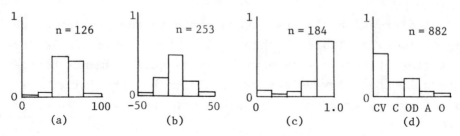

Fig. 12. Histograms of Frequencies

Because classification frequencies are just fractions of a whole,
the sum of them all for any plot is unity:*

$$F_1 + F_2 + \ldots + F_k + \ldots + F_K = 1 \ ,$$

where

F_k = fraction of events falling in the kth class, and

K = number of classes.

In the next chapter we take up the topic of classification frequency estimators.

---

* In Chapter 15 we consider cases involving irrelevancies and ambiguities in
which unity sum may be violated.

QUOTES

ARISTOTLE (c. 335-323 B.C.)

Scientific knowledge is not possible through the act of perception...for perception must be of a particular,whereas scientific
knowledge involves the recognition of the commensurate universal.
So if we were on the moon, and saw the earth shutting out the
sun's light, we should not know the cause of the eclipse: we
should perceive the present fact of the eclipse, but not the
reasoned fact at all, since the act of perception is not of the
commensurate universal. I do not, of course, deny that by
watching the frequent recurrence of this event we might, after
tracking the commensurate universal, possess a demonstration,
for the commensurate universal is elicited from the several
groups of singulars.

...knowledge of things demonstrable cannot be acquired by perception, unless the term perception is applied to the possession
of scientific knowledge through demonstration. Nevertheless,
certain points do arise with regard to connexions to be proved
which are referred for their explanation to a failure in sense-
perception: there are cases when an act of vision would terminate our inquiry, not because in seeing we should be knowing, but
because we should have elicited the universal from seeing; if,
for example, we saw the pores in the glass and the light passing
through, the reason of the kindling would be clear to us because
we should at the same time see it in each instance and intuit
that it must be so in all instances.

> — Aristotle (384-322 B.C.)
> *Posterior Analytics* (c. 335-323 B.C.),Bk. I, Ch.31,
> tr. by G. Mure, *The Basic Works of Aristotle*, ed.
> by. R. McKeon, Random House, NY, 1941,p. 154-155.

OCKHAM (c. 1327)

It ought also to be noted that frequently, in order to acquire
knowledge of the universal proposition, many singulars are required even though the subject of the universal is a species
special; for in most cases a singluar contingent proposition
cannot be evidently known without many apprehensives of single
instances, when it is not easy to know that this herb cured a
certain invalid and that it was not this doctor who cured him.
And so with many other cures, for it is not easy to grasp that
which is experienced, because the same effect in species can
exist through many causes specifically different.

> —William of Ockham (c. 1280-1348)
> *Summa Logicae*, III, pt. 2, Ch. 10 (c. 1327), tr.
> by E.A. Moody, *The Logic of William of Ockham*,
> London, 1935, p. 241.

OCKHAM (c. 1327, cont.)

However often one experiences in himself that he knows something perfectly and intuitively, he never knows something else by means of this knowledge unless he previously has also had knowledge of that other thing.

—William of Ockham (c. 1280-1348)
Ordinatio in Librum Primum Sententiarum, Prol, q. 9, tr. by J.R. Weinberg, *Ockham, Descartes and Hume, Self-Knowledge, Substance and Causality*, The Univ. of Wisconsin Press, Madison, WI, 1977, p. 55. The citation by Weinberg continues: Ockham defines knowing a thing perfectly as "notice of its intrinsic causes, namely of matter and form and knowledge of integral parts suffices for perfect knowledge of a thing, because in this knowledge it suffices to know all that which is of the nature and essence of the thing." References from L. Baudry, *Lexique philosophique de Guillaume d'Ockham*, Paris, 1958, pp. 178-79. Moreover, for Ockham, knowledge of causal efficacy requires experience. See *Summa Logicae*, III, pt 2, Ch. 10; *Ordinatio*, Prol., q. 2 (*Opera Theologica*, I, 90-95); *Expositio super Octo Libros Physicorum*, fol. 122V.

da VINCI (1495-1499)

...to me all sciences seem vain and full of error that are not born of experience, mother of all certainty, and do not terminate in an actual experience; or to put it another way, those of which neither the beginning, nor the middle, nor yet the end is made known to one of the five senses.

—Leonardo da Vinci (1452-1519)
"Paragone," *Codex Urbinas* (1495-1499), 19r-v, excerpted in *The Genius of Leonardo da Vinci*, ed. by André Chastel, The Orion Press, NY, 1961, p. 35.

GILBERT (1600)

Since in the discovery of secret things and in the investigation of hidden causes, stronger reasons are obtained from sure experiments and demonstrated arguments than from probable conjectures and the opinions of philosophical speculators of the common sort...

—William Gilbert (1544-1603)
DeMagnete (Amsterdam, 1600), tr. by P. Mottelay, Dover Pubs., Inc., NY, 1958, p. xlvii.

BACON (1620)

> We have but one simple method of delivering our sentiments, namely, we must bring men to particulars and their regular series and order, and they must for a while renounce their notions, and begin to form an acquaintance with things.
>
> > —Francis Bacon (1561-1626)
> > *Novum Organum; or True Suggestions for the Interpretation of Nature (1620)*, The Colonial Press, NY, 1899, pp. 318-319.

HOBBES (1651)

> For there is no conception in a mans mind, which hath not at first, totally, or by parts, been begotten upon the organs of *Sense*.
>
> > —Thomas Hobbes (1588-1629)
> > *Leviathan* (1651), reprinted by Oxford at the Clarendon Press, London, 1909, p. 11.

NEWTON (1672)

> ...the proper Method for *inquiring* after the properties of things is, to deduce them from Experiments.
>
> > —Isaac Newton (1642-1727)
> > "A Serie's of Quere's propounded by Mr. Isaac Newton, to be determin'd by Experiments, positively and directly concluding his new Theory of Light and Colours," *Phil. Trans., 85,* (July 15, 1672), p. 5004, *Isaac Newton's Papers and Letters on Natural Philosophy,* ed. by I. Cohen, Harvard Univ. Press, Cambridge, MA, 1958, p. 93.

FREUD (1899)

> ...its first structure followed the plan of a reflex apparatus, so that any sensory excitation impinging on it could be promptly discharged along a motor path. But the exigencies of life interfere with this simple function, and it is to them, too, that the apparatus owes the impetus to further development. The exigencies of life confront it first in the form of the major somatic needs. The excitations produced by internal needs seek discharge...An essential component of this experience of satisfaction is a particular perception...the mnemic image of which remains associated thenceforward with the memory trace of the excitation produced by the need.
>
> > —Sigmund Freud (1856-1939)
> > *The Interpretation of Dreams* (1899), ed. and tr. by J. Strachey, Avon Books, Discus Edition, NY, 1965, p. 604.

REFERENCES

Aristotle, *Posterior Analytics*, Book I, Ch. 31, tr. by G. Mure, *The Basic Works of Aristotle*, R. McKeon, ed., Random House, NY, 1941.

Bacon, Francis, *Novum Organum; or True Suggestions for the Interpretation of Nature* (1620), The Colonial Press, NY, 1899.

Baudry, L., *Lexique Philosophique de Guillaume d'Ockham*, Paris, 1958.

Freud, Sigmund, *The Interpretation of Dreams* (1899), tr. and ed. by J. Strachey, Avon Books, Discus Edition, NY, 1965.

Gilbert, William, *DeMagnete* (Amsterdam, 1600), tr. by P. Mottelay, Dover Pubs., Inc., NY, 1958.

Hobbes, Thomas, *Leviathan* (1651), Oxford at the Clarendon Press, London, 1909.

Newton, Isaac, "A Serie's of Quere's propounded by Mr. Isaac Newton, to be determin'd by Experiments, positively and directly concluding his new Theory of Light and Colours," *Phil. Trans, 85,* (July 15, 1672), p. 5004, *Isaac Newton's Papers and Letters on Natural Philosophy*, ed. by I. Cohen, Harvard Univ. Press, Cambridge, MA, 1958.

Ockham, William of, *Summa Logicae*, III, pt. 2, ch. 10, (c. 1327), tr. by E.A. Moody, *The Logic of William of Ockham*, London, 1935.

_____, *Ordination in Librum Primum Sententiarum*, Prol., q. 9, tr. by J.R. Weinberg, *Ockham, Descartes and Hume, Self-knowledge, Substance and Causality*, Univ. of Wisconsin Press, Madison, WI, 1977.

Tolman, Richard C., *The Principles of Statistical Mechanics*, Oxford University Press, London, 1938.

da Vinci, Leonardo, "Paragone," *Codex Urbinas* (1495-1499), 19r-v, excerpted in *The Genius of Leonardo da Vinci*, ed. by Andre Chastel, The Orion Press, NY, 1961.

CHAPTER 7

PROBABILITY, INFORMATION AND ENTROPY

CHAPTER 7

PROBABILITY, INFORMATION AND ENTROPY

A. Probability

Probabilities are estimators of the classification frequencies of yet-to-be-observed outcomes.

For example, when I estimate the frequency of heads on the next 1,000 tosses of a coin to be one-half, I am assigning a value to a probability. This probability is more or less in error depending upon the closeness of the estimate to the actual frequency after it is observed.

When the outcome has not yet occurred, we have a case of *prediction*. When it has already occurred but we have not yet observed it, we have a case of *postdiction*.

It may be useful to impose certain restrictions on probabilities. In seeking reasonable restrictions, we note that:

1. Frequencies are nonnegative numbers.

2. If two outcomes A and B cannot both occur, then the frequency of the logical union "A or B" is equal to the sum of the frequency of A and the frequency of B.

3. The frequency of the logical union of all possible outcomes is unity.

These characteristics are true of all frequencies in definite and unambiguous data sets. (See Chapter 15 for a discussion of other types of data sets.) Since probabilities are frequency estimators, they will generally be less in error if we impose corresponding restrictions on them.

These restrictions are often very helpful, particularly when we are faced with problems of deducing one set of probabilities from another set. However, they do not move us very far along when we are faced with the task of assigning probabilities in the first instance on the basis of observational data.

If we know that there is only a finite number K of outcomes yet to be observed, then the only possible classification frequencies are $0, \frac{1}{K}, \frac{2}{K}, \ldots, \frac{K-1}{K}, 1$. Suppose, for example, we wish to make a prediction about whether or not it will rain tomorrow (for a specific tomorrow). There are only two possibilities:

  1 = rain, and

  0 = no rain.

In this specific case, it would seem foolish to assign the probability of rain a fractional value, say, one-half. We would surely be wrong.

The person involved might be Anne Boleyn, second wife of Henry VIII, and tomorrow may be May 19, 1536. She wants to know whether she should carry an umbrella to her decapitation. She would be totally unimpressed with "long-run" arguments. And it would be beneath her dignity to sheer off half of her umbrella, or to flop it up and down so it covered her during half of the journey from her room in the Tower of London to the executioner's block in the courtyard below. Such may be the behavior of bankers and gamblers wagering fractions of their net worth in the market place and on the gaming tables, but not of the singular Anne.

Knowing (somehow) that the probability is (in some sense) one-half, we might secretly toss a coin, and then give her a definitive prediction based on its outcome.

We might even convince ourselves that our behavior in making predictions this way would be natural. Quantum mechanical events, underlying the behavior or matter, seem to carry with them such secret probabilities (in the form of wave functions), but exhibit only a definitive state selected from those over which the probabilities are distributed.

This may be a proper way of making predictions for ourselves.   However, it is demonstrably not a proper way to provide predictive information to others.

One difficulty is coherency.   Suppose the weather bureau issued all rain forecasts this way.   Then, on some days, everybody would set off on rainy day activities; and on other days, nobody would.

Unwarranted coherence is not the only difficulty.   We would also fault the weather bureau for misrepresenting the amount of its information.   It should issue only what information it has, and let the individual people make their own decisions.   To do otherwise may bias decision-making.   For example, Anne may not wish to take an umbrella unless the probability of rain is at least three-quarters.

Thus, if we properly advise Anne, she will have to toss her own coin.   In this sense, we are merely providing information; *she* is making the prediction.

What is the relation of probabilities to predictions?   Probabilities are frequency estimates that pertain to an imagined indefinite repetition of the event as defined.   Consider an indefinite series of rolls of a particular die.   We may, for example, take the following estimates for the outcome probabilities:

1:	1/6	4:	1/6
2:	1/6	5:	1/6
3:	1/6	6:	1/6

When we collapse to a particular set of estimates for a specific set of say 100, future rolls, we obtain a prediction.   See Fig. 13. For example:

1:	16/100	4:	15/100
2:	17/100	6:	15/100
3:	18/100	7:	19/100

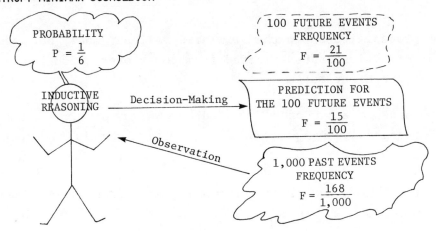

Fig. 13. Frequency, Observation, Probability, Prediction,
Frequency, Observation, Etc.

The procedure of collapsing from probabilities to predictions is a
physical process. It is accomplished by operating a physical model
of the probabilities such as another die we keep handy for this
purpose, or a computerized pseudo-random number generator, or a
scintillation counter/timer monitoring a radioactive source.
Someday, science may establish an international standard for prediction as we already have for mass, length and time. In the meantime, one of these will do.

B. Probabilities of Logical Combinations

Imagine that we observe two events, E_1 and E_2, with outcomes O_1
and O_2. O_1 and O_2 may be the same or they may be different.
Imagine, further, that we have assigned probabilities to these two
outcomes:

$P(O_1)$ = probability of outcome O_1 occurring
$P(O_2)$ = probability of outcome O_2 occurring

We now wish to be able to make probability statements about various
logical combinations of these outcomes. For example:

"O_1 and O_2"   = both O_1 and O_2 occur
"O_1 or O_2"   = either O_1, or O_2, or both occur
"O_1 exor O_2"   = either O_1, or O_2 occurs, but not both
"O_1 nor O_2"   = neither O_1 nor O_2 occurs

"O_1 or not O_2"   = either O_1 occurs, or O_2 does not occur
"O_1 but not O_2" = O_1 occurs and O_2 does not occur

The rules for forming the probabilities of these combined outcomes are available from elementary measure theory:

$$P(O_1 \text{ or } O_2) = P(O_1) + P(O_2) - P(O_1 \text{ and } O_2)$$
$$P(O_1 \text{ exor } O_2) = P(O_1) + P(O_2) - 2P(O_1 \text{ and } O_2)$$
$$P(O_1 \text{ nor } O_2) = 1 - P(O_1) - P(O_2) + P(O_1 \text{ and } O_2)$$
$$P(O_1 \text{ or not } O_2) = 1 - P(O_2) + P(O_1 \text{ and } O_2)$$
$$P(O_1 \text{ but not } O_2) = P(O_1) - P(O_1 \text{ and } O_2)$$

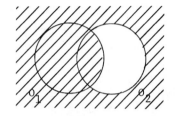   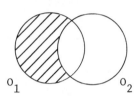

(a) Venn Diagram for "O_1 or not O_2"   (b) Venn Diagram for "O_1 but not O_2"

Fig. 14.   Illustrative Venn Diagrams

Venn diagrams are very helpful graphical analogs of these relations. Consider, for example, the diagram for "O_1 or not O_2" in Fig. 14(a). The possibilities permitted are shown as the cross-hatched region. The excluded (unshaded) area represents the occurrence of O_2 without O_1. The whole space has probability of unity. From this we subtract the probability of O_2. Then we add back the overlap portion, the probability of "O_1 and O_2". Fig. 14(b) gives, as another example, the diagram for "O_1 but not O_2".

There are four distinct regions in a general two-proposition Venn diagram such as those in the examples shown above. Different logical combinations of the two propositions are represented by shading different subsets of these regions. There are, in total, $4^2 = 16$ different combinations, including the proposition-independent combinations of shading the whole space (tautology) and of shading no region (contradiction).

C. Probabilities Subject to Logical Relations

Consider the Venn diagram for the single proposition "O" shown in Fig. 15.

Fig. 15. Venn Diagram for O

Consider also the second proposition "C". This second proposition may bear any of a number of different relations to the original proposition. Five possibilities are shown in Fig. 16.

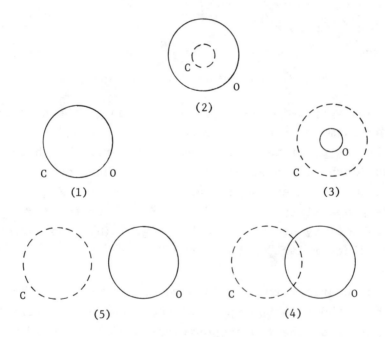

Fig. 16. Venn Diagrams for Illustrative Logical Relations between Propositions "O" and "C"

In the first case, O and C are equivalent. In the second, C is enclosed by O. In the third, O is enclosed by C. In the fourth case they overlap. In the fifth they are disjoint.

Logical combinations are formal abstractions. Logical relations, on the other hand, make a substantive assertion about the relationship between the propositions.

A relation can be specified by asserting that a logical combination is false. Thus, just as there are 16 possible logical combinations of O and C, there are also 16 possible relations between them. For example, the relation "O disjoins C" is specified by asserting that "O and C" is false. The relation "C implies O" is specified by asserting that "C but not O" is false.

Suppose we consider case #2 where C implies O. In this case, the probability of O given C is unity.

If C, and C implies O, then $P(O) = 1$.

Suppose we consider case #5 where O and C are disjoint. Then the probability of O given C is zero.

If C, and C disjoins O, then $P(O) = 0$.

However, in cases #3 and #4, we have an intermediate situation. Roughly speaking:

If C, then $P(O)$ = fraction of C which is enclosed by O.

We use conditional probabilities to express this general case more precisely.

D.   Conditional Probabilities

The *conditional probability* of an outcome O *given* a condition C is defined as the ratio of the probability of the joint occurrence of
O and C to the probability of C.   In logical notation, this is
written:*

$$\text{If } C, \text{ then } P(O) = \frac{P(O \text{ and } C)}{P(C)} .$$

In probability notation, it is shortened to read:**

$$P(O \text{ given } C) = \frac{P(O \text{ and } C)}{P(C)} .$$

It is important to distinguish between "O given C" and "O and C".
In both cases, the reference is to a yet-to-be-observed state in which
outcome O occurs under condition C.   The difference lies in our current knowledge about condition C.   "O given C" assumes that we know
already that condition C will be satisfied.   Thus P(O given C) reflects uncertainty only in the outcome O.   "O and C", on the other
hand, assumes that we are uncertain about whether condition C will
be satisfied.   So P(O and C) reflects uncertainty both in O and
in C.

Another important distinction is between P(O given C) and
P(O or not C).   The probability that

    "O or not C" is true

is the same as the probability that

    C implies O.

---

* P(C) in this formula is the probability of C prior to the supposition "If
C".   Similarly for P(O and C).

** Common mathematical notations for P(O given C) and P(O and C) are $P(O|C)$ and
$P(O \wedge C)$, respectively, although the latter is sometimes written P(O,C).   It requires very little extra space to explicitly insert the words "given", "and",
etc.   So a small inconvenience to the specialist reader has been traded for
what is hoped to be greater convenience to the general reader.

It is given by

P(O or not C) = 1 - P(C) + P(O and C).

As inspection of the Venn diagram in Fig. 14(a) shows, "O or not C" does not necessarily imply C. So this statement is also distinguishable from "O given C".

A third important point is that the notation of conditional probabilities is not restricted to situations where C may *cause* O. It also includes non-causal situations, so long as the probability of O may be different when C occurs than when it does not, e.g., when they both depend upon a third event.

Outcome O is said to be *independent* of condition C when and only when the probability of its occurrence is the same whether or not we specify C as being satisfied.*   Thus**

Iff O is independent of C, then P(O given C) = P(O).

Replacing P(O given C) by its definition, we have:

Iff O is independent of C, then $\dfrac{P(O \text{ and } C)}{P(C)}$ = P(O).

Multiplying by P(C) we obtain:

Iff O is independent of C, then P(O and C)   = P(O)P(C).

Note that the last condition is symmetric under interchange of O and C.   Hence, if O is independent of C, then C is independent of O.

This shows that when O and C are independent, the probability of their joint occurrence is the product of their separate probabilities.   Note that this is true only if they are independent.   In general, they are dependent, and we can only say

P(O and C) = P(O given C) P(C).

---

* Note that just as dependence does not necessarily imply causality, independence does not necessarily imply non-causality.   C may cause O and yet O may be independent of C by there being another cause C' to bring O about in the absence of C.

** "Iff" is read "if and only if".

E.   Bayes' Inverse Probability Theorem

The inverse probability theorem was first proved for a special
case by Bayes in 1763 and later generalized by Laplace.   It states
that the probability for the kth outcome O_k given the condition C
can be expressed as

$$P(O_k \text{ given } C) = \frac{P(O_k) \; P(C \text{ given } O_k)}{\sum_j P(O_j) \; P(C \text{ given } O_j)} \; .$$

Thus, if we know the unconditional probabilities $P(O_k)$ of all
possible outcomes $\{O_k\}$, and each conditional probability of C
given outcome O_k, then we can compute the inverse conditional
probability.

This theorem has had a rather checkered history, though not because
there is any question about its mathematical soundness.   If one
assumes that all of the probabilities $\{P(O_j), P(C \text{ given } O_j)\}$ are
fixed and based on the same evidence,* then Bayes' theorem is a
deductive consequence.   The differences of opinion concern matters
of the applicability of the theorem to real-world situations.   We
will return to this problem in later chapters.

F.   Information

Information, as we define it, consists of statements as to the
values of probabilities.   Two aspects of an informational statement are its domain and its quantity.   A key goal of entropy minimax analysis is to determine the domain as well as the quantity of
information in a given data set.

---

* The latter assumption is usually phrased in terms of "consistency" as defined
by certain axioms, which are not necessarily satisfied for sets of probabilities
based on variable evidence.

By the informational *domain* of a statement we mean its defined scope, which determines the set of events to which it applies. We will take up this matter later (Part III of this volume). For the present, we limit our discussion to the *quantity* of information.

In 1878 Peirce defined the *intensity of a belief* as

$$J(A) = \log \left(\frac{P(A)}{1 - P(A)} \right) .$$

where P(A) is the believed probability of truth of statement A. He chose this definition for three reasons:

o   It depends only on the probability (hence is value-free).

o   It monotonically rises from $-\infty$ to $+\infty$ as P goes from 0 to 1, passing through J = 0 at the equi-uncertainty point P = 0.5.

o   It makes the intensities of beliefs corresponding to independent chances additive:*

$$J(A \text{ and } B) = J(A) + J(B).$$

In 1925 Fisher used the term "information" to refer to a measure of the amount of support a sample gives to an hypothesized value for a parameter in a hypothesized probability distribution. He specifically referred to a measure of how sharply curved the likelihood function is at its peak. The greater the number and closeness (in parameter space) of the data points, the more sharply peaked is the function.

---

* To achieve additivity for independent statements, Peirce defined the joint "chance" associated with two statements A and B as

$$C(A \text{ and } B) = \frac{P((A \text{ and } B) \text{ given } ((A \text{ and } B) \text{ or } (\overline{A} \text{ and } \overline{B})))}{1 - P((A \text{ and } B) \text{ given } ((A \text{ and } B) \text{ or } (\overline{A} \text{ and } \overline{B})))} ,$$

and the joint intensity as

$$J(A \text{ and } B) = \log (C(A \text{ and } B)).$$

The rationale for this definition was a focus on cases in which the two statements agree and hence are either both true or both false.

The definition of information now generally accepted is simpler
than either of these. It was developed in the theory of telegraph
communications. In 1924 Nyquist showed that the speed W of transmission of intelligence over a telegraph circuit with a fixed line
speed (i.e., a fixed rate of sending signal elements) is proportional to the logarithm of a number m of current values used to
encode the message:*

$$W = k \log m.$$

In 1928 Hartley generalized this to all forms of communication,
letting m represent the number of symbols available at each selection of a symbol to be transmitted. He pointed out that the measure of information magnitude should be independent of psychological
value judgment factors, that information becomes "more precise" as
more symbol sequence possibilities are eliminated, and that the
logarithm function is unique for its sole dependence on numbers of
selection options. He further indicated an equiprobability presumption in the definition when he considered the information content of "secondary" characteristics made up of groups of "primary"
symbols, namely the presumption that long characters are not selected more often than their average rate of occurrence.

Shannon adopted Hartley's definition in 1948 as the basis for his
mathematical theory of communication, in which entropy plays a key
role as the expectation value of this measure of information. He
formally incorporated the statistical aspect by defining the quantity of information as proportional to the negative of the logarithm
of probability

$$I = -k \log P,$$

where P = 1/m in the equiprobability case assumed by Nyquist and
Hartley.

---

* The same year, 1924, Küpfmüller independently derived the width of the frequency band for transmission of information and its dependence in the exactness
of the signal transmitted. However, he did not state the logarithm expression
for information.

This definition has the following features:

- o It depends only on the probability P.
- o It monotonically decreases from +∞ to 0 as P goes from 0 to 1.
- o It makes the information quantities of independent statements additive:

    I(A and B) = I(A) + I(B).

Unlike Peirce's intensity of belief, the communication-theoretic definition of information does not assign any particular significance to the value P = 0.5 (which is the equiprobability value only when there are exactly two possibilities).

Because the quantity of information of a statement, using this definition, depends only upon the probability which it asserts, it is sometimes referred to as the *self-information* of the statement.

The property of additivity of the information of independent statements is a result of the following two facts:

- o If two events are independent, then the probability that they will both occur is equal to the product of the probability that one will occur times the probability that the other will occur.

    P(A and B) = P(A) · P(B)
- o The logarithm of the product of two numbers is the sum of the logarithms of the two numbers separately.

    log X·Y = log X + log Y

Combining these, we see that the information in the joint occurrence of two independent events is the sum of the separate quantities of information in the occurrence of the two events.*

    -log P(A and B) = -log P(A) - log P(B)

    I(A and B) = I(A) + I(B)

---

* For notational convenience, we suppress the constant k which fixes the units of information. By using natural logarithms (to the base e = 2.7182818...), we express information in terms of "nats".

The unique feature which distinguishes the Nyquist-Hartley-Shannon definition from Peirce's is the surprisal property. Whereas Pierce defined a measure of the intensity of a belief, the communication theorists defined a measure of the amount of information one would gain if one were to receive a message. Both definitions reference the same probability, but from two different viewpoints.

This same feature distinguishes Fisher's information measure from that of the communication theorists. Fisher's definition measures the state of an existing data sample relative to an hypothesis. The communications theory information measure compares the content of data we receive to our prior state of expectation.

The higher was our prior estimate of the probability for an outcome to occur, the lower will be the information we gain by observing it to occur. At one extreme, were we sure an outcome would occur, then its occurrence would give us no information at all. At the other extreme, were we certain it would not occur, then its occurrence would give us infinite information.

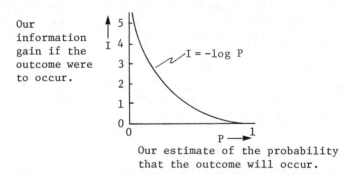

Fig. 17. Future Information versus Present Probability

The information versus probability relationship also holds with past and present translated to present and future. This is illustrated in Fig. 17. It is this application of the concept of information that will prove particularly useful to us for predictive purposes.

G.   Information of Logical Combinations

The definition of information may be applied to compound as well as simple propositions. Thus, just as we may form probabilities of logical combinations, we may also define the corresponding information associated with logical combinations.

For example:

$$I(O_1 \text{ and } O_2) = -\log P(O_1 \text{ and } O_2)$$
$$I(O_1 \text{ or } O_2) = -\log P(O_1 \text{ or } O_2)$$
$$= -\log [P(O_1) + P(O_2) - P(O_1 \text{ and } O_2)]$$

H.   Conditional Information

In Section 7.D we defined conditional probabilities as follows:

P(O given C) = probability of observing outcome O
               given that condition C is satisfied.

In a similar manner, we also define conditional information:

I(O given C) = -log P(O given C)
             = information gained by observing outcome O
               given that condition C is satisfied.

I.   Information Exchange

We now turn to a topic important to the process of communication. This is the concept of information exchange.

Suppose that we compare the unconditional information

$$I(O) = -\log P(O),$$

with the conditional information

$$I(O \text{ given } C) = -\log P(O \text{ given } C).$$

The difference between these is a measure of the extent to which knowledge of condition C affects the information we would gain by observing outcome O. We call this the mutual information exchange between O and C.

$$\Delta I(O \text{ mutual } C) = I(O) - I(O \text{ given } C)$$

$$= \log \frac{P(O \text{ given } C)}{P(O)}.$$

Like Peirce's intensity of belief, mutual information exchange is the logarithm of the ratio of two probabilities. Although it is sometimes referred to simply as "mutual information", the mutual information exchange is not itself information in the surprisal sense. There is no probability of which ΔI is the negative logarithm. Further, unlike information, ΔI can be either positive or negative. For example, if O = not-C, then $\Delta I(O \text{ mutual } C) = -\infty$.

Because of the multiplication theorem

$$P(O \text{ and } C) = P(O \text{ given } C) \, P(C)$$

$$= P(C \text{ given } O) \, P(O),$$

there is complete symmetry between O and C as far as information exchange is concerned.

$$\Delta I(O \text{ mutual } C) = \Delta I(C \text{ mutual } O).$$

Thus, it is completely irrelevant which we take as the condition and which as the outcome. This is, of course, not the case with conditional information.

Because the probability of the joint occurrence of O and C is given by

$$P(O \text{ and } C) = P(O \text{ given } C) \, P(C),$$

we can express the mutual information exchange as

$$\Delta I(O \text{ mutual } C) = \log \frac{P(O \text{ and } C)}{P(O) \, P(C)}.$$

Mutual information exchange is a measure of the interdependence of O and C. When O is independent of C, for example, $\Delta I(O \text{ mutual } C) = 0$.

Positive ΔI reflects positive association of O and C, negative ΔI reflects negative association. See Fig. 18.

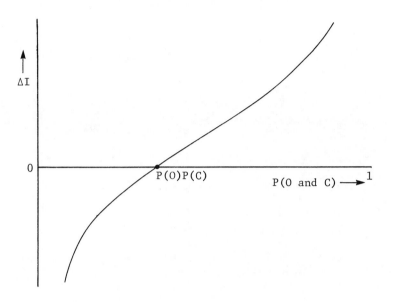

Fig. 18.   Mutual Information Exchange Goes from Minus
           Infinity to Positive Infinity, Passing through
           Zero When O and C Are Independent

If there are two conditions C_1 and C_2, the mutual information exchange for the triplet O, C_1 and C_2 is

$$\Delta I(O \text{ mutual } C_1 \text{ mutual } C_2) = \log \frac{P(O \text{ and } C_1)P(O \text{ and } C_2)P(C_1 \text{ and } C_2)}{P(O)P(C_1)P(C_2)P(O \text{ and } C_1 \text{ and } C_2)} .$$

As McGill showed in 1954, this concept can be extended to any number of conditions.

J. Entropy

To illustrate the concept of entropy, let us list all the possible outcomes in a specific classification scheme for a dependent variable, and list also the probabilities that we assign to each outcome. Each probability has an associated information. We may, as

an example, reformat our earlier Tables 13-16 by estimating probabilities from the data and listing also the associated information
values, giving, as a result, Tables 17-20.*

<table>
<tr><td colspan="3">Table 17</td><td colspan="3">Table 18</td></tr>
<tr><td>Precip.</td><td>Prob.</td><td>Info.</td><td>Price Change</td><td>Prob.</td><td>Info.</td></tr>
<tr><td>80-99</td><td>0.024</td><td>3.730</td><td>above +30</td><td>0.051</td><td>2.976</td></tr>
<tr><td>60-79</td><td>.056</td><td>2.882</td><td>+10 to +30</td><td>.229</td><td>1.474</td></tr>
<tr><td>40-59</td><td>.460</td><td>.777</td><td>-10 to +10</td><td>.463</td><td>.770</td></tr>
<tr><td>10-39</td><td>.428</td><td>.849</td><td>-30 to -10</td><td>.198</td><td>1.619</td></tr>
<tr><td>0-19</td><td>.032</td><td>3.442</td><td>below -30</td><td>.059</td><td>2.830</td></tr>
<tr><td></td><td>1.000</td><td></td><td></td><td>1.000</td><td></td></tr>
</table>

<table>
<tr><td colspan="3">Table 19</td><td colspan="3">Table 20</td></tr>
<tr><td>Fail. Frac.</td><td>Prob.</td><td>Info.</td><td>Cause of Death</td><td>Prob.</td><td>Info.</td></tr>
<tr><td>0.80-1.00</td><td>0.092</td><td>1.286</td><td>Cardiovascular</td><td>0.500</td><td>0.693</td></tr>
<tr><td>.60- .79</td><td>.022</td><td>3.817</td><td>Cancer</td><td>.206</td><td>1.580</td></tr>
<tr><td>.40- .59</td><td>.071</td><td>2.645</td><td>Other disease</td><td>.211</td><td>1.556</td></tr>
<tr><td>.20- .39</td><td>.168</td><td>1.784</td><td>Accident</td><td>.056</td><td>2.882</td></tr>
<tr><td>0- .19</td><td>.647</td><td>.435</td><td>Other</td><td>.027</td><td>3.612</td></tr>
<tr><td></td><td>1.000</td><td></td><td></td><td>1.000</td><td></td></tr>
</table>

Suppose now, that we contemplate making a new observation in each
of these cases. (We have to conjure up special circumstances for
observing the cause of one's own death in the fourth case, but
assume we overcome, somehow, this difficulty.) Prior to making
the observation we ask ourselves: How much information *do* I expect
that I *would* learn from the observation? This question may simply
be one of idle curiosity, or may serve the practical end of helping us to decide whether to bother with the observation at all.

Suppose, further, that the *only* information we have in each case is
that given in Tables 17-20, and that we do not contemplate controlling or influencing the outcome ourselves based on the result of
this analysis. (Informational feedback analyses are more complex
than the simple examples we are considering here.)

---

* For purposes of these examples, we have used sample average estimates of the
probabilities. In subsequent chapters we will explore other estimates.

These are typical examples of problems involving computation of the *expected value* of a *random variable*. The *variable* here is the information, the third column in Tables 17-20. It is *random* in the sense that its value is not known, but rather can be any one of a number of possibilities (here depending upon which outcome happens to occur). Finally, we have the probability of the occurrence of each value, column #2, so we can compute the contribution of each outcome possibility to its expected value. (See Tables 21-24.) The fourth column contains the contribution of each outcome to the expected value of the information, and is simply the product of the information multiplied by the probability (Fig. 19). The total is the expected value of the information, and is referred to as the *entropy*.

Table 21

Precip.	Prob.	Info.	P·I
88-99	0.024	3.730	0.090
60-79	.056	2.882	.161
40-59	.460	.777	.357
20-39	.428	.849	.363
0-19	.032	3.442	.110

Entropy = 1.081

Table 22

Price Change	Prob.	Info.	P·I
above +30	0.051	2.976	0.152
+10 to +30	.229	1.474	.338
−10 to +10	.463	.770	.357
−30 to −10	.198	1.619	.321
below −30	.059	2.830	.167

Entropy = 1.335

Table 23

Fail. Frac.	Prob.	Info.	P·I
0.80-1.00	0.092	2.386	0.219
.60-.79	.022	3.817	.084
.40-.59	.071	2.645	.188
.20-.39	.168	1.784	.300
0-.19	.647	.435	.282

Entropy = 1.073

Table 24

Cause of Death	Prob.	Info.	P·I
Cardiovascular	0.500	0.693	0.347
Cancer	.206	1.580	.325
Other disease	.211	1.556	.328
Accident	.056	2.882	.161
Other	.027	3.612	.098

Entropy = 1.259

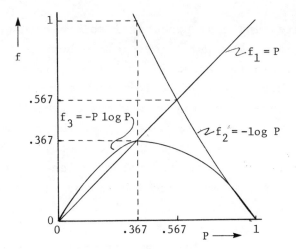

Fig. 19.   Curves Showing Probability, Information, and
the Product of Probability and Information

"Entropie" was a German word coined in 1865 by Clausius to represent a measure of the capacity for change of matter, an idea dating
at least back to Carnot.   Clausius formed his new word by replacing
the last portion of the German "energie" with the Greek "τροπή"
meaning "turning" or "change".*   It was given its English spelling
(paralleling the English "energy") in 1868 by Tait who, along with
Maxwell, originally tried to reverse its definition (to what we
might now call "negentropy"), but who both later returned to using
it with Clausius's original meaning.   Its use to describe physical
systems in thermostatics and thermodynamics was subsequently given
a statistical mechanics interpretation.

Consider, for example, a 3×3×3 meter room filled with air consisting of $6×10^{26}$ nitrogen molecules, $2×10^{26}$ oxygen molecules, and
$7×10^{24}$ other particles.   These molecules interact with each other
and with the walls of the room in a manner described by the laws

---

* "Entropie" was among the "impure" words expunged from the German language during Hitler's Third Reich, being replaced by "Wärmegewicht".

of physics.   Now divide the room into an imaginary grid of twenty-
seven 1×1×1 meter cubes, each labeled with its own serial number,
so that we can specify in which cube each individual molecule is
located at any given instant.   See Fig. 20.

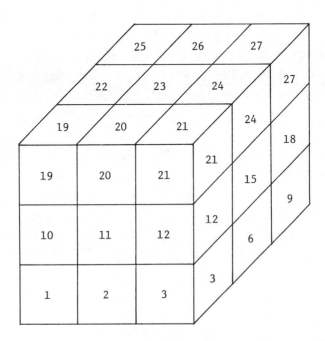

Fig. 20.   Cubic-Shaped Room Divided into 27 Small Cubes for Purposes of
Partially Localizing the Positions of Molecules in the Air

At any given instant, each molecule's past history of motion has
twenty-seven possible outcomes corresponding to its possible presence in each of the twenty-seven imagined cubes.   If we can assign
probabilities to these outcomes, we can then compute the positional entropy of the molecule.

For a dilute gas, it is reasonable to assume that the motions of the
individual molecules are independent.   This enables us to obtain the
total entropy of the air in the room by summing all the individual
molecule entropies.

Even in this highly simplified illustrative situation we can make an important observation about entropy. It is a measure of the *disorder* of the system.

The most ordered possible states are those for which the probability is unity for one of the cubes and zero for the others. Such a state is maximally ordered in two respects. First, our state of knowledge about the positions of the molecules is minimally *confused*, since we know definitely in which cube each molecule resides. Second, the molecules are perfectly *sorted* by the boundaries of these cubes.

In this state, the entropy of the system is zero. See Table 25.

Table 25.   Zero Positional Entropy State
For Room Full of Air Molecules

Cube	Prob.	Info.	P·I
# 1	1	0	0
# 2	0	∞	0
# 3	0	∞	0
'	'	'	'
'	'	'	'
'	'	'	'
#25	0	∞	0
#26	0	∞	0
#27	0	∞	0

Entropy = 0

In this example, the probability of an air particle being in cube
#1 is unity, i.e., it is certainly in this cube. The probability
is zero for the other cubes. The information gained is zero if it
is observed to be in cube #1. It is infinite if observed to be in
one of the other cubes. In each case the product of the probability and the information gain is zero. (Note that the logarithmic
infinity of the information is weaker than the zero of the probability.)

The most disordered states possible are those for which all twenty-
seven outcome probabilities are equal. We are maximally confused
about where the molecules are located. Our cubes have done the
poorest possible job sorting them.

The entropy of this state is computed in Table 26 (to 3-digit
accuracy).

Table 26.   Example of Maximum Entropy State

Cube	Prob.	Info.	P·I
# 1	0.037	3.296	0.122
# 2	0.037	3.296	0.122
# 3	0.037	3.296	0.122
'	'	'	'
'	'	'	'
'	'	'	'
#25	0.037	3.296	0.122
#26	0.037	3.296	0.122
#27	0.037	3.296	0.122

Entropy = 3.296

It is a straightforward mathematical result that the highest possible entropy occurs when all the probabilities are equal. Graphic
illustrations of this fact are shown for the cases of two and three
outcomes in Fig. 21(a) and 21(b).

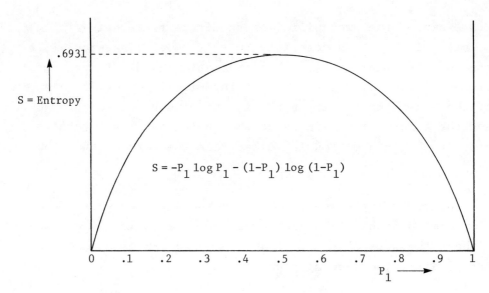

Fig. 21(a).   Entropy Curve for Two Outcome Classes
(Note: $P_2 = 1-P_1$).   The peak of $S = 0.693$
is at $P_1 = P_2 = 1/2$.

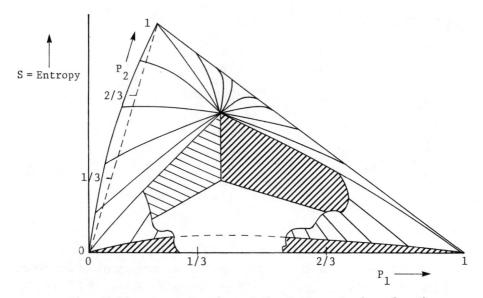

Fig. 21(b).   Cut-Away View of the Entropy Surface for Three
Outcome Classes (Note: $P_3 = 1 - P_1 - P_2$).   The peak
of $S = 1.099$ is at $P_1 = P_2 = P_3 = 1/3$.   The entropy is
zero when and only when one of the probabilities
is unity and the others are zero.

K. Conditional Entropy

We have seen in previous sections how the concept of conditional probability and conditional information are defined. In this section we introduce the concept of conditional entropy.

Consider two entropies

$$S(C) = - \sum_{i=1}^{L} P(C_i) \log P(C_i),$$

and

$$S(O \text{ and } C) = - \sum_{i=1}^{L} \sum_{k=1}^{K} P(O_k \text{ and } C_i) \log P(O_k \text{ and } C_i).$$

$S(C)$ is the expected value of the information we would gain if we were to observe another event and learn what condition it satisfies. $S(O \text{ and } C)$ is the expected value of the information we would gain if we learn the event outcome as well. Obviously, we expect to learn no less if we observe the outcome in addition to the condition.

$$S(O \text{ and } C) \geq S(C)$$

Thus, if we subtract $S(C)$ from $S(O \text{ given } C)$ we obtain a nonnegative result. This difference is the expected excess information from learning the outcome above that already gained by learning the condition. It is referred to as the *conditional entropy* of outcome O *given* condition C:

$$S(O \text{ given } C) = S(O \text{ and } C) - S(C).$$

Because the joint entropy, $S(O \text{ and } C)$, is symmetric in O and C, it has two equivalent decompositions:

$$S(C \text{ given } O) + S(O) = S(O \text{ given } C) + S(C).$$

In subsequent analyses we use the following additional definitions
drawn from communications theory:

$S(O)$ = average transmitted information (about the outcome

$S(C)$ = average received information (about the conditions)

$S(C$ given $O)$ = noise (uncertainty that C is received given that O is sent)

$S(O$ given $C)$ = equivocation (doubtfulness that O was sent given that C was received)

The conditional entropy of O given C is related to the probabilities as follows:

$$S(O \text{ given } C) = - \sum_{i=1}^{L} \sum_{k=1}^{K} P(O_k \text{ and } C_i) \log P(O_k \text{ and } C_i)$$

$$+ \sum_{i=1}^{L} P(C_i) \log P(C_i).$$

From the definition of conditional probability we have

$$P(O_k \text{ and } C_i) = P(O_k \text{ given } C_i) P(C_i).$$

Therefore,

$$S(O \text{ given } C) = - \sum_{i=1}^{L} P(C_i) \sum_{k=1}^{K} P(O_k \text{ given } C_i) \log P(O_k \text{ given } C_i)$$

$$- \sum_{i=1}^{L} P(C_i) \log P(C_i) \sum_{k=1}^{K} P(O_k \text{ given } C_i)$$

$$+ \sum_{i=1}^{L} P(C_i) \log P(C_i).$$

Using the property that probabilities must sum to unity (for a definite unambiguous partition),

$$\sum_{k=1}^{K} P(O_k \text{ given } C_i) = 1 \quad \text{for all } i=1,\ldots,L,$$

we conclude

$$S(O \text{ given } C) = -\sum_{i=1}^{L} P(C_i) \sum_{k=1}^{K} P(O_k \text{ given } C_i) \log P(O_k \text{ given } C_i).$$

This formula will be used in Chapter 12 in the minimization aspect of entropy minimax.

L. Entropy Exchange

Another inequality of general usefulness is

$$S(O \text{ given } C) \leq S(O).$$

This relation states that knowledge of condition C can only reduce or leave unchanged the information we expect to gain by observation of outcome O. The amount of the reduction is the expected differential value of learning the outcome O while not knowing condition C, above the value of learning the outcome while knowing the condition. This quantity is referred to as the mutual entropy exchange between O and C.

$$\Delta S(O \text{ mutual } C) = S(O) - S(O \text{ given } C)$$

$$= S(O) + S(C) - S(O \text{ and } C).$$

Note that entropy exchange, like information exchange, is symmetric, that is

$$\Delta S(O \text{ mutual } C) = \Delta S(C \text{ mutual } O).$$

Therefore

$$\Delta S(O \text{ mutual } C) = S(C) - S(C \text{ given } O).$$

Using the previously obtained expressions for $S(O)$ and $S(O \text{ given } C)$, and the partition formula

$$P(O_k) = \sum_{i=1}^{L} P(O_k \text{ given } C_i) \, P(C_i),$$

it follows that

$$\Delta S(O \text{ mutual } C) = \sum_{i=1}^{L} P(C_i) \sum_{k=1}^{K} P(O_k \text{ given } C_i) \log \frac{P(O_k \text{ given } C_i)}{P(O_k)} .$$

$\Delta S(O \text{ mutual } C)$ is the average information exchange between O and C, i.e., the expected value of the amount of information each gives *about* the other. Suppose O represents a message transmitted through a noisy channel, and C represents the received message.   Then $\Delta S(O \text{ mutual } C)$ represents the average amount of information provided by the received signal C about the message O. It is equal to the average amount of information required to specify the message minus the amount still required to do so after reception of the signal.

Since entropy exchange is symmetric, $\Delta S(O \text{ mutual } C)$ is also equal to the information which knowledge of the message at the transmitter end of the noisy channel would provide about what the signal will be at the reception end. See Fig. 22.

How does the foregoing mathematical development relate to the world of empirical predictions? In this world, the outcome O represents the "reality" we wish to predict. The condition C represents the information available to us from our experience base. We adopt the view that reality is communicated to us through an imperfect system, namely our limited sensory receptors for collecting data and our limited conceptual apparatus for processing these data. Information lost by these limitations results in equivocation when predicting the future and in reference to random noise when explaining past observations.   In summary:

        Physical picture (events, observations, ideas):
           Message + Noise = Data = Signal + Equivocation
        Conceptual picture (future, present, past):
           Message - Equivocation = Knowledge = Signal - Noise
        Analysis: Knowledge = Data - Noise - Equivocation

One of the key objectives of entropy minimax analysis is to minimize the equivocation.   This will be the subject of Chapter 12. However, we must first clear up the matter of how to compute the probabilities.   This is the subject of the next chapter.

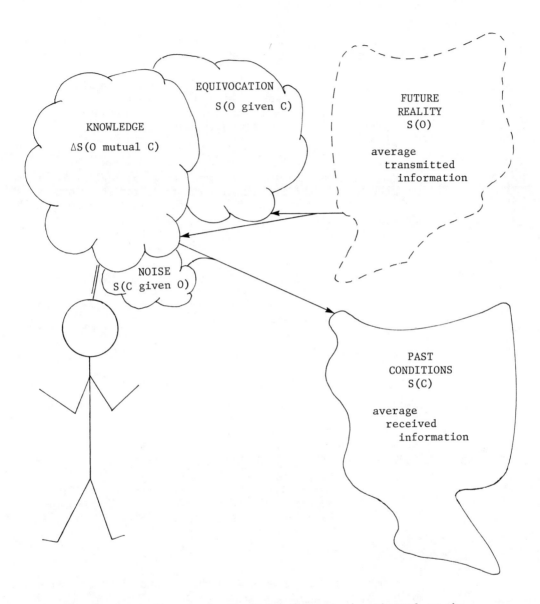

Fig. 22. The Observer is the Noisy Transmission Channel as the
Future Communicates with the Past

QUOTES

CARDANO (c. 1564)

I am as able to throw 1, 3 or 5 as 2, 4 or 6. The wagers are therefore laid in accordance with this equality if the die is honest, and if not, they are made so much the larger or smaller in proportion to the departure from true equality.

* * * *

In three casts of two dice the number of times that at least one ace will turn up three times in a row falls far short of the whole circuit, but its turning up twice differs from equality by about 1/12. *The argument is based upon the fact that such a succession is in conformity with a series of trials and would be inaccurate apart from such a series.*

> —Girolamo Cardano (1501-1576)
> *De Ludo Aleae* (c. 1564), in *Opera Omnia, Vol. I*, Amsterdam, 1663, tr. by Gould, quoted by I. Hacking, *The Emergence of Probability*, Camb. Univ. Press, Camb., 1975, p. 54.

GALILEO (c. 1632)

...9 and 12 can be made up in as many ways as 10 and 11...

...it is known from long observation that dice players consider 10 and 11 to be more advantageous than 9 and 12...a very simple explanation, namely that some numbers are more easily and more frequently made than others, which depends on their being able to be made up with more variety of numbers.

> — Galilei Galileo (1564-1642)
> memorandum on tossing three dice (c. 1632), quoted by I. Hacking, *The Emergence of Probability*, Camb. Univ. Press, Camb., 1975, p. 52.

HOBBES (1650)

...though a man have always seen the day and night to follow one another hitherto, yet can he not hence conclude they shall do so, or that they have done so eternally: *experience concludeth nothing universally.* If the signs hit twenty times for one missing, a man may lay a wager of twenty to one of the event but may not conclude it for a truth.

> —Thomas Hobbes (1588-1679)
> *Humane Nature, or the fundamental Elements of Policie* (1650), quoted by I. Hacking, *The Emergence of Probability*, Camb. Univ. Press, Camb., 1975, p. 48.

LEIBNIZ (1678)

...probability is degree of possibility...

...equally easy, that is to say, equally possible...

> —Gottfried Wilhelm Leibniz (1646-1716)
> *De incerti aestimatione* (Sept. 1678), excerpts tr.
> in I. Hacking, *The Emergence of Probability*, Camb.
> Univ. Press, London, 1975, p. 127.

BERNOULLI (1713)

...since this type of calculation is quite uncommon, and there
is frequent occasion for employing it, I shall here succinctly
expound its principle or method...

The basic principle of my method is this: in games of hazard,
the chance or the expectation of each player in gaining any
point must be estimated in the degree to which he can ultimately
realize that same chance or expectation when playing with equal
stakes.

PROPOSITION I

If I expect a or b and one may happen as easily as the
other, my expectation may be said to have the value of $\frac{a+b}{2}$.

$$\star \quad \star \quad \star \quad \star$$

COROLLARY

Hence it is evident that, if something or a is hidden in
one hand, and nothing in the other, the expectation of
each player separately is half of that something, or ½ a.

...the value of the expectation always signifies something
intermediary between the best that we hope and the worst
that we fear.

$$\star \quad \star \quad \star \quad \star$$

COROLLARY

It is evident also, if in one or several boxes there is
nothing hidden, that my expectation will similarly be a
third of what is contained in the other boxes: or a
fourth, if there are four, or a fifth, if there are five:
and so on.

> —Jacob Bernoulli (1654-1705)
> *Ars Conjectandi* (posthumous, Basel, 1713), tr. by
> H.E. Wedeck, in *Classics in Logic*, ed. by
> D.D. Runes, Philosophical Library, Inc., NY,
> 1962, pp. 126, 128, 130.

DE MOIVRE (1738)

...if we constitute a Fraction whereof the Numerator be the
number of Chances whereby an Event may happen, and the Denominator the number of all the Chances whereby it may either happen
or fail, that Fraction will be a proper designation of the
Probability of happening.

> —Abraham De Moivre (1667-1754)
> *The Doctrine of Chances* (1738), reprinted by
> Chelsea Pub. Co., NY, 1967, p. 1-2.

BAYES (1763)

5. The *probability of any event* is the ratio between the value
at which an expectation depending on the happening of the event
ought to be computed, and the value of the thing expected upon
it's happening.

6. By *chance* I mean the same as probability.

*   *   *   *

If a figure be described upon any base *AH* (Vid. Fig.) having for
it's equation $y = x^p r^q$; where y, x, r are respectively the ratios
of an ordinate of the figure insisting on the base at right
angles, of the segment of the base inter-
cepted between the ordinate and *A* the beginning of the base, and of the other segment
of the base lying between the ordinate and
the point *H*, to the base as their common
consequent. I say then that if an unknown
event has happened p times and failed q in
$p + q$ trials, and in the base *AH* taking any
two points as f and t you erect the ordinates
fC, *tF* at right angles with it, the chance
that the probability of the event lies
somewhere between the ratio of *Af* to *AH* and
that of *At* to *AH*, is the ratio of *tFCf*, that
part of the before-described figure which is
intercepted between the two ordinates, to
ACFH the whole figure insisting on the base *AH*.

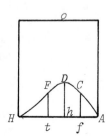

> —Thomas Bayes (1702-1761)
> "An Essay Towards Solving a Problem in the Doctrine
> of Chances" (posthumous, 1763), *The Philosophical
> Transactions of the Royal Society of London, 53*,
> London, 1763, pp. 376, 394.

LAPLACE (1814)

Probability is relative, in part to [our] ignorance, in part to our knowledge...

The theory of chance consists in reducing all the events of the same kind to a certain number of cases equally possible, that is to say, to such as we may be equally undecided about in regard to their existence, and in determining the number of cases favorable to the event whose probability is sought. The ratio of this number to that of all the cases possible is the measure of this probability, which is thus simply a fraction whose nunumerator is the number of favorable cases and whose denominator is the number of all the cases possible.

> —Pierre Simon de Laplace (1749-1827)
> *A Philosophical Essay on Probabilities* (1814),
> tr. by F. Truscott and F. Emory, Dover Pubs., Inc.,
> NY, 1951, pp. 6-7.

GAUSS (1821)

Certain causes of errors depend, for each observation, on circumstances which are variable and independent of the result which one obtains: the errors arising from such sources are called *irregular* or *random*...

On the other hand, there exist causes which in all observations of the same nature produce an identical error, or depend on circumstances essentially connected with the result of the observation. We shall call the errors of this category *constant* or *regular*.

...if one repeats indefinitely the measurement of a single angle, the errors arising from an imperfect division of the circular scale will belong to the class of constant errors.

> —Karl F. Gauss (1777-1855)
> "Theory of the Combination of Observations Which Leads to the Smallest Errors" (1821), *Gauss' Work (1803-1826) on the Theory of Least Squares,* tr. by H.F. Trotter, Tech. Rept. No. 5, Dept. of Army, Project No. 5B99-01-004, Statistical Techniques Research Group, Section of Mathematical Statistics, Dept. of Mathematics, Princeton Univ., Princeton, NJ, 1957, pp. 1-2.

POISSON (1837)

> In many different fields, empirical phenomena appear to obey a
> certain general law, which can be called the Law of Large Numbers.
> This law states that the ratios of numbers derived from the ob-
> servation of a very large number of similar events remain prac-
> tically constant, provided that these events are governed partly
> by constant factors and partly by variable factors whose varia-
> tions are irregular and do not cause a systematic change in a
> definite direction. Certain values of these relations are charac-
> teristic of each given kind of event. With the increase in
> length of the series of observations the ratios derived from such
> observations come nearer and nearer to these characteristic con-
> stants. They could be expected to reproduce them exactly if it
> were possible to make series of observations of an infinite length.

> > —Siméon D. Poisson (1781-1840)
> > *Recherches sur la Probabilité des Jugements en*
> > *Matière Criminelle et en Matière Civile*, Paris,
> > Bachelier, Imprimeur-Libraire, 1837, p. 7, tr. in
> > R. von Mises, *Probability, Statistics and Truth*,
> > Allen and Unwin, London, 1957, pp. 104-105.

ELLIS (1842)

> If the probability of a given event be correctly determined, the
> event will on a long run of trials tend to recur with frequency
> proportional to their probability. This is generally proved
> mathematically. It seems to me to be true *a priori*...I have
> been unable to sever the judgment that one event is more likely
> to happen than another from the belief that in the long run it
> will occur more frequently.

> > —Robert Leslie Ellis (1817-1859)
> > "On the Foundations of the Theory of Probabilities,"
> > (paper read Feb. 1842), *Trans. Cambridge Phil. Soc.*,
> > *8*, 1844, pp. 1-6. Quoted by R. Carnap, *Logical*
> > *Foundations of Probability*, Univ. of Chicago Press,
> > Chicago, IL, 1950, pp. 185-186.

RANKINE (1854)

$$F = \int \frac{dH}{Q} = \int \frac{1+\phi \cdot Q}{Q}\, dQ + \int \frac{dP}{dQ}\, dV$$

This function...I shall call a *Thermo-dynamic function*...

* * * *

$$Q = \mathbf{k}(\tau - \chi)$$

where Q is the actual heat in unity of weight of a substance, τ
its temperature, measured from the absolute zero of gaseous ten-
sion, χ the temperature of absolute cold, measured from the same

RANKINE (1854, cont.)

point, and $\mathbb{k}$ the real specific heat of the substance, expressed in terms of motive power.

* * * *

The expression for the Thermo-dynamic function denoted by F takes the form

$$F = \int \frac{1+\frac{1}{\mathbb{k}} \cdot f' \cdot \tau}{\tau - \chi} \cdot d\tau + \frac{1}{\mathbb{k}} \int \frac{d\mathrm{P}}{d\tau} \cdot dV ;$$

but a more convenient thermo-dynamic function, bearing the same relation to temperature as reckoned from the point of absolute cold, which the function F does to actual heat, is formed by multiplying the latter by the real specific heat $\mathbb{k}$, thus: —

$$\Phi = \mathbb{k}F = \int \frac{\mathbb{k} + f' \cdot \tau}{\tau - \chi} d\tau + \int \frac{d\mathrm{P}}{d\tau} dV$$

> —William J.M. Rankine (1820-1872)
> "On the Geometrical Representation of the Expansive Action of Heat, and the Theory of Thermo-dynamic Engines," *Philosophical Transactions of the Royal Society of London, 144,* 1854, pp. 126, 147, 149.

CLAUSIUS (1865)

If one looks for a characteristic name for S, one can, similar to the way in which it is said that the quantity U is the *heat- and work-content* of matter, say that the quantity S is the *capacity for change* of matter. Here, I feel it would be better to take the name for an important scientific quantity from the old language, so that it can be used unchanged in all new languages. So I propose that the quantity S be named the change, the *Entropie* of matter, after the Greek words ἡ τροπη. I have intentionally formed the word *Entropie* similar to the word *Energie,* since both quantities which these words designate are of such a close meaning one to the other that a clear analogy in the naming seems to me to be appropriate.

> —Rudolf J.E. Clausius (1822-1888)
> "Ueber verschiedene für die Anwendung bequeme Formen der Hauptgleichungen der mechanischen Wärmetheorie," *Annalen der Physik und Chemie, CXXV,* Leipzig, J.A. Barth, Publ., 1865, p. 390, tr. by J. Monk and R. Christensen.

VENN (1866)

Here becomes apparent the full importance of the distinction so
frequently insisted on, between the actual irregular series before us and the substituted one of calculation, and the meaning
of the assertion...that it was in the case of the latter only
that strict scientific inferences could be made. For how can we
have a 'limit' in the case of those series which ultimately exhibit irregular fluctuations? When we say, for instance, that
it is an even chance that a given person recovers from the
cholera, the meaning of this assertion is that in the long run
one half of the persons attacked by that disease do recover.
But if we examined a sufficiently extensive range of statistics,
we might find that the manners and customs of society had produced such a change in the type of the disease or its treatment,
that we were no nearer approaching towards a fixed limit than
we were at first. The conception of an ultimate limit in the
ratio between the numbers of the two classes in the series
necessarily involves an absolute fixity of the type. When
therefore nature does not present us with this absolute fixity,
as she seldom or never does except in games of chance (and not
demonstrably there), our only resource is to introduce such a
series, in other words, as has so often been said, to substitute
a series of the right kind.

> —John Venn (1834-1923)
> *The Logic of Chance* (1866), Chelsea Pub. Co., NY,
> 4th ed., 1962, pp. 164-165.

GIBBS (1873)

"...entropy as defined by Clausius is synonymous with the
thermodynamic function as defined by Rankine."

> —J. Willard Gibbs (1839-1903)
> "A Method of Geometrical Representation of the
> Thermodynamic Properties of Substances by Means of
> Surfaces," *Trans. of the Conn. Academy, II,* (Dec.
> 1873), pp. 382-404, reprinted in *The Scientific
> Papers of J. Willard Gibbs,* Vol. I, *Thermodynamics,*
> Dover Pubs., Inc., NY, 1961, p. 52.

PEIRCE (1878)

Probability is the ratio of the favourable cases to all the
cases. Instead of expressing our result in terms of this ratio,
we may make use of another--the ratio of favourable to unfavourable cases. This last ratio may be called the *chance* of an event.

* * * *

Any quantity which varies with the chance might, therefore, it
would seem, serve as a thermometer for the proper intensity of
belief. Among all such quantities there is one which is pecul-
iarly appropriate. When there is a very great chance, the feeling of belief ought to be very intense. Absolute certainty, or
an infinite chance, can never be attained by mortals, and this
may be represented appropriately by an infinite belief. As the
chance diminishes the feeling of believing should diminish,
until an even chance is reached, where it should completely van-
ish and not incline either toward or away from the proposition.
When the chance becomes less, then a contrary belief should
spring up and should increase in intensity as the chance diminishes, and as the chance almost vanishes (which it can never
quite do) the contrary belief should tend toward an infinite intensity. Now, there is one quantity which, more simply than any
other, fulfills these conditions; it is the *logarithm* of the
chance. But there is another consideration which must, if admitted, fix us to this choice for our thermometer. It is that
our belief ought to be proportional to the weight of evidence,
in this sense, that two arguments which are entirely independent,
neither weakening nor strengthening each other, ought, when they
concur, to produce a belief equal to the sum of the intensities
of belief which either would produce separately. Now, we have
seen that the chances of independent concurrent arguments are
to be multiplied together to get the chance of their combination,
and therefore the quantities which best express the intensities
of belief should be such that they are to be *added* when the
chances are multiplied in order to produce the quantity which
corresponds to the combined chance. Now the logarithm is the
only quantity which fulfills this condition.

—Charles Sanders Peirce (1839-1914)
"The Probability of Induction," *Popular Science
Monthly, 12,* April 1878, pp. 708, 708-709.

GIBBS (1902)

> It is the relative numbers of systems which fall within different
> limits, rather than the absolute numbers, with which we are most
> concerned. It is indeed only with regard to relative numbers
> that such discussions as the preceding will apply with literal
> precision, since the nature of our reasoning implies that the
> number of systems in the smallest element of space which we
> consider is very great. This is evidently inconsistent with a
> finite value of the total number of systems, or of the density-
> in-phase. Now if the value of D is infinite, we cannot speak of
> any definite number of systems within any finite limits, since
> all such numbers are infinite. But the ratios of these infinite
> numbers may be perfectly definite. If we write N for the total
> number of systems, and set
>
> $$P = \frac{D}{N} \, ,$$
>
> P may remain finite, when N and D become infinite.
> The integral
>
> $$\int \cdots \int P \; dp_1, \; \ldots, \; dq_n$$
>
> taken within any given limits, will evidently express the ratio
> of the number of systems falling within those limits to the
> whole number of systems. This is the same thing as the *proba-
> bility* that an unspecified system of the ensemble (i.e., one of
> which we only know that it belongs to the ensemble) will lie
> within the given limits.
>
> —J. Willard Gibbs (1839-1903)
> *Elementary Principles in Statistical Mechanics*
> (1902), Dover Pubs., Inc., NY, 1960, p. 16.

GIBBS (1906)

> "On entropy as mixed-up-ness."
>
> —J. Willard Gibbs (1839-1903)
> Unpublished fragments of a supplement to the
> "Equilibrium of Heterogeneous Substances"
> (posthumous, 1906), *The Scientific Papers of
> J. Willard Gibbs*, Longmans, Green and Company,
> London, 1906, reprinted by Dover Pubs., Inc., NY,
> 1961, p. 418.

KEYNES (1921)

Let our premises consist of any set of propositions h, and our conclusion consist of any set of propositions a, then, if a knowledge of h justifies a rational belief in a of degree α, we say that there is a *probability-relation* of degree α between a and h.

*   *   *   *

...*a logical relation between two sets of propositions* in cases where it is not possible to argue demonstratively from one to the other.

*   *   *   *

I believe, therefore, that the practice of underwriters weakens rather than supports the contention that all probabilities can be measured and estimated numerically.

Another set of practical men, the lawyers, have been more subtle in this matter than the philosophers. A distinction, interesting for our present purpose, between probabilities, which can be estimated within somewhat narrow limits, and those which cannot, has arisen in a series of judicial decisions respecting damages.

*   *   *   *

Whether or not such a thing is theoretically conceivable, no exercise of the practical judgment is possible, by which a numerical value can actually be given to the probability of every argument. So far from our being able to measure them, it is not even clear that we are always able to place them in an order of magnitude. Nor has any theoretical rule for their evaluation ever been suggested.

The doubt, in view of these facts, whether any two probabilities are in every case even theoretically capable of comparison in terms of numbers, has not, however, received serious considera-tion. There seems to me to be exceedingly strong reasons for entertaining the doubt.

*   *   *   *

...new evidence would give us a new probability, not a fuller knowledge of the old one...

> —John Maynard Keynes (1883-1946)
> *A Treatise on Probability* (1921), Macmillan & Co.,
> Ltd., London, 1957, pp. 4, 8, 24, 27-28, 31.

NYQUIST (1924)

The speed at which intelligence can be transmitted over a telegraph circuit with a given line speed, i.e., a given rate of
sending of signal elements, may be determined approximately by
the following formula...

$$W = K \log m$$

Where W is the speed of transmission of intelligence,
    m is the number of current values,
  and, K is a constant.

> —Harry Nyquist (1889-1976)
> "Certain Factors Affecting Telegraph Speed," *Bell
> Syst. Tech. Journ.*, *3*, 1924, pp. 332-333.

FISHER (1925)

The idea of an infinite hypothetical population is, I believe,
implicit in all statements involving mathematical probability.
If, in a Mendelian experiment, we say that the probability is
one half that a mouse born of a certain mating shall be white,
we must conceive of our mouse as one of an infinite population
of mice which might have been produced by that mating. The
population must be infinite for in sampling from a finite population the fact of one mouse being white would affect the
probability of others being white, and this is not the hypothesis
which we wish to consider; moreover, the probability may not
always be a rational number. Being infinite the population is
clearly hypothetical, for not only must the actual number produced by any parents be finite, but we might wish to consider
the possibility that the probability should depend on the age
of the parents, or their nutritional conditions. We can, however, imagine an unlimited number of mice produced upon the
conditions of our experiment, that is, by similar parents, of
the same age, in the same environment. The proportion of white
mice in this imaginary population appears to be the actual meaning to be assigned to our statement of probability.

*   *   *   *

It is a common case for a sample of n observations to be distributed into a finite number of classes, the numbers "expected"
in each class being functions of one or more unknown parameters,

FISHER (1925, cont.)

if p is the probability of an observation falling into any one class, the amount of information in the sample is

$$\sum \left\{ \frac{1}{m} \left(\frac{\partial m}{\partial \theta} \right)^2 \right\}$$

where $m = np$, is the expectation in any one class.

—Ronald A. Fisher (1890-1962)
"The Theory of Statistical Estimation," *Proceedings of the Cambridge Philosophical Society, 22,* 1925, pp. 700, 710. [The symbol "S" used by Fisher to denote summation over the sample has been replaced by the more familiar "Σ".]

HARTLEY (1927)

In any given communication the sender mentally selects a particular symbol and by some bodily motion, as of his vocal mechanism, causes the attention of the receiver to be directed to that particular symbol. By successive selections a sequence of symbols is brought to the listener's attention. At each selection there are eliminated all of the other symbols which might have been chosen. As the selections proceed more and more possible symbol sequences are eliminated, and we say that the information becomes more precise.

* * * *

For a particular system let the amount of information associated with n selections be

$$H = Kn, \tag{4}$$

where K is a constant which depends on the number s of symbols available at each selection. Take any two systems for which s has the values s_1 and s_2 and let the corresponding constants be K_1 and K_2. We then define these constants by the condition that whenever the numbers of selections n_1 and n_2 for the two systems are such that the number of possible sequences is the same for both systems, then the amount of information is also the same for both; that is to say, when

$$s_1^{n_1} = s_2^{n_2}, \tag{5}$$

$$H = K_1 n_1 = K_2 n_2, \tag{6}$$

from which

$$\frac{K_1}{\log s_1} = \frac{K_2}{\log s_2} . \tag{7}$$

HARTLEY (1927, cont.)

This relation will hold for all values of s only if K is connected with s by the relation

$$K = K_0 \log s \qquad\qquad (8)$$

where K_0 is the same for all systems.

* * * *

A difficulty, more apparent than real, arises from the fact that a given number of secondary or character selections may necessitate widely different numbers of primary selections, depending on the particular characters chosen. This would seem to indicate that the values of information deduced from the primary and secondary symbols would be different. It may easily be shown, however, that this does not necessarily follow.

If the sender is at all times free to choose any secondary symbol, he may make all of his selections from among those containing the greatest number of primary symbols. The secondary symbols will then all be of equal length, and, just as for the uniform code, the number of primary symbols will be the product of the number of characters by the maximum number of primary selections per character. If the number of primary selections for a given number of characters is to be kept to some smaller value than this, some restriction must be placed on the freedom of selection of the secondary symbols. Such a restriction is imposed when, in computing the average number of dots per character for a non-uniform code, we take account of the average frequency of occurrence of the various characters in telegraph messages. If this allotted number of dots per character is not to be exceeded in sending a message, the operator must, on the average, refrain from selecting the longer characters more often than their average rate of occurrence.

* * * *

We have then to examine the ability of such a continuous function to convey information. Obviously over any given time interval the magnitude may vary in accordance with an infinite number of such functions. This would mean an infinite number of possible secondary symbols, and hence an infinite amount of information. In practice, however, the information contained is finite for the reason that the sender is unable to control the form of the function with complete accuracy, and any distortion of its form tends to cause it to be confused with some other function.

> —Ralph V.L. Hartley (1888-1970)
> "Transmission of Information," presented at the
> Intl. Congress of Telegraphy and Telephone, Lake
> Como, Italy, Sept. 1927, *Bell Syst. Tech. Jour.*,
> *7*, 1928, pp. 536, 540, 541, 542-543.

von MISES (1928)

> ...a collective is a mass phenomenon or a repetitive event, or, simply, a long sequence of observations, for which there are sufficient reasons to believe that the relative frequency of the observed attribute would tend to a fixed limit if the observations were indefinitely continued. This limit will be called *the probability of the attribute considered within the given collective.*

> > —Richard von Mises (1883-1953)
> > *Probability, Statistics and Truth* (1928), tr. by J. Neyman, D. Scholl and E. Rabinowitsch, 2nd rev. ed. by H. Geiringer, George Allen and Unwin, Ltd., London, 1957, p. 15.

JEFFREYS (1939)

> Our fundamental idea will not be simply the probability of a proposition p, but the probability of p on data q. Omission to recognize that a probability is a function of two arguments, both propositions, is responsible for a large number of serious mistakes; in some hands it has led to correct results, but at the cost of omitting to state essential hypotheses and giving a delusive appearance of simplicity to what are really very difficult arguments. *It is no more valid to speak of the probability of a proposition without stating the data than it would be to speak of the value of x + y for given x, irrespective of the value of y.*

> > *   *   *   *

> If there is no reason to believe one hypothesis rather than another, the probabilities are equal. In terms of our fundamental notions of the nature of inductive inference, *to say that the probabilities are equal is a precise way of saying that we have no ground for choosing between the alternatives.*

> > —Harold Jeffreys (1891-   )
> > *Theory of Probability* (1939), Oxford at the Clarendon Press, London, 1961, pp. 15, 33.

SHANNON (1948)

If the number of messages in the set is finite then this number
or any monotonic function of this number can be regarded as a
measure of the information produced when one message is chosen
from the set, all choices being equally likely. As was pointed
out by Hartley the most natural choice is the logarithmic function. Although this definition must be generalized considerably
when we consider the influence of the statistics of the message
and when we have a continuous range of measures, we will in all
cases use an essentially logarithmic measure.

$*$ $*$ $*$ $*$

Suppose we have a set of possible events whose probabilities of
occurrence are $p_1, p_2, \ldots, p_n$. These probabilities are known
but that is all we know concerning which event will occur. Can
we find a measure of how much "choice" is involved in the selection of the event or of how uncertain we are of the outcome?

$*$ $*$ $*$ $*$

Quantities of the form $H = -\Sigma \, p_i \log p_i$...play a central role in
information theory as measures of information, choice and uncertainty. The form of H will be recognized as that of entropy as
defined in certain formulations of statistical mechanics where
p_i is the probability of a system being in cell i of its phase
space. H is then, for example, the H in Boltzmann's famous H
theorem. We shall call $H = -\Sigma \, p_i \log p_i$ the entropy of the set
of probabilities $p_1, \ldots, p_n$.

—Claude E. Shannon (1916-   )
"A Mathematical Theory of Communication," *The Bell
System Technical Journal, 27,* July 1948, pp. 379,
392, 393-394.

CARNAP (1950)

Some authors believe they have given a solution to the problem of
probability, in our terminology, an explication for probability,
by merely constructing an axiom system for probability without
giving an interpretation; for a genuine explication, however, an
interpretation is essential.

$*$ $*$ $*$ $*$

CARNAP (1950, cont.)

As soon as we go over from the field of formal mathematics to that of knowledge about the facts of nature, in other words, to empirical science, which includes applied mathematics, we need more than a mere calculus or axiom system; an interpretation must be added to the system.

> —Rudolf Carnap (1891-1970)
> *Logical Foundations of Probability,* The Univ. of Chicago Press, Chicago, IL, 1950, pp. 16, 18.

HEISENBERG (1958)

...the transition from the "possible" to the "actual" takes place during the act of observation...

* * * *

The discontinuous change in the probability function, however, takes place with the act of registration, because it is the discontinuous change of our knowledge in the instant of registration that has its image in the discontinuous change of the probability function.

> —Werner Heisenberg (1901-1976)
> *Physics and Philosophy, The Revolution in Modern Science* (1958), Harper & Bros., NY, 1962, pp. 54,55.

CARNAP (1966)

...I would say both Keynes and Jeffreys were pioneers who worked in the right direction. My own work on probability is in the same direction. I share their view that logical probability is a logical relation. If you make a statement affirming that, for a given hypothesis, the logical probability with respect to given evidence is .7, then the total statement is an analytic one. This means that the statement follows from the definition of logical probability (or from the axioms of a logical system) without reference to anything outside the logical system, that is, without reference to the structure of the actual world.

> —Rudolf Carnap (1891-1970)
> *Philosophical Foundations of Physics,* Basic Books, Inc., NY, 1966, p. 32.

REFERENCES

Aigrain, P., "A Theory of Communication," *Ann. Telecomm.*, *4*, Dec. 1949, p. 406.

Ash, Robert, *Information Theory*, J. Wiley & Sons, NY, 1965.

Bayes, Thomas, "An Essay Towards Solving a Problem in the Doctrine of Chances" (posthumous, 1763), *Philosophical Transactions of the Royal Society of London*, *53*, London, 1763, pp. 370-418.

Bernoulli, Jacob (James), *Ars Conjectandi* (posthumous, Basel, 1713), tr. by H.E. Wedeck, *Classics in Logic*, D.D. Runes, ed., Philosophical Library, Inc., NY, 1962.

Boltzmann, Ludwig, *Lectures on Gas Theory* (1896-98), Univ. of Calif. Press, Berkeley, CA, 1964.

Boole, George, *An Investigation of The Laws of Thought* (1854), Dover Pubs., Inc., NY, 1961.

Carnap, Rudolf, *Logical Foundations of Probability*, Univ. of Chicago Press, Chicago, IL, 1950.

_____, *Philosophical Foundations of Physics*, Basic Books, Inc., NY, 1966.

Christensen, R.A., "Induction and the Evolution of Language," Physics Dept., Univ. of Calif., Berkeley, CA, July 19, 1963. [*Chapter 7 of Volume II.*]

_____, "Inductive Reasoning and the Evolution of Language," Physics Dept., Univ. of Calif., Berkeley, CA, Dec. 1964. [*Chapter 8 of Volume II.*]

Clausius, Rudolf J.E., "Ueber die bei einem stationären elektrischen Strome in dem Leiter gethane Arbeit und erzeugte Wärme," *Annalen der Physik und Chemie, LXXXVII*, Leipzig, J.A. Barth, Pub., 1852, pp. 415-426.

_____, "Ueber verschiedene für die Anwendung bequeme Formen der Hauptgleichungen der mechanischen Wärmetheorie," *Annalen der Physik und Chemie, CXXV*, Leipzig, J.A. Barth, Publ., 1865, p. 390.

Clavier, A.G., "Evaluation of Transmission Efficiency According to Hartley's Expression of Information Content," *Elec. Comm. 15*, Dec. 1948, p. 414.

Darwin, C.G. and R.H. Fowler, "Fluctuations in an Assembly in Statistical Equilibrium," *Proc. Cambridge Philos. Soc.*, *21*, 1922, p. 391.

De Moivre, Abraham, *The Doctrine of Chances* (1738), reprinted by Chelsea Pub. Co., NY, 1967.

Edwards, A.W.F., *Likelihood, An Account of the Statistical Concept of Likelihood and Its Application in Scientific Inference*, Camb. Univ. Press, London, 1972.

Ellis, Leslie, "On the Foundations of the Theory of Probabilities," *Trans. Camb. Phil. Soc.*, *8*, 1844, pp. 1-6.

Fisher, R.A., "Theory of Statistical Estimation," *Proceedings of the Cambridge Philosophical Society*, *22*, 1925, pp. 700-725.

Gabor, D., "Theory of Communication," *Jour. Intro. Elec. Engr.*, *93*, Pt. 3, 1964, p. 429.

Gauss, Karl F., "Theory of the Combination of Observations Which Leads to the Smallest Errors" (1821), in *Gauss' Work (1803-1826) on the Theory of Least Squares*, tr. by H.F. Trotter, Technical Rept. No. 5, Dept. of Army, Project No. 5B99-01-004, Statistical Techniques Research Group, Section of Mathematical Statistics, Dept. of Mathematics, Princeton Univ., Princeton, NJ, 1957.

Gibbs, J. Willard, "A Method of Geometrical Representation of the Thermodynamic Properties of Substances by Means of Surface," *Trans. of the Conn. Academy*, *II*, (Dec. 1873), reprinted in *The Scientific Papers of J. Willard Gibbs*, *Vol. I, Thermodynamics*, Dover Pubs., Inc., NY, 1961, pp. 33-54.

_____, *Elementary Principles in Statistical Mechanics* (1902), Dover Pubs., Inc., NY, 1960.

_____, unpublished fragments of a supplement to "Equilibrium of Heterogeneous Substances" (1906), *The Scientific Papers of J. Willard Gibbs*, *Vol. I, Thermodynamics*, Dover Pubs., Inc., NY, 1961, p. 418-434.

Goldman, S., "Some Fundamental Considerations Concerning Noise Reduction and Range in Radar Communications," *Proc. of I.R.E.*, *36*, Nov. 1948, p. 584.

Hacking, Ian, *The Emergence of Probability*, Cambridge Univ. Press, London, 1975.

Hartley, R.V.L., "Transmission of Information," *Bell System Tech. Journal*, *7*, 1928, pp. 535-563.

Heisenberg, Werner, *Physics and Philosophy, The Revolution in Modern Science* (1958), Harper & Bros., NY, 1962.

Jeffreys, Harold, *Theory of Probability* (1939), Oxford at the Clarendon Press, London, 1961.

Kac, Marc, *Statistical Independence in Probability, Analysis and Number Theory*, The Math. Assoc. of America, Providence, RI, 1959.

Keynes, John Maynard, *A Treatise on Probability* (1921), Macmillan & Co., Ltd., London, 1957.

Khinchin, A.I., *Mathematical Foundations of Information Theory*, tr. by R.A. Silverman and M.D. Friedman, Dover Pubs., Inc., NY, 1957.

Kolmogorov, A., *Foundations of the Theory of Probability* (1933), tr. by Nathan Morrison, Chelsea Publ. Co., NY, 1950.

Kullback, Solomon, *Information Theory and Statistics*, J. Wiley & Sons, Inc., NY, 1959.

Küpfmüller, K., "Transient Phenomena in Wave Filters," *Elektrische Nachrichten-Technik*, *1*, 1924, pp. 141-152.

Laplace, P.S., *A Philosophical Essay on Probabilities* (1814), Dover Pubs., Inc., NY, 1951.

Leibniz, Gottfried Wilhelm, *De incerti aestimatione* (Sept. 1678), excerpts tr. in I. Hacking, *The Emergence of Probability*, Camb. Univ. Press, London, 1975.

McGill, W.J., "Multivariate Information Transmission," *Transactions PGIT, 1954 Symposium on Information Theory*, PGIT-4, pp. 93-111.

McMullen, Charles W., *Communication Theory Principles*, The Macmillan Co., NY, 1968.

Mill, John Stuart, *A System of Logic* (1843), Longmans, Green & Co., London, 1879.

von Mises, Richard, *Probability, Statistics and Truth* (1928), tr. by N. Neyman, D. Scholl and E. Rabinowitsch, 2nd rev. ed. by H. Geiringer, George Allen and Unwin, Ltd., London, 1957.

Nyquist, H., "Certain Factors Affecting Telegraph Speed," *Bell Syst. Tech. Journal, 3,* 1924, p. 324-346.

Peirce, Charles Sanders, "The Probability of Induction," *Popular Science Monthly, 12,* April 1878, pp. 705-718.

Peusner, Leonardo, *Concepts in Bioenergetics*, Prentice-Hall, Inc., Englewood Cliffs, NJ, 1974.

Poisson, S.D., *Recherches sur la Probabilité des Jugements en Matière Criminelle et en Matière Civile,* Paris, Bachelier, Imprimeur-Libraire, 1837.

Rankine, William J.M., "On the Geometrical Representation of the Expansive Action of Heat, and the Theory of Thermo-dynamic Engines," *Phil. Trans. of the Roy. Soc. of London, 144,* London, 1854, pp. 115-175.

Reichenbach, Hans, *The Theory of Probability*, Univ. of Calif. Press, Berkeley, CA, 1953.

Shannon, C.E., "The Mathematical Theory of Communication," *Bell System Technical Journal, 27,* 1948, pp. 379-423, 623-656.

_____, "Communication in the Presence of Noise," *Proc. I.R.E., 37,* 1949, p. 10.

Shannon, C.E. and W. Weaver, *The Mathematical Theory of Communication,* Univ. of Illinois Press, Urbana, IL, 1949.

Tait, Peter Guthrie, *Sketch of Thermodynamics*, Edmonston and Douglas, Edinburgh, 1868.

Tuller, W.G., "Theoretical Limits on the Rate of Transmission of Information," *Proc. of I.R.E., 37,* 1949, p. 468.

Venn, John, *The Logic of Chance* (1866), Chelsea Pub. Co., NY, 4th ed., 1962.

Weston, J.D., "A Note on the Theory of Communication," *Phil. Mag., 40,* 1949, p. 449.

Woodward, P.M., *Probability and Information Theory, with Applications to Radar,* Pergamon Press, NY, 1953.

Zipf, G.K., "The Repetition of Words, Time Perspective and Semantic Balance," *J. of Gen. Psych., 32,* 1945, p. 127.

_____, *Human Behavior and the Principle of Least Effort,* Addison-Wesley Press, Inc., Cambridge, MA, 1949.

CHAPTER 8

MAXIMUM ENTROPY PROBABILITIES

CHAPTER 8

MAXIMUM ENTROPY PROBABILITIES

A.   The Principle of Equiprobabilities

The astute reader will have noticed that, in the previous chapter,
we asserted that the mathematical definition of probability is
by itself inadequate to determine probabilities from observational
data, but that, up to this point, we also have skirted the issue.
We now turn to this question.

We begin by considering the situation in which our knowledge is
limited to being able to list the possible outcomes; we have no
sample of past observations, and are given no auxiliary information
about the situation.

The following examples illustrate such situations:

   o   We come upon a scientist timing radioactive decay
       events by recording the clock times of individual
       Geiger counter clicks. He poses the following prob-
       lem: "Consider the next three clicks after you
       say 'Go', and subdivide the interval between the
       first and third clicks into six sub-intervals. Es-
       timate the probability that the second click will
       occur during each of these sub-intervals." Fig. 23.

   o   We come upon a group of gamblers playing craps. One
       of them says: "I am going to roll a single die.
       Estimate the probabilities for each of its six faces."
       Fig. 24.

   o   We come upon a statistician shaking the proverbial
       urn. He says: "In this urn, I have six balls, each
       colored differently: red, orange, yellow, green,
       blue and violet. I will reach in and take one out.
       Estimate the probability of each color." Fig. 25.

It is not at all clear that we really do not have a sample of
past observations or auxiliary information in these cases.

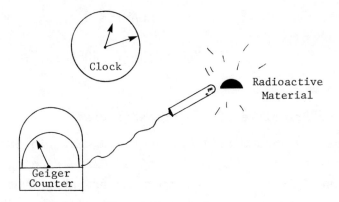

Fig. 23.   Scientist Recording Time of Geiger Counter Clicks
When Sensor Held near a Radioactive Source

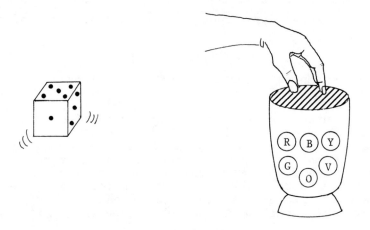

Fig. 24.   Gambler Tossing Die   Fig. 25.   Statistician Reaching
                                                  into Urn Containing Six
                                                  Differently Colored Balls

Consider the die-throwing example.   Simply by knowing that it is
a "die" which is being thrown might be taken to imply that we have
quite a fund of past examples, namely past instances of "die-tosses", though not necessarily of *that particular die*.   However, all particular events are singular.   What does it matter whether it is the

die, or the *toss*, or both, that distinguishes a particular case? We do have past experience with other cases of *die-tossing*.

It might even be argued that it is impossible to pose a real-life situation in which we know the possible outcomes but have no past experience or auxiliary information. To communicate the verbal classifications of the elements of the situation is to reference it to our past experience and general background information.

Be this as it may, we will, for discussion purposes, *pretend* that we know only the possible outcomes. We are dealing with a situation which is so abstract that we know only the *number* of outcome classes and are able to individually identify each of them and attach some label thereto.

Thus, all three of the above examples, from this viewpoint, are identical. Each example has six possible outcomes and each outcome is distinguishable from the other five so the events can be uniquely labeled as to outcome class.*

Now suppose we apply the *principle of equiprobabilities* to this situation. This is the principle that probabilities should be taken as equal when there is no information to the contrary. We have explicitly defined the situation to exclude any such "information to the contrary." Thus, we have an ideal case to which to apply this principle. The result is that we simply assign the value 1/6 to each of the six probabilities:

$$P_1 = P_2 = P_3 = P_4 = P_5 = P_6 = 1/6.$$

_____

* Alternative analyses apply when permanent distinguishability, as assumed in Bernoulli's scheme of repeated trials with single replacement, described by the multinomial distribution, is not assumed. Repeated trials without replacement, described by the multidimensional hypergeometric distribution, for example, leads to Fermi-Dirac statistics. Repeated trials with double replacement (Polya's urn scheme in the form of a Bayes-Laplace analysis from which the rule of succession is derived) lead to Bose-Einstein statistics. See de Finetti, Section 10.3, for a discussion.

The principle of equiprobabilities appears to suffer from a serious case of arbitrariness on two counts:

1.   Just because we have no reason to believe probabil-
     ities P_A and P_B are different, why should we assume
     that they are the same? Are we saying that we *do*
     have reason to believe they are equal? An argument
     from ignorance can lead to any conclusion. Why not
     assume, for instance, that P_A is twice P_B?

2.   Even if we were able to justify the assignment of equal
     probabilities in the absence of reasons to the con-
     trary, why apply them to these rather arbitrarily
     chosen outcome classes?

The principle of maximum entropy focuses on the first of these
problems, the principle of minimum entropy on the second.   Entropy
minimization will be taken up in Chapter 12.   We turn now to
entropy maximization.

B.   The Principle of Maximum Entropy

The privileged status of ignorance with respect to nonequality
can be expressed in information-theoretic terms.   We begin with
the following definition of *induced probabilities:*

Probabilities are said to be *induced* from a set of observational data, for given background information, if they represent all of, but no more than, the available information.

The "but no more than" component of this definition is the principle of maximum entropy.*   This principle may be stated as:

The induced probabilities are those probabilities for which
the entropy is a maximum, consistent with given constraints
and available data.

Recall that we defined probability as an estimator of the classification frequency of yet-to-be-observed outcomes.   Its values
must lie on the interval from zero to unity.   We divide this

---

* E.T. Jaynes, "Information Theory and Statistical Mechanics," *The Physical
Review*, *106*, 1957, pp. 620-630, and *108*, 1957, pp. 171-190.

interval into K equal length segments; K is chosen sufficiently large for us to be satisfied to know simply on which segment the frequency lies.   See Fig. 26.

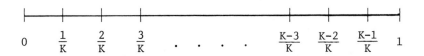

Fig. 26.   Frequency Range Partitioned into K Equal Segments

Let P_k represent the probability that the frequency is on the kth segment.   The entropy of this segmentation is given by

$$S = - \sum_{k=1}^{K} P_k \log P_k.$$

Since the actual frequency must be on one and only one of the segments*, it can be shown that the maximum possible value of this entropy is the logarithm of K

$$S = \log K,$$

and that this maximum occurs when and only when all the probabilities are the same and given by

$$P_k = \frac{1}{K} \qquad \text{for all } k = 1, 2, \ldots, K.$$

For example, imagine that we wish to compute the entropy of the six die-tossing outcome possibilities assuming equiprobabilities. This gives

$$S = -[\tfrac{1}{6} \log \tfrac{1}{6} + \tfrac{1}{6} \log \tfrac{1}{6} + \tfrac{1}{6} \log \tfrac{1}{6} + \tfrac{1}{6} \log \tfrac{1}{6} + \tfrac{1}{6} \log \tfrac{1}{6} + \tfrac{1}{6} \log \tfrac{1}{6}]$$

$$= \log 6$$

$$= 1.792.$$

---

* We define them as $0 \le x < \frac{1}{K}, \frac{1}{K} \le x < \frac{2}{K}, \ldots, \frac{K-2}{K} \le x < \frac{K-1}{K}, \frac{K-1}{K} \le x \le 1.$

This is the maximum possible entropy for six outcome classes. Any other set of six probabilities summing to unity would yield a lower value. For example, suppose we had chosen the following set of probabilities:

$$P_1 = 0.3 \qquad\qquad P_4 = 0.1$$

$$P_2 = 0.2 \qquad\qquad P_5 = 0.1$$

$$P_3 = 0.1 \qquad\qquad P_6 = 0.2$$

Then the entropy would be

$$S = -[0.3 \log 0.3 + 0.2 \log 0.2 + 0.1 \log 0.1$$
$$+ 0.1 \log 0.1 + 0.1 \log 0.1 + 0.2 \log 0.2]$$
$$= 1.465.$$

Recall that entropy is, by definition, the expected value of information. Since equiprobability implies maximum entropy (under conditions of no constraints other than unity sum), it further implies maximum expected value of information. The information referred to is, of course, the information we would gain by future observation. Thus, the maximum entropy probabilities define the state of maximum ignorance. By adopting these equiprobability values as our estimates, we maximize the information we expect to gain from future observational data. Adopting any other set of probabilities would be tantamount to saying that we already have some information* about the situation, contradicting our earlier assumption. For example, we would be implying that we had information which leads us to believe that the die is loaded in favor of ones, twos and sixes.

---

* There is, of course, the matter of whether what we mean by "information" in stating the problem is what is measured as the magnitude of "information" by this mathematical analysis. If this is made so by assumption, we might be accused of circular reasoning. But this accusation does not bother us. All mathematical reasoning is tautological with respect to its definitions and assumptions. It is still of interest to explore its logical (deductive) implications. The problem of justifying a principle such as basing beliefs on no more than available information will be deferred to Part V of this volume.

In the same way that we have analyzed the problem of the die, a parallel analysis can be conducted with respect to the problem of drawing a ball out of the urn and the problem of timing the second Geiger counter click.

There appears to be, however, an additional complication in the Geiger counter case. We may divide the interval between the first and second clicks into sub-intervals of unequal duration. For example, we may make them become successively larger. See Fig. 27.

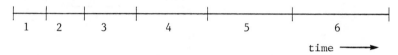

Fig. 27. Frequency Range Partitioned into Unequal Segments

Our information-theoretic reasoning leads to equiprobable sub-intervals regardless of how unequal we make the sub-interval lengths along the underlying time axis.

It may be argued that, by consciously making unequal sub-intervals, we introduce information, contrary to our stated assumptions. An equivalent situation can arise, however, without our being conscious of it at all. For example:

o We might be using a poorly-designed clock that slows down as the spring is unwound. Fig. 28(a).

o The scientist might be on a space ship that is moving with a velocity near light-speed away from us on Earth. His radioactive process slows down relative to our clock, in accord with the special theory of relativity. Fig. 28(b).

o The scientist might be on a space ship approaching the gravitational field of a black hole. In this case also, his radioactive process slows down relative to our clock, in accord with the general theory of relativity. Fig 28(c).

As it is possible to imagine a great number of different physical setups, each yielding a different principle for segmenting the

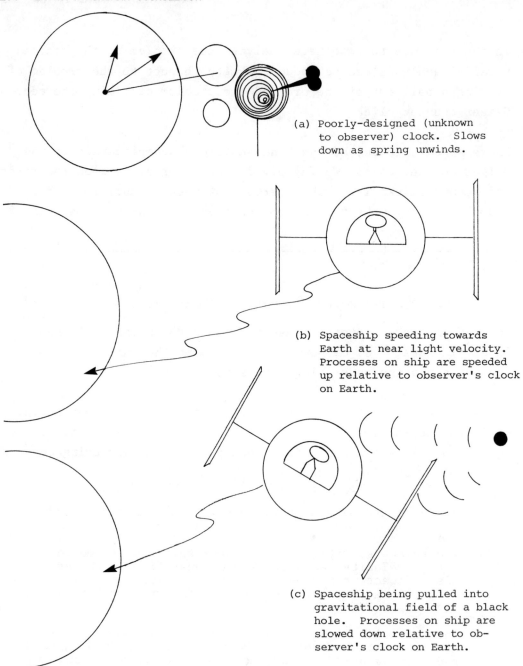

(a) Poorly-designed (unknown
    to observer) clock. Slows
    down as spring unwinds.

(b) Spaceship speeding towards
    Earth at near light velocity.
    Processes on ship are speeded
    up relative to observer's clock
    on Earth.

(c) Spaceship being pulled into
    gravitational field of a black
    hole. Processes on ship are
    slowed down relative to ob-
    server's clock on Earth.

Fig. 28. Examples of Situations in Which Radioactive Decays
          During Equal Intervals of "Time" Are not Equally
          Probable, Despite Absence of Any Consciously
          Introduced Information by the Observer

time axis, how are we to proceed? The answer lies in the conventions of our language which suggest the underlying variable along
which to take equal length segments. These language conventions
contain background information imparted to us by our predecessors.
This background information is manifest in the meanings of our
words and the grammar of our sentences.

When we say that we have defined an "underlying variable" along
which we will assume equiprobabilities for equal length segments,
what we mean is that we have reached the limits of our ignorance.

We are aware of no further information. It is, of course, entirely
possible that we are wrong. We may not be aware, for example, of
the poor design of the clock. But this says only that there may
be aspects of physical reality beyond our knowledge. With respect
to the information available to us, however, we are logically consistent in assigning equal probabilities to equal intervals as
measured by this clock.

Finally, it should be noted that the problem of how to subdivide
a dependent variable also arises in situations in which the underlying variable is discrete rather than continuous. For example,
the statistician with the urn would confront us with the problem
if he amended his request to say: "Incidentally, I forgot to tell
you whether the blue ball was dark blue or light blue. In fact,
I'm not sure of the answer myself. Could you please give me a
probability for each of the *seven* colors?"

We now have seven distinguishable discrete possibilities. Or have
we? Are "light blue" and "dark blue" both *possibilities* in the same
sense that "yellow" and "red" are? If we choose to distinguish
them differently, are we not thereby manufacturing artificial *information* which we do not in fact possess? And is not the source
of this information the conventions of language about the words
"balls", "painted", and so forth? See Fig. 29(a).

(a)
Do we have six or seven
possibilities when we do
not know whether the blue
ball is dark blue or light
blue?

(b)
Why might not "red ball"
and "yellow ball" be two
fundamentally different
and noncomparable things?

Red ball glued to
bottom of urn.

(c)
Why cannot "red" and "yellow"
be considered an irrelevant distinction with respect to the
"truly" underlying variable explaining the ball selection probabilities?

Blue ball much
larger than the
others.

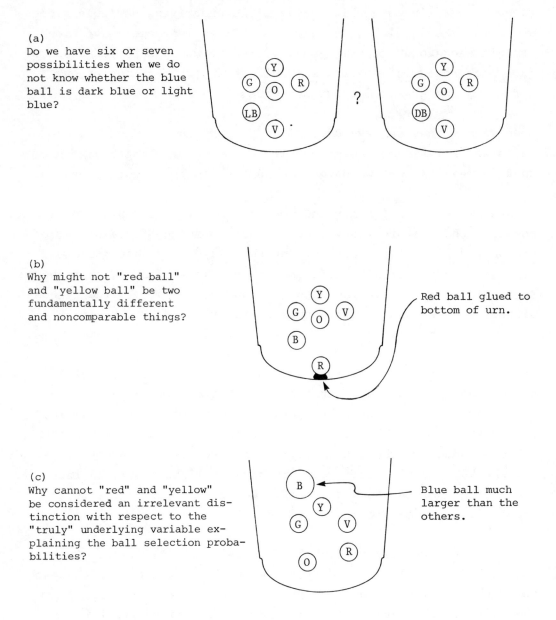

Fig. 29.   Example of Difficulties with the Definition of "Sameness"

Granting that there are difficulties with this introduction of an extra kind of blue, it is less obvious but nonetheless true that there are equivalent difficulties with the original situation. We can, for instance, ask why a "red ball" and a "yellow ball" should be considered as alternatives of the *same* thing, namely *being a ball*. Why might not "red ball" and "yellow ball" be two fundamentally different and noncomparable things, like "red ball" and "die toss resulting in prime number"? See Fig. 29(b). Alternatively, why cannot "red" and "yellow" be considered irrelevant distinctions with respect to some other underlying variable explaining the ball selection probabilities? See Fig. 29(c).

Questions of this kind are fundamental to observational science. Only by comparing our observations to the limits of our ignorance and thereby pushing to new depths in our search for the underlying equiprobability variables can we gain new information about the world in which we live. The search for the proper variables along which to display our observations is a key step in the scientific discovery of patterns in nature.

C. Bayes' Uniform Distribution Postulate

Although less well-known than his famous inverse probability theorem, Bayes also formulated a postulate about frequency distributions over probability which we discuss here because of its connections to entropy minimax theory.

Consider the experiment described in Chapter 7 in which each event consists of 100 rolls of a die and the dependent variable is the frequency of ones. We want to partition the range of this variable, which extends from 0 to 1, into a number of sub-intervals and estimate the probability for each sub-interval.

Once the range has been segmented, the maximum entropy principle will yield the desired probabilities. Given the existence of no data and of no constraints other than unity sum, the result is simply equiprobabilities. However, this argument will hold whether we make the segments of equal or unequal length.

Bayes postulated that the proper segmentation of a frequency range for purposes of applying the equiprobability principle is one which gives the segments equal lengths.

It is not at all obvious that this is necessarily the proper way. It might be, for example, that regions very near extremes of 0 or 1 should be subdivided more finely or more coarsely.

Bayes gave no rationale for his choice, but it can be seen to be consistent with the uniformity of the possible effect of any subsequent observation on the frequency.

To illustrate this concept, imagine that, starting with no data, we observe n rolls of which x turn up ones. The frequency is then x/n. Now suppose we contemplate making one extra roll. If it is a one, our frequency rises to $(x+1)/(n+1)$. If it is not a one, it drops to $x/(n+1)$. The difference, $(x+1)/(n+1) - x/(n+1) = 1/(n+1)$, is independent of x, i.e., independent of the current frequency. Thus, our current uncertainty of what the frequency would be after another observation is uniform over the range of frequency $0 \leq x/n \leq 1$ and equal to $1/(n+1)$, depending only upon n and not at all upon x.

Suppose we measure the distribution of our ignorance in terms of the ability of future observations to reduce our uncertainty. Then Bayes' postulate is consistent with taking the underlying equiprobability variable to be one over which our ignorance is distributed uniformly:

$$f(p) = 1 \quad \text{for } 0 \leq p \leq 1.$$

In entropy minimax analysis, the matter of chopping up independent variable space is treated as a part of the problem of induction; but the matter of chopping up the dependent variable space is treated

as depending upon the purpose for which the analysis is being conducted. Here we are treating frequency as the dependent variable.
The proper way of partitioning it therefore depends upon the purpose of our analysis.

Our purpose may be to reduce our uncertainty about how much the next
event will affect the frequency. Suppose we have already observed n
events. Then, whatever the frequency is after n observations, the
frequency after the next observation will be in one of n+1 intervals
of length $\Delta p = \frac{1}{n+1}$. In the limit of arbitrarily large samples, this
is approximated by equal infinitesimal intervals dp.

Although this role-reversal for probability and frequency was formulated to explain the use of equal intervals, it also has implications for the frequency limit hypothesis. With probability treated
solely as a model of frequency and not as a separate reality, it
has no meaning as the limit of a sequence of frequencies.

D. Laplace's Rule of Succession

Laplace's rule of succession is remembered most for its application
to the question: Will the sun rise tomorrow?

Imagine that we have observed the sun to rise n times in which
n = 1,825,000, corresponding to 5,000 years of recorded history.
Then, according to the rule of succession, the probability that it
will rise tomorrow is given by

$$P = \frac{n+1}{n+2}$$

$$= \frac{1,825,001}{1,825,002}$$

$$= 0.9999995.$$

To a being for whom the 1,825,000 observations are precisely all the
available information, this is indeed the induced probability.

A more general form of this rule is the statement that if a particular outcome has been observed to occur on x of n observations
of an event, and to fail to occur on n-x observations, then the probability of its occurrence is given by*

$$P = \frac{x+1}{n+2} .$$

In the preceding section, we showed how Bayes' postulate of uniform
distribution over probability follows from the principle of entropy
maximization.   In this section, we show how Laplace derived his
rule of succession from Bayes' postulate.   This links the rule of
succession to entropy maximization.

Let p be the unconditional probability of occurrence of the outcome
in question, p = P(O).   Then 1-p is the probability of nonoccurrence.   The probability of observing exactly x occurrences for n
independent events is given by the binomial distribution

$$P(x{:}n \text{ given } p) = \binom{n}{x} p^x (1-p)^{n-x}.$$

We wish to compute the conditional probability of the outcome given
the observed sample: P(O given x:n).   From Bayes' theorem, we have

$$f(p \text{ given } x{:}n) dp = \frac{P(x{:}n \text{ given } p) f(p) dp}{\int_0^1 P(x{:}n \text{ given } p') f(p') dp'} .$$

At this point we apply entropy maximization via Bayes' postulate

$$f(p) = 1 \qquad \text{for } 0 \le p \le 1.$$

Then

$$f(p \text{ given } x{:}n) dp = \frac{P(x{:}n \text{ given } p) dp}{\int_0^1 P(x{:}n \text{ given } p') dp'} .$$

---

* P.S. Laplace, *Mém. de l'Acad. R.D. Sci.*, *6*, Paris, 1774, p. 621.

Finally, we use this distribution over probability given our observed sample to compute the expected value of the probability.

$$P(0 \text{ given } x:n) = \int_0^1 p \, f(p \text{ given } x:n) dp$$

$$= \frac{\int_0^1 p \binom{n}{x} p^x (1-p)^{n-x} dp}{\int_0^1 \binom{n}{x} p^x (1-p)^{n-x} dp}$$

$$= \frac{x+1}{n+2} \, .$$

This completes the derivation.

The last step is obtained via successive integrations by parts. It can alternatively be obtained from the fact that

$$\int_0^1 z^{a-1} (1-z)^{b-1} dz = B(a,b),$$

where $B(a,b)$ is the beta function

$$B(a,b) = \frac{\Gamma(a)\Gamma(b)}{\Gamma(a+b)} \, ,$$

and the gamma function is an extension of factorials from the integer to the real domain

$$\Gamma(n) = (n-1)! \text{ for } n = \text{integer} > 0.$$

E.   Induced Probabilities for Flat Distributions

In this section we derive a generalization of Laplace's rule of succession to an arbitrary partition of outcome space. The formula which we will derive is given by*

$$P = \frac{x+t}{n+t+f},$$

* R.A. Christensen, "Induction and the Evolution of Language," Physics Dept., Univ. of Calif., Berkeley, CA, 1963, (*Volume II*, p. 137); *Foundations of Inductive Reasoning*, Berkeley, CA, 1964, Sec. 7.E.

where t is the number of ways in which the outcome can occur, and f is the number of ways in which it can fail to occur.

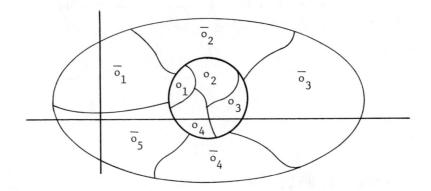

Fig. 30.   Partition of Outcome Space

We begin by subdividing the outcome O into t nonoverlapping ways of occurring o_1, o_2,...,o_t.   See Fig. 30.   Thus, O may be represented by a set of "or" statements:

$$O = o_1 \text{ or } o_2 \text{ or } o_3 \text{ or},...,\text{or } o_t$$

$$= \sum_{k=1}^{t} o_k.$$

We designate the set of outcomes that are not O by $\overline{O}$ and subdivide $\overline{O}$ into the f distinct ways of O failing to occur.   Thus, $\overline{O}$ may be represented as

$$\overline{O} = \overline{o}_1 \text{ or } \overline{o}_2 \text{ or } \overline{o}_3,...,\text{or } \overline{o}_f$$

$$= \sum_{k=1}^{f} \overline{o}_k.$$

Let the grand union of these possibilities be denoted as

$$R = o_1 \text{ or } o_2 \text{ or}, \ldots, \text{or } o_t \text{ or } \overline{o}_1 \text{ or } \overline{o}_2 \text{ or}, \ldots, \overline{o}_f$$

$$= \sum_{j=1}^{t+f} r_j \, ,$$

where

$$r_1 = o_1$$
$$\vdots$$
$$r_t = o_t$$
$$r_{t+1} = \overline{o}_1$$
$$\vdots$$
$$r_{t+f} = \overline{o}_f .$$

Now consider a particular way o of O occurring, and arrange R such that this is the last of the $K = t + f$ options.

$$R = r_1 \text{ or } r_2, \ldots, \text{or } r_{K-1} \text{ or } r_K,$$

where

$$r_K = o.$$

Now consider the conditional probability $P(\overline{o}$ given $(o \text{ or } r_j))$ where $j < K$. Using the definition of conditional probability and the fact that the possibilities r_j, $r_2, \ldots, r_{K-1}$, o form a complete partition, we have (for $j < K$)

$$P(\overline{o} \text{ given } (o \text{ or } r_j)) = \frac{P(\overline{o} \text{ and } (o \text{ or } r_j))}{P(o \text{ or } r_j)}$$

$$= \frac{P(\overline{o} \text{ and } r_j)}{P(o \text{ or } r_j)}$$

$$= \frac{P(r_j)}{P(o \text{ or } r_j)} \, ,$$

where the second line uses the property of intersections. Therefore, for j < K

$$P(r_j) = P(\bar{o} \text{ given } (o \text{ or } r_j)) P(o \text{ or } r_j).$$

Using the fact that $\{r_j\}$, for j < K, forms a complete partition of
$\bar{o}$, we have

$$p(\bar{o}) = \sum_{j=1}^{K-1} p(r_j).$$

Substituting for $P(r_j)$:

$$P(\bar{o}) = \sum_{j=1}^{K-1} P(\bar{o} \text{ given } (o \text{ or } r_j)) P(o \text{ or } r_j).$$

We can also rewrite the conditional probability $P(o \text{ given } (o \text{ or } r_j))$
as:

$$P(o \text{ given } (o \text{ or } r_j)) = \frac{P(o \text{ and } (o \text{ or } r_j))}{P(o \text{ or } r_j)}$$

$$= \frac{P(o)}{P(o \text{ or } r_j)},$$

again using the property of intersections. Therefore, with rearrangement of the terms, we obtain

$$P(o \text{ or } r_j) = \frac{P(o)}{P(o \text{ given } (o \text{ or } r_j))}.$$

Substituting for $P(o \text{ or } r_j)$ in the expression for $P(\bar{o})$, we have:

$$P(\bar{o}) = \sum_{j=1}^{K-1} P(\bar{o} \text{ given } (o \text{ or } r_j)) \frac{P(o)}{P(o \text{ given } (o \text{ or } r_j))}$$

$$= P(o) \sum_{j=1}^{K-1} \frac{1 - P(o \text{ given } (o \text{ or } r_j))}{P(o \text{ given } (o \text{ or } r_j))}$$

$$= P(o) \left[1 - K + \sum_{j=1}^{K-1} \frac{1}{P(o \text{ given } (o \text{ or } r_j))} \right].$$

Probabilities for the union of all possibilities sum to unity

$$P(o) + P(\bar{o}) = 1.$$

Substituting $1-P(o)$ for $P(\bar{o})$, we get

$$P(o) + P(o) \left[1-K + \sum_{j=1}^{K-1} \frac{1}{P(o \text{ given } (o \text{ or } r_j))} \right] = 1.$$

Rearranging terms,

$$P(o) = \frac{1}{2-K + \sum_{j=1}^{K-1} \frac{1}{P(o \text{ given } (o \text{ or } r_j))}}.$$

Now we are ready to use the principle of maximum entropy in the form of Laplace's rule of succession. Let x_k be the number of observations of a particular way o_k of O occurring. Let x_j be the number of observations of the jth other alternative (one of the f ways of O not occurring or of the remaining t-1 ways of O occurring).

From Section 8.D above, if we are given that that which actually occurs is either o_k or r_j, then, since o_k and r_j are mutually exclusive, Laplace's rule of succession says that

$$P(o \text{ given } (o_k \text{ or } r_j)) = \frac{x_k+1}{x_k+x_j+2}.$$

For our case with K = t+f, this gives

$$P(o_k) = \frac{1}{2-(t+f) + \sum_{j \neq k} \frac{x_k+x_j+2}{x_k+1}}.$$

Since

$$\sum_{j \neq k} x_j = n - x_k,$$

this becomes

$$P(o_k) = \frac{x_k + 1}{n + t + f}.$$

Now, since $\{o_i\}$ is a complete partition of O, we have

$$P(O \text{ given } x:n) = \sum_{k=1}^{t} p(o_k),$$

and since

$$\sum_{k=1}^{t} x_k = x,$$

we obtain the final result

$$P(O \text{ given } x:n) = \frac{x + t}{n + t + f}.$$

As we will see for an analogous situation in the next section, we could have obtained this formula with less work by assuming a beta distribution. But then we would need to justify the assumption. Here we have obtained the result using only entropy maximization.

F. Induced Probabilities for Curved Distributions

The flat distribution over probability shown in Fig. 31 arises from entropy maximization in the special case in which there is no auxiliary information. A curved distribution, $f(p)$, can result from entropy maximization when there is auxiliary information in the form of constraints.

For example, Fig. 32 shows a situation in which the range of p is subdivided more finely at the extremes. In this case the distribution is concave. Fig. 33 shows a convex distribution for a situation in which the range is more finely divided in the vicinity of a particular probability.

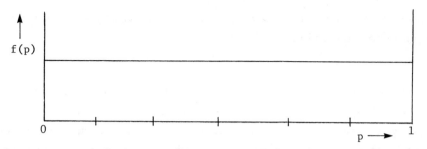

Fig. 31.   Situation Satisfying Bayes' Postulate in which our
Ignorance Is Uniformly Distributed over Probability

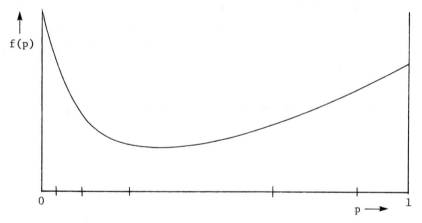

Fig. 32.   A Nonuniform Situation in which Our Ignorance
is Skewed toward Some Particular Probability

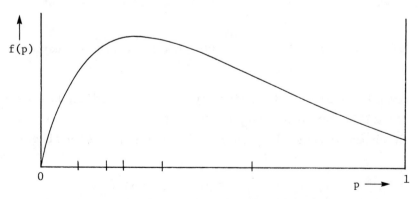

Fig. 33.   A Nonuniform Situation in which Our Ignorance
Is Skewed toward Extreme Probabilities

Our auxiliary information may be about some variable v which depends upon the probability

$$v = v(p).$$

An example is the temperature of a volume of gas, where the probability describes the motions of individual molecules entering and exiting the volume through an orifice. Another example is the daily "take" by a roulette wheel, in which the probability describes the betting odds.

What we will generally know about such variables depends, of course, upon the particular situation. The implications of several general classes of auxiliary information have been studied. Examples are given in Table 27.

Table 27. Examples of Types of Auxiliary Information

Name and symbol	Mathematical formulation
known mean (μ)	$\int_0^1 f(p)\, v(p)\, dp$ = constant
known variance (σ^2)	$\int_0^1 f(p)\, [v(p) - \mu]^2\, dp$ = constant
known mean log (S_+)	$-\int_0^1 f(p)\, \log v(p)\, dp$ = constant
known mean log (S_-)	$-\int_0^1 f(p)\, \log [1-v(p)]\, dp$ = constant

The determination of the function $f(p)$ is handled in the maximum entropy formalism by using Lagrange multipliers to find that value of $f(p)$ which maximizes the renormalized distributional entropy

$$S_R = -\int_0^1 f(p)\, \log f(p)\, dp,$$

subject to the constraints. (See Sec. 14.C on entropy renormalization.)

For all physically sensible constraints, we have the frequency-correspondence principle: Maximum entropy probabilities approach sample average estimators in the limit of large samples.

Listed below are results for a few illustrative types of constraints:

Known Mean

If we maximize entropy subject to the constraint of a known value for the mean, we obtain the exponential distribution[*]

$$f(p) = \frac{1}{\mu} e^{-v/\mu} \frac{dv}{dp} .$$

Known Mean and Variance

If our auxiliary information includes both the mean and the variance, we obtain the normal distribution:[**]

$$f(p) = \frac{1}{\sigma\sqrt{2\pi}} e^{-\frac{1}{2}(\frac{v-\mu}{\sigma})^2} \frac{dv}{dp} .$$

Known Mean and Mean log

If we assume, instead, a known mean and mean log, we obtain the gamma distribution:[***]

$$f(p) = \frac{\mu}{\sigma^2 \Gamma(\mu^2/\sigma^2)} (\mu v/\sigma^2)^{(\mu^2/\sigma^2 - 1)} e^{-\mu v/\sigma^2} \frac{dv}{dp} .$$

---

[*]   *Cf.* C.E. Shannon, "A Mathematical Theory of Communication, Part III," *Bell System Technical Journal*, *27*, 1948, p. 631.

[**]   C.E. Shannon, *supra*, p. 629.

[***]   M. Tribus, R. Evans and G. Crellin, *The Use of Entropy in Hypothesis Testing*, Thayer School of Engr., Dartmouth Univ., Hanover, NH, 1963, pp. 8-13.

Known S_+ and S_-

If the auxiliary information fixes the value of S_+ and S_-, entropy maximization yields a beta distribution:*

$$f(p) = \frac{v^a (1-v)^b}{B(a,b)} \frac{dv}{dp} ,$$

where

$$a = \frac{\mu^2 (1-\mu)}{\sigma^2} - \mu , \text{ and}$$

$$b = \frac{\mu (1-\mu)^2}{\sigma^2} - (1-\mu) .$$

G.   Maximum Entropy A Priori Weights

The effects of specific forms of auxiliary information upon the induced probabilities are determined by inserting them into a generalized succession rule

$$P(\text{o given } x{:}n, \{\theta\}) = \frac{\int_0^1 p^{x+1} (1-p)^{n-x} f(p) \, dp}{\int_0^1 p^x (1-p)^{n-x} f(p) \, dp} ,$$

where $\{\theta\}$ is the set of parameters fixed by the auxiliary information.

Suppose, for example, that we have auxiliary information about a variable v which is proportional to p

$$v = Kp .$$

Then, for known mean μ the maximum entropy distribution is

$$f(p) = \frac{K}{\mu} e^{-Kp/\mu} .$$

---

* M. Tribus, *et al.*, *op. cit.*

The probability of obtaining the outcome in a specified way is given by

$$P(o \text{ given } x:n,K,\mu) = \frac{\int_0^1 p^{x+1} (1-p)^{n-x} e^{-Kp/\mu} dp}{\int_0^1 p^x (1-p)^{n-x} e^{-Kp/\mu} dp}.$$

The integrals may be evaluated with the help of the relation

$$\int_0^1 p^{\nu-1}(1-p)^{\mu-1} e^{zp} dp = B(\nu,\mu) \; \Phi(\nu,\mu+\nu;z),$$

where Φ is the degenerate hypergeometric function

$$\Phi(\alpha,\gamma;z) = 1 + \frac{\alpha}{\gamma}\frac{z}{1!} + \frac{\alpha(\alpha+1)}{\gamma(\gamma+1)}\frac{z^2}{2!} + \frac{\alpha(\alpha+1)(\alpha+2)}{\gamma(\gamma+1)(\gamma+2)}\frac{z^3}{3!} + \cdots$$

and B is the beta function defined in Section 8.D. This gives

$$P(o \text{ given } x:n,K,\mu) = \frac{B(x+2,n-x+1)}{B(x+1,n-x+1)} \frac{\Phi(x+2,n+3;-K/\mu)}{\Phi(x+1,n+2;-K/\mu)}$$

$$= \frac{x+1}{n+2} \frac{\Phi(x+2,n+3;-K/\mu)}{\Phi(x+1,n+2;-K/\mu)}.$$

The functions B and Φ may be obtained from standard tables.

As a second case, we take an auxiliary information variable proportional to p, but now assume known S_+ and S_-. In this case the maximum entropy distribution is given by

$$f(p) = \frac{(Kp)^a (1-Kp)^b}{B(a,b)} K.$$

The probability of obtaining the outcome in a specific way is now

$$P(o \text{ given } x{:}n,K,S_+,S_-) = \frac{\displaystyle\int_0^1 p^{x+1+a} (1-p)^{n-x} (1-Kp)^b \, dp}{\displaystyle\int_0^1 p^{x+a} (1-p)^{n-x} (1-Kp)^b \, dp}.$$

The integrals in this case may be evaluated with the help of the relation

$$\int_0^1 p^{\lambda-1}(1-p)^{\mu-1}(1-Kp)^{-\nu} \, dp = B(\lambda,\mu) \, F(\nu,\lambda;\lambda+\mu;K),$$

where F is Gauss' hypergeometric function

$$F(\alpha,\beta;\gamma;z) = 1 + \frac{\alpha \cdot \beta}{\gamma} \frac{z}{1!} + \frac{\alpha(\alpha+1)\beta(\beta+1)}{\gamma(\gamma+1)} \frac{z^2}{2!}$$

$$+ \frac{\alpha(\alpha+1)(\alpha+2)\beta(\beta+1)(\beta+2)}{\gamma(\gamma+1)(\gamma+2)} \frac{z^3}{3!} + \cdots$$

This yields the result

$$P(o \text{ given } x{:}n,K,S_+,S_-) = \frac{B(x+2+a,n-x+1)}{B(x+1+a,n-x+1)} \frac{F(-b,x+2+a;n+3;K)}{F(-b,x+1+a;n+2;K)}$$

$$= \frac{x+1+a}{n+2+a} \frac{F(-b,x+2+a;n+3;K)}{F(-b,x+1+a;n+2;K)}.$$

The functions B and F may be obtained from standard tables. For the special case of $K = 1$ (known mean probability), this reduces to

$$P(o \text{ given } x{:}n,1,S_+,S_-) = \frac{x+1+a}{n+2+a+b},$$

as is easily seen by the fact that

$$(1-p)^{n-x} (1-p)^{b} = (1-p)^{(n+a+b)-(x+a)}.$$

The case of known S_+ and S_- for $v = p$ has an interesting interpretation.   Note that for this case

$$S_+ = \int_0^1 f(p) \ I(p) \ dp, \text{ and}$$

$$S_- = \int_0^1 f(p) \ I(1-p) \ dp.$$

Thus, when $v = p$, S_+ is equivalent to the average information content of prior observations of occurring in way o, and S_- is equivalent to the average information content of prior observations of not occurring in way o.   The numbers of "virtual" prior observations of these two types are a and b, respectively.

The effect of the auxiliary information upon the induced probabilities is determined by using the appropriate expression for P(o given x:n) in the derivation of P(O given x:n) for arbitrary t and f.   For example, with $v = p$ and S_+, S_- known, we get

$$P(O \text{ given } x{:}n,1,S_+,S_-) = \frac{x+t+a}{n+t+f+a+b},$$

where now a and b are the aggregated number of virtual prior observations in the auxiliary information.

We will, in subsequent chapters, generally make use of this latter form to represent background information, recognizing that more complex forms may be needed for special situations.

H.   Weight Normalization

Suppose that the outcome O is identified as the kth in the classification $O_1, O_2, \ldots, O_k, \ldots, O_K$.

Define the following quantities:

n = total number of events observed,

n_k = number of events observed with outcome O_k,

t_k = number of ways O_k can occur,

y_k = number of virtual O_k events in a deterministic model,

a_k = number of virtual O_k events in other background information, and

$w_k = t_k + y_k + a_k$.

Then the probability of outcome O_k is given by *

$$P_k = \frac{n_k + w_k}{n + w},$$

where

$$n = \sum n_k, \text{ and}$$

$$w = \sum w_k.$$

---

* R.A. Christensen, "Seminar on Entropy Minimax Method of Pattern Discovery and Probability Determination," Carnegie-Mellon Univ., Tech. Rept. No. 40.3.75, April 17, 1971, (*Volume III*, pp. 54-56); "Entropy Minimax Method of Pattern Discovery and Probability Determination," Arthur D. Little, Inc., Cambridge, MA, March 7, 1972, (*Volume III*, pp. 69-76).

The virtual event counts y_k for a mechanistic model can be obtained from the probability q_k and associated uncertainty u_k given by the model:

$$y_k = \frac{q_k(1-q_k)}{u_k^2}.$$

Assuming no specific form of additional background information is available, we let the "other" virtual events represent all types of events (in addition to the type for which we are making the probability estimate). In the terminology of Chapter 11, these other virtual events represent the whole of "feature space," not just that portion constituting the pattern currently under investigation. Thus a_k will be proportional to the frequency of the kth outcome among all events in the entire sample distributed throughout feature space.

$$a_k = \alpha f_k.$$

The proportional constant α fixes the weight normalization,

$$w = \sum_k (t_k + y_k + a_k).$$

The procedure for determining this normalization will be described in Chapters 20 and 23. Before that, however, we complete our examination of distributions along the dependent variable, and then turn to the topics of the independent variables and patterns in feature space.

QUOTES

BERNOULLI (1713)

...for each of the games that follow there is no greater probability that fortune should favor the same persons whom it favored
before than those who have been the most unfortunate of all...

> —Jacob Bernoulli (1654-1705)
> *Ars Conjectandi* (posthumous, Basel, 1713), tr. by
> H.E. Wedeck, in *Classics in Logic*, ed. by D.D.
> Runes, Philosophical Library, Inc., NY, 1962,
> p. 136.

GAUSS (1809)

If, any hypothesis H *being made, the probability of any determin-
ate event* E *is* h, *and if, another hypothesis* H' *being made ex-
cluding the former and equally probable in itself, the probability of the same event is* h': *then I say, when the event* E *has
actually occurred, that the probability that* H *was the true hypothesis, is to the probability that* H' *was the true hypothesis,
as* h *to* h'.

> —Karl Friedrich Gauss (1777-1885)
> *Theory of the Motion of the Heavenly Bodies Moving
> about the Sun in Conic Sections* (1809), tr. by
> C. Davis, Dover Pubs., Inc., NY, 1963, p. 255.

LAPLACE (1814)

Each of the causes to which an observed event may be attributed
is indicated with just as much likelihood as there is probability
that the event will take place, supposing the event to be constant.
The probability of the existence of any one of these causes is a
fraction whose numerator is the probability of the event resulting from this cause and whose denominator is the sum of the similar probabilities relative to all the causes; if these various
causes, considered *à priori*, are unequally probable, it is necessary, in place of the probability of the event resulting from
each cause, to employ the product of this probability by the
possibility of the cause itself. This is the fundamental principle of this branch of the analysis of chances which consists in
passing from events to causes.

> —Pierre Simon de Laplace (1749-1827)
> *A Philosophical Essay on Probabilities* (1814), tr.
> by F. Truscott and F. Emory, Dover Pubs., Inc.,
> NY, 1951, pp. 15-16.

GAUSS (1821)

If the causes of error are such that there is no reason for two equal errors of opposite signs to have unequal probabilities, one will have

$$\varphi(x) = \varphi(-x).$$

> —Karl F. Gauss (1777-1855)
> "Theory of the Combination of Observations Which Leads to the Smallest Errors," in *Gauss' Work (1803-1826) on the Theory of Least Squares*, tr. by Hale F. Trotter, Tech. Report No. 5, Dept. of the Army, Proj. No. 5B99-01-004, Statistical Techniques Research Group, Sec. of Math. Stats., Dept. of Math., Princeton Univ., Princeton, NJ, 1957, p. 5.

CARNOT (1824)

...the maximum of motive power resulting from the employment of steam is also the maximum of motive power realizable by any means whatever.

Since every re-establishment of equilibrium in the caloric may be the cause of the production of motive power, every re-establishment of equilibrium which shall be accomplished without production of this power should be considered as an actual loss...The necessary condition of the maximum is, then, *that in the bodies employed to realize the motive power of heat there should not occur any change of temperature which may not be due to a change of volume.* Reciprocally, every time that this condition is fulfilled the maximum will be attained. This principle should never be lost sight of in the construction of heat-engines; it is its fundamental basis.

* * * *

The motive power of heat is independent of the agents employed to realize it; its quantity is fixed solely by the temperatures of the bodies between which is effected, finally, the transfer of the caloric.

We must understand here that each of the methods of developing motive power attains the perfection of which it is susceptible. This condition is found to be fulfilled if, as we remarked above, there is produced in the body no other change of temperature than that due to change of volume, or, what is the same thing in other words, if there is no contact between bodies of sensibly different temperatures.

> —Sadi Carnot (1796-1832)
> *Reflections on the Motive Power of Fire, and on Machines Fitted to Develop That Power* (Paris, 1824), tr. by R. Thurston, in *Reflections on the Motive Power of Fire*, ed. by E. Mendoza, Dover Pubs., Inc., NY, 1960, pp. 12-13, 20.

BOOLE (1854)

> If the *à priori* probabilities of the causes are equal, ...*whatever the nature of the connexion among the causes*, the *à posteriori* probability of each cause will be proportional to the probability of the observed event *E* when the cause is known to exist.

> * * * *

> It is, however, to be observed, that in all those problems the probabilities of the *causes* involved are supposed to be known *à priori*. In the absence of this assumed element of knowledge, it seems probable that arbitrary constants would *necessarily* appear in the final solution.

> * * * *

> It has been said, that the principle involved in the above and in similar applications is that of the equal distribution of our knowledge, or rather of our ignorance--the assigning to different states of things of which we know nothing, and upon the very ground that we know nothing, equal degrees of probability. I apprehend, however, that this is an arbitrary method of procedure.

> > —George Boole (1815-1864)
> > *An Investigation of the Laws of Thought* (1854), Dover Pubs., Inc., NY, 1961, pp. 357, 363, 370.

CLAUSIUS (1865)

> 1) The energy of the world is constant.
> 2) The entropy of the world tends toward a maximum.

> > —Rudolf J.E. Clausius (1822-1888)
> > "Ueber verschiedene für die Anwendung bequeme Formen der Hauptgleichungen der mechanischen Wärmetheorie," *Annalen der Physik und Chemie, CXXV,* Leipzig, 1865, p. 400, tr. by R. Christensen.

MAXWELL (1871)

> ...if we conceive of a being whose faculties are so sharpened that he can follow every molecule in its course, such a being, whose attributes are still as essentially finite as our own, would be able to do what is at present impossible to us...

> * * * *

> In dealing with masses of matter, while we do not perceive the individual molecules, we are compelled to adopt what I have described as the statistical method of calculation, and to abandon the strict dynamical method, in which we follow every motion by the calculus.

> > —James Clerk Maxwell (1831-1879)
> > *Theory of Heat,* Longmans, Green & Co., London, 1871, pp. 301, 309.

GIBBS (1876)

...if the system is in a state in which its entropy is greater than in any other state of the same energy, it is evidently in equilibrium, as any change of state must involve either a decrease of entropy or an increase of energy, which are alike impossible for an isolated system.

> —J. Willard Gibbs (1839-1903)
> "On the Equilibrium of Heterogeneous Substances," *Trans. of the Conn. Academy, III*, Oct. 1875 - May 1976, pp. 108-248, in *The Scientific Papers of J. Willard Gibbs*, Vol. I, *Thermodynamics*, Dover Pubs., Inc., NY, 1961, pp. 58-59.

BOLTZMANN (1896-1898)

The fact that in nature the entropy tends to a maximum shows that for all interactions (diffusion, heat conduction, etc.) of actual gases the individual molecules behave according to the laws of probability in their interactions, or at least that the actual gas behaves like the molecular-disordered gas which we have in mind.

The second law is thus found to be a probability law.

> —Ludwig Boltzmann (1844-1906)
> *Lectures in Gas Theory* (1896-1898), tr. by S. Brush, Univ. of Calif. Press, Berkeley, CA, 1964, pp. 74-75.

KEYNES (1921)

We must first determine what parts of our evidence are relevant on the whole by a series of judgments of relevance, not easily reduced to rule...If this relevant evidence is *of the same form* for both alternatives, then the Principle authorises a judgment of indifference.

* * * *

...the Principle of Indifference is not applicable to a pair of alternatives, if we know that either of them is capable of being further split up into a pair of possible but incompatible alternatives...

* * * *

...if the original alternatives each comprise a definite number of indivisible and indifferent sub-alternatives, we can compute their probabilities...it is easy in most cases to discover another set of alternatives which do satisfy the condition, and which will often serve our purpose equally well. Suppose, for instance, that a point lies on a line of length $m.l.$, we may write the alternative 'the interval of length l on which the point lies is the xth interval of that length as we move along the line from left to

KEYNES (1921, cont.)

> right' ≡ $\phi(x)$; and the Principle of Indifference can then be applied safely to the m alternatives, $\phi(1)$, $\phi(2)$..., $\phi(m)$, the number m increasing as the length l of the intervals is diminished. There is no reason why l should not be of any definite length however small.
>
> —John Maynard Keynes (1883-1946)
> *A Treatise on Probability* (1921), Macmillan & Co., London, 1957, pp. 56, 61, 62.

FISHER (1922)

> ...fundamental problems have been ignored and fundamental paradoxes left unresolved. This anomalous state of statistical science is strikingly exemplified by a recent paper (1) entitled "The Fundamental Problem of Practical Statistics," in which one of the most eminent of modern statisticians presents what purports to be a general proof of Bayes' postulate, a proof which, in the opinion of a second statistician of equal eminence, "seems to rest upon a very peculiar--not to say hardly supposable-- relation." (2)
>
> *   *   *   *
>
> There would be no need to emphasize the baseless character of the assumptions made under the title of inverse probability and Bayes' Theorem in view of the decisive criticism to which they have been exposed at the hands of Boole, Venn, and Chrystal, were it not for the fact that the older writers, such as Laplace and Poisson, who accepted these assumptions, also laid the foundations of the modern theory of statistics, and have introduced into their discussions of this subject ideas of a similar character.
>
> *   *   *   *
>
> We do not, and cannot, know, from the information supplied by a sample, anything about the probability that p should lie between any named values.
>
> —Ronald A. Fisher (1890-1962)
> "On the Mathematical Foundations of Statistics," *Trans. of the Royal Society of London, 222A,* May 1922, pp. 310-311, 326, 327.

(1) K. Pearson (1920). "The Fundamental Problem of Practical Statistics," 'Biom.,' xiii, pp. 1-16.

(2) F.Y. Edgeworth (1921). "Molecular Statistics," 'J.R.S.S.,' lxxxiv., p. 83.

KHINCHIN (1943)

Some authors[1] try to generalize the notion of entropy by considering it as being a phase function which, depending on the phase,
can assume different values for the same set of thermodynamical
parameters, and try to prove that entropy so defined must
increase, with overwhelming probability. However, such a proof
has not yet been given and it is not at all clear how such an
artificial generalization of the notion of entropy could be
useful to the science of thermodynamics.

* * * *

"the entropy of a system is proportional to the logarithm of
the probability of the corresponding state" (Boltzmann's
postulate). This statement, which is absolutely meaningless
in the case of an isolated system, obtains, as we see, some
meaning for a system in the larger system. This can be accomplished however, only by using the above described generalization
of the notion of entropy which is introduced "ad hoc". In fact,
one must not forget that this notion is used in connection with
the second law of thermodynamics which loses meaning when the
generalized definition of entropy is used. All existing attempts
to give a general proof of this postulate must be considered as
an aggregate of logical and mathematical errors superimposed on
a general confusion in the definition of the basic quantities.[2]
In the most serious treatises on that subject (for example:
R.H. Fowler "Statistical Mechanics" Cambridge 1936) the authors
refuse to accept this postulate, indicating that it cannot be
proved, and cannot be given a sensible formulation even on the
basis of the exact notions of thermodynamics.

> —Aleksandr I. Khinchin (1894-1959)
> *Mathematical Foundations of Statistical Mechanics*
> (1943), tr. by G. Gamow, Dover Pubs., Inc., NY,
> 1949, pp. 139, 142.

---

1 Comp. Borel, Mécanique statistique classique, Paris 1925.

2 Comp. the "proof" in "Thermodynamik" by M. Planck, and the
   corresponding critique in "Statistical Mechanics" by R.H.
   Fowler.

BRILLOUIN (1953)

Information can be changed into negentropy and vice versa...

Any experiment by which an information is obtained about a physical system corresponds *in average* to an increase of entropy in the system or in its surroundings. This average increase is always larger than (or equal to) the amount of information obtained.

The smallest possible amount of negentropy required in an observation is of the order of k. [Boltzmann's constant]

...the principle of negentropy of information...applies in different branches of physics, technology, and even in some very general problems of human knowledge.

> —Leon Brillouin (1889-1969)
> "The Negentropy Principle of Information,"
> *J. of Applied Physics, 24,* No. 9, Sept. 1953,
> p. 1153.

JAYNES (1957)

...in making inferences on the bases of partial information we must use that probability distribution which has maximum entropy subject to whatever is known. This is the only unbiased assignment we can make; to use any other would amount to arbitrary assumption of information which by hypothesis we do not have.

> —Edwin T. Jaynes (1922-   )
> "Information Theory and Statistical Mechanics.I,"
> *Phys. Rev., 106,* No. 4, May 15, 1957, p. 623.

RENYI (1960)

For instance we may ask what is the amount of information concerning a random variable ξ obtained from observing an event E, which is in some way connected with the random variable ξ. If $\mathscr{P}$ denotes the original (unconditional) distribution of the random variable ξ and $\mathcal{Q}$ the conditional distribution of ξ under the condition that the event E has taken place, we shall denote a measure of the information concerning the random variable ξ contained in the observation of the event E by $I(\mathcal{Q}|\mathscr{P})$.

* * * *

RENYI (1960, cont.)

In the case when the random variable ξ takes on only a finite number of different values $x_1, x_2, \ldots, x_n$ and we put $P\{\xi = x_k\} = p_k$ and $P\{\xi = x_k | E\} = q_k$ for k=1, 2, $\ldots n$, then...

$$I_1(\mathcal{Q}|\mathcal{P}) = \sum_{k=1}^{n} q_k \, \log_2 \frac{q_k}{p_k}$$

—Alfred Renyi (1921-1970)
"On Measures of Entropy and Information," *Proc. of the Fourth Berkeley Symposium on Mathematical Statistics and Probability, Vol. I, June 20 - July 30, 1960,* Univ. of Calif. Press, Berkeley, CA, 1961, pp. 553, 554.

TRIBUS (1961)

"Frequency" refers to a set of measurements that have been made; "probability" refers to measurements which have not yet been made (or the results of which have not yet been communicated). The problem of statistical inference (as Hume pointed out) is to find a method for assigning probabilities that is minimally biased; that is, which uses the available information and leaves the mind unbiased with respect to what is not known.

Of course, if in the past we have observed the frequency of events and found them to be of magnitude f_i, then for the future we should assign, as a *rational procedure*, $p_i = f_i$. But what should we do if the frequencies of occurrence are not measure-able, as in the case with atoms and molecules, as well as many other problems of statistical inference? The answer, according to Jaynes, is this (Jaynes' "Principle of Minimum Prejudice"): "Assign that set of values to the p_i which is consistent with the given information and which maximizes the uncertainty."

—Myron Tribus (1921-   )
"Information Theory as the Basis for Thermostatics and Thermodynamics," *Journal of Applied Mechanics,* March 1961, p. 2.

The entropy of an isolated system will not spontaneously decrease.

Since entropy and uncertainty are synonymous, the above state-ment is equivalent to saying that the only thing that can happen to our knowledge of an isolated system is that it must grow more vague with time.

—Myron Tribus (1921-   )
Thermostatics and Thermodynamics, D. Van Nostrand Co., Inc., Princeton, NJ, 1961, pp. 145-146.

BRILLOUIN (1962)

Whenever we make an observation on a physical system, we must have all sorts of sources of negentropy. We use this negentropy and we increase the total entropy of the laboratory containing the system under observation and the measuring instruments. As a result, we obtain a certain amount of information about the system. The increase in entropy is, however, always larger than the information gained. This result represents an extension of Carnot's principle...

> —Leon Brillouin (1889-1969)
> *Science and Information Theory*, Academic Press, NY, 1962, p. 293.

GOOD (1965)

The probabilities that govern the ordinary sampling are here called probabilities of Type I, and those governing the selection of populations from the superpopulation as probabilities of Type II.

* * * *

Even without a sample, there are occasions when the Bayes' postulate is unreasonable. The actuaries, G.F. Hardy* and Lidstone,** suggested that a more flexible and convenient class of initial densities was given by the beta form, proportional to $p^{\alpha}(1-p)^{\beta}$, where $\alpha > -1$ and $\beta > -1$.

...Given a sample of r successes in N trials, the final Type II expectation and variance of p are

$$\mathcal{E}_{II}(p \mid r,s) = \frac{\alpha+r+1}{\alpha+\beta+N+2}$$

and

$$\text{var}_{II}(p \mid r,s) = \frac{(\alpha+r+1)(2\alpha+2\beta+r+1)}{(\alpha+\beta+N+2)(\alpha+\beta+N+3)}$$

> —Irving John Good (1916-   )
> *The Estimation of Probabilities: An Essay on Modern Bayesian Methods* (1965), MIT Press, Camb., MA, 1968, pp. 9, 17.

---

* Hardy, G.F., in correspondence in Insurance Record, 1889, reprinted in *Trans. Fac. Actuaries, 8,* (1920).

** Lidstone, G.J., "Note on the General Case of the Bayes-Laplace Formula for Inductive or *A Posteriori* Probabilities," *Trans. Fac. Actuar., 8,* 182-192 (1920).

REFERENCES

Bernoulli, Jacob, *Ars Conjectandi* (posthumous, Basel, 1713), tr. by H.E. Wedeck, in *Classics in Logic*, ed. by D.D. Runes, Philosophical Library, Inc., NY, 1962.

Boltzmann, Ludwig, *Lectures on Gas Theory* (1896-1898), tr. by S. Brush, Univ. of Calif., Berkeley, CA, 1964.

Boole, George, *In Investigation of the Laws of Thought* (1854), Dover Pubs., Inc., NY, 1961.

Brillouin, Leon, "Maxwell's Demon Cannot Operate: Information and Entropy I," *J. Applied Physics, 22*, 1951, pp. 334-337. (2) "Physical Entropy and Information II," *J. Applied Physics, 22,* 1951, pp. 338-343. (3) "The Negentropy Principle of Information," *J. Applied Physics, 24*, 1953, pp. 1152-1163.

_____, *Science and Information Theory*, Academic Press, NY, 1962.

Broad, C.D., "On the Relation Between Induction and Probability," *Mind, 27,* 1918, pp. 389-404.

Carnap, Rudolf, *Two Essays on Entropy*, Univ. of Calif. Press, Berkeley, CA, 1977.

Carnot, Lazare, *Fundamental Principles of Equilibrium and Movement*, Chapelet, Imprimeur, Paris, 1803.

Carnot, Sadi, *Reflections on the Motive Power of Fire, and on Machines Fitted to Develop That Power* (Paris, 1824), tr. by R. Thurston, in *Reflections on the Motive Power of Fire*, ed. by E. Mendora, Dover Pubs., Inc., NY, 1960.

Christensen, R.A., "Induction and the Evolution of Language," Physics Dept., Univ. of Calif., Berkeley, CA, July 19, 1963. [*Chapter 7 of Volume II.*]

_____, *Foundations of Inductive Reasoning*, Berkeley, CA, 1964.

_____, "Inductive Reasoning and the Evolution of Language," Physics Dept., Univ. of Calif., Berkeley, CA, Dec. 1964. [*Chapter 8 of Volume II.*]

_____, "A General Approach to Pattern Discovery," Tech. Report No. 20, Computer Center, Univ. of Calif., Berkeley, CA, June 29, 1967 (revised Nov. 15, 1967). [*Chapter 1 of Volume 3.*]

_____, "Seminar on Entropy Minimax Method of Pattern Discovery and Probability Determination," presented at Carnegie-Mellon Univ. and MIT, Feb. and March 1971, Tech. Report No. 40.3.75, Carnegie-Mellon Univ., Pittsburgh, PA, April 7, 1971. [*Chapter 3 of Volume III.*]

_____, "Entropy Minimax Method of Pattern Discovery and Probability Determination," Arthur D. Little, Inc., Camb., MA, March 7, 1972. [*Chapter 4 of Volume III.*]

_____, "Crossvalidation: Minimizing the Entropy of the Future," *Info. Proc. Letters, 4*, No. 3, Dec. 1975, pp. 73-76. [*Chapter 10 of Volume III.*]

Clausius, Rudolf J.E., "Ueber die bewegende Kraft der Wärme, und die Gestze, welche sich daraus für die Wärmelehre selbst ableiten lassen," *Annalen der Physik und Chemie, LXXIX*, Leipzig, J.A. Barth, Publ., March and April 1850, pp. 368-397; pp. 500-524.

Clausius, Rudolf, J.E., "Ueber die bei einem stationären electrischen Strome in dem Leiter gethane Arbeit und erzeugte Wärme," *Annalen der Physik und Chemie, LXXXVII*, Leipzig, J.A. Barth, Publ., 1852, pp. 415-426.

_____, "Ueber verschiedene für die Anwendung bequeme Formen der Hauptgleichungen der mechanischen Wärmetheorie," *Annalen der Physik und Chemie, CXXV*, Leipzig, J.A. Barth, Publ., 1865, pp. 353-400.

de Finetti, Bruno, *Theory of Probability*, J. Wiley & Sons, Inc., NY, 1975.

Fisher, Ronald A., "On the Mathematical Foundations of Statistics," *Phil. Trans. of the Royal Society of London*, Ser. A, *222*, May 1922, pp. 309-368.

Gauss, Karl F., *Theory of the Motion of the Heavenly Bodies Moving about the Sun in Conic Sections* (1809), tr. by C. Davis, Dover Pubs., Inc., NY, 1963.

_____, Theory of the Combination of Observations Which Leads to the Smallest Errors," in *Gauss' Work (1803-1826) on the Theory of Least Squares*, tr. by H.F. Trotter, Tech. Rept. No. 5, Dept. of Army, Project No. 5B99-01-004, Statistical Techniques Research Group, Section of Mathematical Statistics, Dept. of Math., Princeton Univ., Princeton, NJ, 1957.

Gibbs, J. Willard, "On the Equilibrium of Heterogeneous Substances," *Trans. of the Conn. Academy, III*, Oct. 1875 - May 1976, pp. 108-248, in *The Scientific Papers of the J. Willard Gibbs*, Vol. I, *Thermodynamics*, Dover Pubs., Inc., NY, 1961.

_____, *Elementary Principles in Statistical Mechanics*, Yale Univ., Press, New Haven, CT, 1902.

Good, Irving John, "The Population Frequencies of Species and the Estimation of Population Parameters," *Biometrika, 40*, 1953, pp. 237-264.

_____, *The Estimation of Probabilities: An Essay on Modern Bayesian Methods* (1965), MIT Press, Cambridge, MA, 1968.

Jaynes, Edwin T., "Information Theory and Statistical Mechanics," *The Physical Review, 106*, 1957, pp. 620-630, and *108*, 1957, pp. 171-190.

_____, "Information Theory and Statistical Mechanics," *Statistical Physics*, ed. by K. Ford (1962 Brandeis Lectures), W. Benjamin, Inc., NY, 1963, pp. 181-218.

_____, "Prior Probabilities," *IEEE Trans. on Systems Science and Cybernetics, SSC-4*, 1968, pp. 227-241.

Jeffreys, Harold, *Theory of Probability*, Oxford Univ. Press, London, 1961.

Keynes, John M., *A Treatise on Probability* (1921), Macmillan & Co., London, 1957.

Khale, Ludwig Martin, *Elementa Logicae Probabilium Methodo Mathematica in Usam Scientiarum et Vitae Adornata*, Halle, 1735.

Khinchin, A.E., *Mathematical Foundations of Statistical Mechanics*, tr. by G. Gamow, Dover Pubs., Inc., NY, 1949.

Laplace, Pierre Simon de, *Mém. de l'Acad. R.d. Sci., 6*, Paris, 1774, p. 621.

_____, *A Philosophical Essay on Probabilities* (1814), tr. by F. Truscott and F. Emory, Dover Pubs., Inc., NY, 1951.

Levine, Raphael D. and Myron Tribus (eds.), *The Maximum Entropy Formalism*, MIT Press, Camb., MA, 1979. Includes papers by Myron Tribus, Edwin T. Jaynes, Richard T. Cox, Robert B. Evans, N. Agmon, Y. Alhassid, Raphael D. Levine, Walter M. Elsasser, James C. Kech, C. Alden Mead, Baldwin Robertson, Rolf Landauer, John G. Pierce, Bernard O. Koopman, Jerome Rothstein, Edward H. Kerner, and George J. Chaitin.

Maxwell, James Clerk, *Theory of Heat*, Longmans, Green & Co., London, 1871.

Patrick, Edward A., "Proof That *A Posteriori* Density Maximizes Entropy," *Fundamentals of Pattern Recognition*, Prentice-Hall, Inc., Englewood Cliffs, NJ, 1972, pp. 156-160.

Planck, Max, *A Survey of Physics* (1923), tr. by R. Jones and D. Williams, reprinted as *A Survey of Physical Theory*, Dover Pubs., Inc., NY, 1960.

Rankine, William J.M., "On the Geometrical Representation of Expansive Action of Heat and the Theory of Thermo-dynamic Engines," *Phil. Trans.*, *144*, 1854, pp. 115-175.

Rao, C. Radhakrishna, *Linear Statistical Inference and Its Applications* (1965), J. Wiley & Sons, Inc., NY, 2nd ed., 1973.

Reichert, Thomas A., "The Amount of Information Stored in Proteins and Other Short Biological Sequences," *Proc. of the Sixth Berkeley Symposium on Mathematical Statistics and Probability*, Univ. of Calif. Press, Berkeley, CA, 1971, pp. 197-309.

Reichert, Thomas A., Donald N. Cohen and Andrew K.C. Wong, "An Application of Information Theory to Genetic Mutations and the Matching of Polypeptide Sequences," *J. Theor. Biol.*, *41*, 1973, pp. 245-261.

Reichert, Thomas A. and Andrew K.C. Wong, "Toward a Molecular Taxonomy," *J. Molec. Evol.*, *1*, 1971, pp. 99-111.

Renyi, Alfred, "On Measures of Entropy and Information," *Proc. of the Fourth Berkeley Symposium on Mathematical Statistics and Probability*, Vol. I, June 20 - July 30, 1960, Univ., of Calif. Press, Berkeley, CA, 1961.

Rothstein, Jerome, "Informational Generalization of Entropy in Physics," *Quantum Theory and Beyond* (ed. by T. Bastin), Cambridge Univ. Press, London, 1971, pp. 291-305.

Shannon, C.E., "A Mathematical Theory of Communication, Part III," *Bell System Technical Journal*, *27*, 1948, pp. 623-656.

Tribus, Myron, "Information Theory as the Basis for Thermostatics and Thermodynamics," *Jour. Appl. Mech.*, *28*, March 1961, pp. 1-8.

_____, *Thermostatics and Thermodynamics*, D. Van Nostrand Co., Inc., Princeton, NJ, 1961.

_____, "The Use of the Maximum Entropy Estimate in the Estimation of Reliability," *Recent Developments in Information and Decision Processes*, ed. by R.E. Mashol and Paul Gray, Macmillan Co., NY, 1962, pp. 102-140.

Tribus, Myron, R.B. Evans and G. Crellin, *The Use of Entropy in Hypothesis Testing*, Thayer School of Engineering, Dartmouth, Hanover, NY, 1963.

Tolman, Richard C., *The Principles of Statistical Mechanics*, Oxford Univ. Press, London, 1938.

van der Waals, J.D., Jr., "Ueber die Erklarung der Naturgesetze auf Statische-mechanischer Grundlage," *Physik. Zeischr. XII*, 1911, pp. 547-549.

Wrinch, Dorothy and Harold Jeffreys, "On Certain Fundamental Principles of Scientific Inquiry," *Phil. Mag.*, *42*, 1921, pp. 369-390 and *45*, 1923, pp. 368-374.

CHAPTER 9

FREQUENCY DISTRIBUTIONS
AND GOODNESS-OF-FIT TESTS

CHAPTER 9

FREQUENCY DISTRIBUTIONS AND GOODNESS-OF-FIT TESTS

A.   Frequency Distributions

The histograms in Figs. 5, 6 and 7 of Chapter 6 represent a more
or less "natural" ordering of the events. This is because the
dependent variables are *quantitative* in these cases, and "nearness"
along them is generally related to "nearness" in terms of the purposes for which we want to predict their values.*

The shape of the histogram in Fig. 8, on the other hand, is arbitrary. Because the dependent variable is *qualitative* in this case,
the ordering is arbitrary. We can, of course, arrange the Fig. 8
events in a sequence of increasing "utility" according to some
preference-order scheme, but this will depend upon the particular
application intended.

The shapes of histograms defined along a quantitative variable can
be described in terms of similarity to one or another curve defined
by a mathematical function. These functions are called *frequency
distributions* .

Frequency distributions are commonly employed in a much stronger
way than simply to describe the general shape of an empirical
histogram. Often a research worker will attempt to justify the
assumption that an observed histogram *is* a particular instance
of one or another of a set of candidate frequency distributions to
some appropriate degree of accuracy.

---

* One can, of course, imagine quantitative variables which would have to be
rearranged in order to satisfy the triangle inequality for an underlying
preference-order scheme.

The usual argument can be delineated into the following ten steps:

Step 1:   Define the variable to be predicted.

Step 2:   Define the sample of data.

Step 3:   Define a categorization... $< x_{k-1} < x_k < x_{k+1} < \dots$ of the dependent variable so that the sample of N events can be sorted into K categories and the frequencies computed.   They are given by

$$F_k = \frac{\text{\# events in } k^{th} \text{ category}}{\text{\# events in sample}} \quad , \quad k=1,\dots,K.$$

Step 4:   Define a candidate frequency distribution, $f(x)$, along the dependent variable.

$f(x)dx$ = probability of observed value lying between x and x+dx.

Step 5:   Compute the expected frequency for each category based on the candidate distribution.

$$f_k = \int_{x_{k-1}}^{x_k} f(x)\,dx$$

Step 6:   Define a test-statistic measure of error

$$E = \text{function of } \{F_k\} \text{ and } \{f_k\}$$

such that:

$E = 0$ if and only if $F_k = f_k$ for all $k=1,\dots,K$, and

$E < E'$ if and only if $\{F_k\}$ is, in some sense, *closer* to $\{f_k\}$ than is $\{F_k'\}$.

Step 7:   Compute the theoretical distribution for the test-statistic, assuming the candidate distribution $f(x)$ for the underlying variable.

$g(E)dE$ = probability of test-statistic being between E and E+dE assuming $f(x)$ to be the distribution of x.

If the exact distribution $g(E)$ is unknown, estimate it either by Monte-Carlo calculation or by investigation of its asymptotic shape at large sample sizes and its distance from the asymptote as a function of sample size.

Step 8:   Compute the observed histogram frequencies.

$$\hat{F}_k = \frac{\text{\# observed events in } k^{\text{th}} \text{ range}}{\text{\# events in observed sample}} \, .$$

Step 9:   Compute the value of the test-statistic for the actually observed sample.

$$\hat{E} = \text{function of } \{\hat{F}_k\} \text{ and } \{f_k\}.$$

Step 10: Compute the probability of obtaining a larger than observed test-statistic assuming the candidate distribution.

$$P(E > \hat{E}) = \int_{\hat{E}}^{\infty} g(E) \, dE.$$

Goodness-of-fit reasoning then concludes as follows: Assume that the data actually are distributed according to the candidate function. Then, if we draw a large number of samples, we expect to obtain one which fits the candidate more "poorly" than the observed sample roughly a fraction $P(E > \hat{E})$ of the time. We adopt the policy of rejecting candidate distributions for which $P(E > \hat{E})$ is below some previously chosen "significance" level, say 0.10, 0.05, or 0.01. For each significance level, we can compute the critical region $E_{\text{crit}} < E < \infty$ for which we will reject the candidate as being too "poorly" matched by the data. (The concept of "poorness" used here is, of course, defined by the choice of test-statistic E.)

The following example illustrates this type of reasoning:

Step 1:   *Define the variable*

          Lifetime (age to nearest integer years) of people who died by accident in the United States in 1977.

Step 2:   *Define the sample*

          n = 100 persons drawn by a specified random selection procedure from death records.

Step 3: *Define categories for the variable*

1: $0 \leq x \leq 5$ yrs	6: $45 < x \leq 54$
2: $5 < x \leq 14$	7: $55 < x \leq 64$
3: $15 < x \leq 24$	8: $65 < x \leq 74$
4: $25 < x \leq 34$	9: $75 < x \leq 84$
5: $35 < x \leq 44$	10: $85 < x$

Step 4: *Define candidate frequency distribution of the variable*

$$\text{Rectangular:} \quad f(x) = \begin{cases} 0.11 & 0 \leq x \leq 90 \\ 0 & \text{otherwise} \end{cases}$$

Step 5: *Compute expected frequencies based on candidate distribution*

$f_1 = 0.06$	$f_6 = .11$
$f_2 = .11$	$f_7 = .11$
$f_3 = .11$	$f_8 = .11$
$f_4 = .11$	$f_9 = .11$
$f_5 = .11$	$f_{10} = .06$

Step 6: *Define a test-statistic of "poorness" of fit*

$$T = n \sum_{k=1}^{K} \frac{(F_k - f_k)^2}{f_k}$$

Step 7: *Derive the distribution of this test-statistic for the candidate distribution of the variable*

$$g(T) \approx \frac{T^{\frac{\nu}{2} - 1} e^{-T/2}}{2^{\nu/2} \, \Gamma\left(\frac{\nu}{2}\right)} \quad \text{for large } n_k = f_k n$$

where $\nu = K-1$

Step 8: *Compute observed frequencies*

$\hat{F}_1 = 0.04$	$\hat{F}_6 = .09$
$\hat{F}_2 = .06$	$\hat{F}_7 = .09$
$\hat{F}_3 = .25$	$\hat{F}_8 = .09$
$\hat{F}_4 = .14$	$\hat{F}_9 = .09$
$\hat{F}_5 = .09$	$\hat{F}_{10} = .06$

Step 9:   *Compute the observed value of the test-statistic for the candidate distribution*

$$T = n \sum_{k=1}^{10} \frac{(F_k - f_k)^2}{f_k}$$

$$= 23.6$$

Step 10: *Compute the probability of the test-statistic having the observed value or greater, assuming the candidate distribution*

$$P(T > \hat{T}) = \int_{\hat{T}}^{\infty} g(T) \, dT$$

$$= 0.005$$

If we have adopted, for example, a 0.01 significance level criterion for rejecting the hypothesis that the sample is drawn from a population with the candidate distribution, then we would reject the rectangular distribution in this case.

Goodness-of-fit reasoning may appear to be based on Bayes' inverse probability theorem, but this is not so. It involves a separate assumption which is not derivable from Bayes' theorem.

For example, assume that:

O_1 = population is rectangular distributed

O_2 = population is gamma distributed

O_3 = population is Weibull distributed

.
.
.

C = test-statistic is greater than or equal to that for the empirical sample $(T > \hat{T})$

Now let us test hypothesis O_1. Bayes' theorem says:

$$P(O_1 \text{ given } T > \hat{T}) = \frac{P(O_1) P(T > \hat{T} \text{ given } O_1)}{P(O_1) P(T > \hat{T} \text{ given } O_1) + P(O_2) P(T > \hat{T} \text{ given } O_2) + \ldots}.$$

All that the test-statistic analysis gives us is the value of $P(T > \hat{T}$ given $O_1)$, i.e., the probability of getting as "poor" a fit as the observed, assuming the rectangular distribution (for the particular type of "poorness" defined by the test-statistic).

What we want is the inverse probability $P(O_1$ given $T > \hat{T})$. We cannot get it because we do not know $P(O_1)$ and the P's associated with all the other distributions. Depending upon the values of these unknown probabilities, the desired quantity $P(O_1$ given $T > \hat{T})$ can be *anywhere* in the interval from 0 to 1.

Thus, it is clear that when we reject a candidate distribution because the test-statistic gives a value of $P(T > \hat{T})$ below a selected "significance level", we are not reaching a purely mathematical conclusion. We are, rather, employing a statistical decision-making policy. This is the xenophobic policy of rejecting hypotheses on the basis of which the (pre-defined) class in which our actual observation happens to fall satisfies us as being sufficiently improbable. This is discussed further in Section 16.G.

B.   Candidates for Frequency Distributions

The shopping list of available mathematical distribution functions is very long. Some examples are given in Table 28.*

The equations describing a few of the more common of these functions are given in Appendices 1 and 2 along with sample plots and listings of some of their properties.

Most distribution functions depend upon one or more parameters as well as the variable in question. The problem of estimating these parameters is taken up in the next chapter. Here we assume that the parameter values are given and that our interest is in choosing among the many different available functional forms.

---

* Listed roughly in the order of appearance in volumes by Johnson and Kotz referenced at end of chapter.

Table 28.   Examples of Distribution Functions

o Binomial
o Singly truncated binomial
o Doubly truncated binomial
o Poisson
o Singly truncated Poisson
o Doubly truncated Poisson
o Compound Poisson
o Displaced Poisson
o Hyper-Poisson
o Poisson Gram-Charlier
o Kolmogorov
o Negative binomial
o Geometric
o Truncated geometric
o Polya-Eggenberger
o Inverse Polya-Eggenberger
o Miller
o Poisson-Pascal
o Pearson Type
o Hypergeometric
o Positive hypergeometric
o Negative hypergeometric
o Compound hypergeometric
o Noncentral hypergeometric
o Generalized hypergeometric
o Extended hypergeometric
o Logarithmic series
o Truncated logarithmic series
o Thomas
o Polya-Aeppli
o Neyman Type
o Log-Zero Poisson
o Woodbury
o Zeta
o Zipf
o Yule
o Harmonic
o Factorial
o Arfwedson's
o Borel-Tanner
o Ising-Stevens
o Morse
o Student's
o Normal
o Half-normal
o Singly truncated normal
o Doubly truncated normal
o Compound normal
o Lognormal
o Logarithmic Gram-Charlier
o Wald
o Truncated Wald
o Random walk
o Cauchy

o Generalized Cauchy
o Half-Cauchy
o Folded Cauchy
o Gamma
o Truncated gamma
o Compound gamma
o Generalized gamma
o Erlang
o Helmert's
o Exponential
o Pareto
o Double Pareto
o Generalized Pareto
o Lomax
o Champernowne
o Power-function
o Weibull
o Compound Weibull
o Log-Weibull
o Extreme value
o Gumbel
o Logistic
o Generalized logistic
o Hyperbolic secant
o Generalized hyperbolic secant
o Laplace
o Beta
o Compound beta
o Non-central beta
o Arc-sine
o Generalized arc-sine
o Uniform
o Triangular
o Binomial-uniform
o F
o Non-central F
o Bradford
o Student's t
o Pseudo-t
o Non-central t
o Chi-squared
o Non-central chi-squared
o Rayleigh
o Bose-Einstein
o Fermi-Dirac
o Wakeby
o Leipnik
o Kolmogorov-Smirnov
o Circular normal
o Wrapped-up normal
o Gompertz
o Planck
o Continuous Poisson
o Fisher-Stevens

C.   Criteria for Distribution Selection

Key considerations in choosing among candidate functions include:

o   What criteria do we adopt for selecting a goodness-of-fit test? What goodness-of-fit test do we select? How well does the shape of the candidate distribution function match the histogram of the observed data according to this test?

o   Are there any properties of the process producing the sample data which impose restrictions on the properties of the distribution function? If so, how well does the candidate function comply with these restrictions?

o   Is the function sufficiently simple for convenient mathematical manipulation? Do we know its basic properties? Are critical value tables for it available?

o   Does the candidate function have special properties such as asymptotic normality? Are these properties desirable in the given situation?

The permissible range of the variable is one important aspect of the shape of a distribution to consider. For example:

o   Unbounded distributions include: normal, Cauchy, extreme value, Laplace, logistic, chi-squared, Student's-t.

o   Bounded below only distributions include: log-normal, exponential, Weibull, F-distribution, gamma, binomial, negative binomial, Poisson, compound Poisson, geometric.

o   Bounded below and above distributions include: uniform, rectangular, triangular, beta.

Some unbounded distributions decay to zero sufficiently fast to reasonably approximate bounded ones. For example, under certain circumstances, the normal distribution approximates the binomial.

Several basic aspects of the shape of a distribution can be obtained from its first few moments.

o   Mean: The first moment indicates the value about which the bulk of the distribution is centered.

o   Variance: The second moment about the mean indicates how widely the distribution is dispersed about its mean. The positive square root of the variance is called the standard deviation, denoted by σ. Another measure of distribution spread or thickness is given for continuous distributions (see McConnell) by

$$\tau = e^{S_R},$$

where S_R is the renormalized entropy (see Sec. 14.C)

$$S_R = -\int f(x) \log f(x) \, dx,$$

and for discrete distributions by

$$\tau = e^S,$$

where

$$S = -\sum_x f(x) \log f(x).$$

We call τ the standard width.*

o Skewness: The third moment indicates the degree of symmetry about the mean. A distribution which is perfectly
   symmetric about its mean has zero skewness. Positive skewness indicates an asymmetrical distribution towards the
   right, negative towards the left. Distributions with zero
   skewness include the normal, uniform rectangular, triangular, Laplace, logistic, and Student's-t. Distributions with
   positive skewness include the log-normal, exponential, extreme
   value, chi-squared, gamma, negative binomial, Poisson, geometric and F-distribution. Distributions which may have
   either positive or negative skewness (depending upon a parameter value) include the binomial and the beta.

o Kurtosis: The fourth moment indicates the peakedness of
   the distribution compared to a normal distribution. The perfectly bell-shaped normal distribution has zero kurtosis.
   Distributions which are truncated at finite limits have negative kurtosis, while those with tails decaying more slowly
   than the normal have positive values. Distributions with positive kurtosis include the log-normal, exponential, extreme
   value, Laplace, logistic, chi-squared, Student's-t, gamma,
   negative binomial, Poisson, and geometric. Distributions with
   negative kurtosis include the uniform, rectangular, and triangular. Distributions which may have either positive or negative kurtosis (depending upon a parameter) include the binomial and the beta.

---

* For the normal distribution, $\tau = 4.1327\sigma$ (see McConnell). For the rectangular, $\tau = 3.4641\sigma$; for the isosceles triangular, $\tau = 4.0385\sigma$; for the Maxwell-
Boltzmann, $\tau = 4.0209\sigma$; for the exponential, $\tau = 2.7183\sigma$; for the logistic,
$\tau = 4.0738\sigma$; for the extreme value, $\tau = 3.7749\sigma$; and for the Laplace, $\tau = 3.8442\sigma$.
For the Cauchy distribution, $f(x) = (\pi b)^{-1}\left[1+(\frac{x-a}{b})^{-2}\right]^{-1}$, the standard width is
$\tau = 4\pi b$, while the standard deviation is infinite. For the log-normal distribu-
tion, $\tau = \sigma \sqrt{\dfrac{2\pi e}{e^{c^2}(e^{c^2}-1)}}$, where c is a shape-determining parameter. τ measures
the range of values of x spanned by $f(x)$. It is thus unchanged if segments of the dis-
tribution are rearranged along the x-axis. Appendix 1 shows two-parameter distributions plotted with common mean and entropy, rather than common mean and variance, compared to a hypothetical normally distributed sample. A slightly greater mismatch of
extremes is traded for a better match of the informational bulk of the distributions.
Fitting of variance, skewness and kurtosis can be used to fix shape-determining parameters and to choose among distributions.

D. Test-Statistics

Just as there is a long shopping list of distribution functions to
consider when fitting an observed histogram, there is also a long
shopping list of test-statistics to consider when measuring the
goodness-of-fit.

Examples whose properties have been studied and which are described
in generally available references are listed in Table 29.[*]

Table 29.   Examples of Test-Statistics

Bernoulli's Binomial Test
De Moivre's Z Test for Measurements
Poisson's Isolated Occurrences Test
Arbuthnott's Sign Test for the Median Difference
Cox-Stuart's S_2 Sign Test for Trend in Location
Cox-Stuart's S_3 Sign Test for Trend in Location
Cox-Stuart's S_3 Sign Test for Trend in Dispersion
Noether's Binomial Test for Cyclical Trend
Mosteller's Test of Predicted Order
Savur's Population Quantities Test
Pearson's Chi-Squared Test
Fisher's "Exact" Test
Westenberg's Median Test
Cox-Stuart's Median Test for Linear Trend
Westenberg's Test for Interquartile Range
Cochran's "Median" Test for Correlation
McNemar's Test for a Difference between Correlated Proportions
Fisher's Matched Pairs Test
Wilcoxon's Sum of Ranks Test
Wilcoxon's Signed Ranks Test
Wilcoxon's Stratified Test
Lord's Range Test
Friedman's Test
Kruskal-Wallis's Test
Walsh's Test for Location of the Median
Fisher's Unmatched Data Test
Mann-Whitney's Test
Pitman's Correlation Test

---

[*] Listed roughly in the order of presentation by James V. Bradley, *Distribution-Free Statistical Tests*, WADD Tech. Rept. 60-661, Wright-Patterson Air Force Base, Ohio, August 1960. See also J. Durbin, *Distribution Theory for Tests Based on the Sample Distribution Function*, SIAM, Philadelphia, PA, 1973; and R. Langley, *Practical Statistics*, Dover Pubs., Inc., NY, 1970.

Table 29 (cont.)

Spearman's Rank Difference Correlation Coefficient
Wald-Wolfowitz's Test for Serial Correlation
Kendall's Rank Order Correlation Test
Mann's Inversion Test for Linear Trend
Mann's K Test
Wald-Wolfowitz's Total Number of Runs Test
Mood-Mosteller's Length of the Longest Run Test
Ramachandran-Ranganathar's Sum of Squared Run Lengths
Dixon's Test
David's Chi-Square "Smooth" Test
Olmstead's Length of Longest Run Up or Down Test
Wilks-Epstein's Exceedances Test
Wilks-Epstein's Includances Test
Wilks-Epstein's Univariate Tolerance Limit Test
Smirnov's Maximum Absolute Deviation Test
Pyke's Statistic
Smirnov-Cramer-von Mises's W^2 Test
Birnbaum-Tang's Statistic
Kuiper's Statistic
Watson's Statistic
Ajne's Statistic
Anderson-Darling's Weighted Maximum Deviation Test
Tsao's Truncated Maximum Absolute Deviation Test
Marshall's Test
Kendall's Rank Serial Correlation
Wilcoxon's Two-Sample Rank Test
White's Two-Sample Rank Test
Kruskal-Wallis's H Test
Mann-Whitney's Multi-Sample Test
Pitman's Multi-Sample Test
Mood-Brown's Multi-Sample Median Test
Wilson's Median Mulit-Sample Test
Cochran's Q Test
Bloomquist's Median Test
Tukey's Divergent Population Tests
Moran's Transitivity of Preference Test
Moore-Wallis's Difference Sign Test
Foster-Stuart's D Test for Trend
Foster-Stuart's d Test for Trend
Cox-Stuart's S_1 Sign Test for Trend
David's Combinatorial Test
Olmstead-Tukey's "Corner" Test for Peripheral Association
Mood's Rank Test for Dispersion
Lehmann's "Quadruple" Test for Dispersion
Rosenbaum's Test for Dispersion
Mann's T Test
Daniel's Test
Terry's Test
Lehmann's Most Powerful Test
van der Waerden's X Test

Different test-statistics can, in general, give very different assessments of goodness-of-fit.*   Thus, as a guard against contrivedness, we need to consider criteria for selecting a test-statistic.

Assume we are testing hypothesis H_1: the data are drawn from a population with a specific candidate frequency distribution. Compare this to the following null hypothesis H_0: the population frequency distribution is that of a population on which we have no data, i.e., the maximum entropy probabilities are $\{w_k/w\}$, where w_k and w are the virtual counts described in Section 8.H and Chapter 20.

The significance level α_0 of the null hypothesis for a specific test-statistic T is defined as

$$\alpha_0 = \int_{T_0}^{\infty} f(T)\,dT,$$

where $f(T)$ is the distribution of this statistic and T_0 is its value for the null hypothesis. This equation defines the functional dependence of T_0 upon α_0, $T_0 = T_0(\alpha_0)$.

The significance level of the hypothesis H_1 is similarly defined as

$$\alpha_1 = \int_{T_1}^{\infty} f(T)\,dT,$$

where H_1 is assumed true, and T_1 is the value of the test-statistic on the data.

---

* For a striking example, see Janos Galambos, *The Asymptotic Theory of Extreme Order Statistics*, J. Wiley & Sons, Inc., NY, 1978, p. 90.

Desirable properties for a test-statistic include those listed in Table 30.

Table 30. Examples of Criteria
for Selecting Test-Statistics

o Unique perfect match characterization: The test-statistic has this property if there is a defined value (usually zero) corresponding to a perfect match between the candidate frequency distribution and the observed histogram.

o Monotonic in extent of mismatch: The test-statistic has this property if it deviates more and more from the perfect-match value as the fit becomes progressively "worse" (in some sense). For example, suppose all the $\hat{F}_k$ exactly match the corresponding f_k, $k=1,\ldots,K-1$, except one, say $k=j$. Then the test-statistic is monotonic in the difference $\hat{F}_k - f_j$ for $\hat{F}_j > f_j$, and similarly monotonic for $\hat{F}_j < f_j$. (Selection of the concept of "worse" is, ultimately, a decision-theory problem. See Chapter 25.)

o High power to discriminate against the null hypothesis H_o: This power is defined as the probability

$$P(\alpha_o) = \int_{T_o(\alpha_o)}^{\infty} f(T) dT,$$

assuming H_1 to be true.

o Unbiased: The significance level α_o of the null hypothesis H_o is higher if H_o is true than if it is false.

o Consistent: The discriminatory power increases with sample size.

o Distribution-Free: For goodness-of-fit testing it is typical to use tests which are calculable without explicitly referencing any specific distribution. This enables the use of general tabulations for critical values of the test-statistic, rather than the computation of them for each distribution separately.

o Convenient: The test-statistic has this property if its distribution is known (exactly or approximately) for the candidate distribution, and is either easy to compute or available in tabulated form.

These seven requirements are generally insufficient to uniquely determine a test-statistic. However, because it can be difficult to derive the distribution of a test-statistic for an assumed candidate frequency distribution, it is a frequent, though less than optimal, practice to accept the first test-statistic one can find meeting the requirements. Fortunately, there are some test-statistics which have well known distributions for a wide variety of candidate functions, so one usually does not have to look too far.

The most commonly used test-statistic is Pearson's T. It was proposed as a test-statistic by Pearson in 1900, based on its theoretical properties derived by Helmert in 1875. We shall describe it here, leaving others to be discussed in the next chapter.

Let n_k be the number of observed events in the kth category, and $n = \Sigma n_k$ by the total number of observed events. Let P_k be the probability of the kth category. Then Pearson's T is defined as

$$T = \sum_{k=1}^{K} \frac{(n_k - nP_k)^2}{nP_k} .$$

For a wide variety of candidate distribution functions underlying the probabilities $\{P_k\}$, Pearson's T is asymptotically chi-squared distributed.

$$f(T) \approx f_{\chi^2}(T) \qquad \text{for large } n_k ,$$

where the chi-squared distribution is defined as

$$f_{\chi^2}(T) = \frac{T^{\frac{\nu}{2}-1} e^{-\frac{T}{2}}}{2^{\frac{\nu}{2}} \Gamma(\frac{\nu}{2})} ,$$

with

$$\nu = \text{number of degrees of freedom.}$$

Frequency distributions for which Pearson's T test-statistic is asymptotically chi-squared distributed include:

- Multinomial distribution, with $\nu = K-1$.

- Binomial distribution, using maximum likelihood estimates of p and n, with $\nu = K-3$. (If n is known, then $\nu = K-2$.)

- Poisson distribution, using maximum likelihood estimate for μ, with $\nu = K-2$.

- Normal distribution, using maximum likelihood estimates of μ and σ, with $\nu = K-3$.

- General maximum likelihood estimation, with $\nu = K-1-\ell$, where ℓ = number of parameters estimated.

The chi-squared approximation to the true $f(T)$ distribution is generally quite good even when the expected number of events per class, nP_k, is small. The minimum expected number needed for T to have reasonable closeness to χ^2 behavior is generally accepted as 5 for $K \geq 5$ and as 6 or greater for $2 \leq K \leq 4$. Cochran claims the approximation to be good so long as not more than 20% of the classes have expectation between 1 and 5. For lower expectation cases, the T-test could, of course, still be used, but its distribution would have to be computed separately (such as by a Monte-Carlo technique) rather than being taken from χ^2 tables.

Setting $F_k = n_k/n$, Pearson's T can also be written as

$$T = n \sum_{k=1}^{K} \frac{(F_k - P_k)^2}{P_k}.$$

If $\{P_k\}$ are the class probabilities, then $f_{\chi^2}(T')dT'$ is (approximately) the probability that $T' < T < T' + dT'$. Suppose we formulate two hypotheses:

Null Hypothesis H_o: $P_k = w_k/w$, and

Test Hypothesis H_1: $P_k = f_k$,

where w_k and w were defined in Subsection 8.H. Assume we choose the following as test-statistics for H_o and H_1, respectively:

$$T_o = n \sum_{k=1}^{K} \frac{(F_k - w_k/w)^2}{w_i/w}, \text{ and}$$

$$T_1 = n \sum_{k=1}^{k} \frac{(F_k - f_k)^2}{f_k}.$$

If H_o is assumed true, then

$$\alpha(T_c) = \int_{T_c}^{\infty} f_{\chi^2}(T_o) dT_o$$

is the probability that $T_o > T_c$. We refer to α as the confidence level and T_c as the critical value. This equation also defines the functional dependency of T_c upon α.

$$T_c = T_c(\alpha).$$

If H_1 is assumed true, then

$$\beta(T_a) = \int_{-\infty}^{T_a} f_{\chi^2}(T_1) dT_1$$

is the probability that $T_1 < T_a$. We refer to β as the acceptance level and T_a as the acceptance value.

Suppose, for example, that we adopt, as our decision-making policy, a procedure of rejecting the null hypothesis at the $\alpha = 0.05$ level and accepting a test hypothesis at the $\beta = 0.07$ level. (There is no reason why the two should be the same, especially considering the fact that they refer to two different distributions.)

There are four possible outcomes of our decision process:

Consider H_1,
Reject H_o: In this case we might provisionally accept H_1, recognizing, of course, that it is not necessarily the best hypothesis. There will, in general, be another hypothesis which is better. But at least we know of this H_1 and it meets our criteria for decision-making.

Don't Consider H_1,
Reject H_o: In this case we must look further for another hypothesis H_1 to consider.

Don't Consider H_1,
Don't Reject H_o: In this case we might consider accepting the null hypothesis for purposes of decision-making.

Consider H_1,
Don't Reject H_o: In this case we are unable to distinguish between the two. One way to resolve the matter is to reassess the levels α and β. Another is to seek a test-statistic with greater discriminatory power.

The discriminatory power of a test-statistic is easy to define. However, for nonparametric estimation such as this, it is difficult to compute. Even if we do not compute it, the definition may give us an idea of what we are looking for.

The power of a test-statistic T to discriminate hypothesis H_1 from H_o is the probability that H_o will be rejected given that H_1 is true. In other words, it is the probability that $T_o > T_c$, given H_1. Formally, the power is given by

$$\text{Power} = \sum_{\{n_k\} \ni T_o > T_c} n! \prod_{k=1}^{K} \frac{f_k^{n_k}}{n_k} .$$

The difficulty is that this probability cannot be estimated by
using

$$\int_{T_c}^{\infty} f_{\chi^2}(T_o)\,dT_o \; ,$$

because when we assume H_1 rather than H_o, $f_{\chi^2}(T_o)$ no longer approximates the distribution for T_o. Rather, $f_{\chi^2}(T_1)$ approximates
the distribution for T_1. If T_o and T_1 only differed by a parameter value, this could be used to obtain the power. But in our
case T_o and T_1 are related through the entire set of numbers $\{n_k\}$.

E.   Distributions over Probability

Frequency distributions are used to describe histograms for quantitative variables of all kinds. Of special importance is the class
of histograms over the outcome probabilities themselves. Frequency distributions associated with this type of histogram are used to
represent uncertainties in the predictions for a specific set of outcomes. These might be thought of as "second level" probabilities.
Actually, they are just probabilities defined for a different event.
This event is an *ensemble* of N of the original events. The dependent variable for this ensemble-event is the outcome frequency for
the N original events.

As an example, consider the die-tossing experiment of Section 7.A.
In each such experiment, we roll a die 100 times. Now define six
dependent variables: the frequency of ones, the frequency of twos,
etc. Let us focus on the variable "frequency of ones". We can now
adopt a classification scheme for this variable. An example of a
5-class scheme is depicted in Fig. 34.

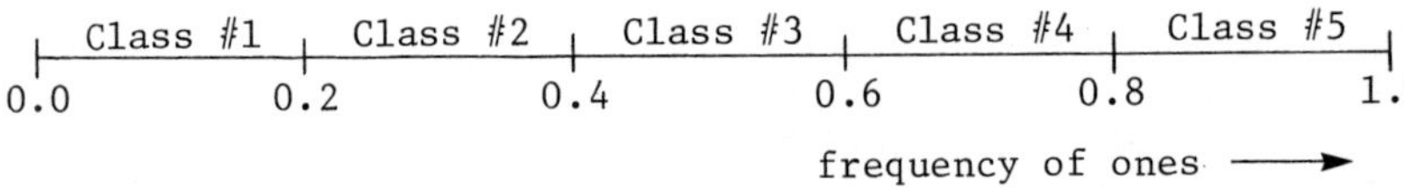

Fig. 34.   Five-Class Partition of Frequency Range

After collecting data by conducting a number of such experiments, we plot a histogram representing the results in this classification scheme. Since our dependent variable (the frequency of ones) is quantitative, the ordering of the outcomes is meaningful. Thus, it makes sense to ask what we might reasonably assume to be the underlying distribution function. For a finite number of independent repetitions of the experiment, the binomial distribution is generally appropriate. For a continuous variable with upper and lower bounds, the beta distribution is available. This distribution can be single-peaked near the mean, single-peaked at an extremum, or bi-peaked at both extrema. Fig. 35 shows an example which is single-peaked near the mean.

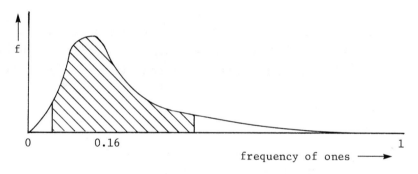

Fig. 35.   Example of Beta Distribution

(Note that a normal distribution would not be appropriate since our range is limited to values from 0 to 1, while normal distributions extend from minus infinity to positive infinity.)

The distribution shown in Fig. 35 is centered about a prediction of 0.16. Of special interest is the concentration of the area under this curve in the vicinity of our 0.16 prediction. One way to express this is in terms of the upper and lower limits of a segment which includes a specified percentage (say 70%) of the area under the curve (with equal amounts on each side of the prediction). We write this as:

$$\text{Prediction} = 0.16 \, {}^{+0.08}_{-0.06} \quad @ \; 70\% \; \text{confidence (symmetric).}$$

Note that these confidence intervals represent more information than simply the outcome probability for an individual die-toss. First, they are centered on a prediction which is one of a finite set of possible results of the 100 tosses. Second, their width is a function not only of the assumed die-toss probabilities ($\frac{1}{6}, \frac{1}{6}, \frac{1}{6}, \frac{1}{6}, \frac{1}{6}, \frac{1}{6}$) but also of the number of die-tosses (n = 100) involved in the prediction.

Of course, if we are providing predictive information to others, rather than making the prediction ourselves, we would not ourselves settle upon a prediction; but, rather, simply specify the probability. For example, to 3-significant figures, we might specify

$$\text{Probability} = 0.167 \, {}^{+0.075}_{-0.063} \quad @ \; 70\%.$$

The confidence interval portion of this statement presupposes that we have selected a number N of independent events to which we expect the probability to be applied for prediction purposes. If we do not know this, and are not willing to assume such a number, then all we can say is

$$\text{Probability} = 0.167,$$

and leave it to the person who finally makes the prediction to compute the confidence interval.

There is, however, an interpretation we can give to the confidence interval which can be useful even without committing ourselves to the ultimate purposes for which the information will be used.

This is to use the interval to represent the amount of data upon which our probability estimate is based. We do this by adopting the notational convention that a statement of the form

$$\text{Probability} = 0.167 \; ^{+0.075}_{-0.063}$$

refers to an imagined indefinite repetition of the actual experience upon which the prediction is based. Unless otherwise stated, the usual convention is that the upper and lower bounds refer to a symmetric confidence interval of approximately 70%.*

---

* If the distribution were normal, an interval of plus or minus one standard deviation would correspond to 68% confidence. Even for non-normal distributions, a symmetric 68% confidence interval is often referred to as being two standard deviations wide, although this is not strictly true. A normal distribution interval three standard deviations wide would be an 87% confidence interval, and a four standard deviation width would correspond to 95%. In terms of standard width units, $\tau = e^{SR}$, one-half standard width corresponds to a 70% confidence interval, two-thirds standard width to 83%, three-quarters standard width to 88%, and one standard width to 96%.

QUOTES

GAUSS (1809)

Therefore, that will be the most probable system of values of the unknown quantities p,q,r,s, *etc., in which the sum of the squares of the differences between the observed and computed values of the functions* V,V',V", *etc., is a minimum,* if the same degree of accuracy is to be presumed in all the observations. This principle, which promises to be of most frequent use in all applications of the mathematics to natural philosophy, must, everywhere be considered an axiom with the same propriety as the arithmetical mean of several observed values of the same quantity [and] is adopted as the most probable value.

This principle can be extended without difficulty to observations of *unequal* accuracy.

> —Karl F. Gauss (1777-1855)
> *Theory of the Motion of the Heavenly Bodies Moving about the Sun in Conic Sections* (Göttingen, 1809), tr. by C.H. Davis, Dover Pubs., Inc., NY, 1963, p. 260.

GAUSS (1821)

Let us denote by $\varphi(x)$ the relative probability of an error x: One should understand by that, because of the continuity of the errors, that $\varphi(x)dx$ is the probability that the error is contained between the limits x and x+dx. It is not in general possible to assign *a priori* the form of the function φ, and one may even state that this function will never be known in practice.

* * * *

The integral

$$\int_a^b \varphi(x)dx$$

expresses the probability that the error, still unknown, falls between the limits a and b.

* * * *

GAUSS (1821, cont.)

...the integral

$$\int_{-\infty}^{\infty} x\varphi(x)dx,$$

that is to say the average value of x...; similarly, the integral

$$\int_{-\infty}^{\infty} x^2\varphi(x)dx,$$

that is to say the average value of x^2, seems very suitable for defining and measuring, in a general way, the uncertainty of a system of observations...

*   *   *   *

Among the infinite number of functions...it seems natural to choose the simplest, which is, without doubt, the square of the error...

Laplace has considered the question in an analogous manner, but he adopted the absolute value of the error itself as a measure of the loss. This hypothesis, unless we deceive ourselves, is no less arbitrary than ours; should one in fact consider an error as more or less regrettable than an error of half the size repeated twice, and should one, in consequence, give it an importance double or more than double that of the second? The question is not clear and it is one on which mathematical arguments have no bearing; each must resolve it to his own liking. One cannot deny, however, that Laplace's hypothesis departs from the law of continuity and is consequently less suited to analytic study; ours, on the contrary, recommends itself by the generality and simplicity of its consequences.

> —Karl F. Gauss (1777-1855)
> "Theory of the Combination of Observations Which Leads to the Smallest Errors" (Göttingen, 15 Feb. 1821), *Gauss' Work (1803-1826) on the Theory of Least Squares*, tr. by H. Trotter, Tech. Rept. No. 5, DOA No. 5B99-01-004, Statistical Techniques Research Group, Dept. of Mathematical Statistics, Dept. of Math., Princeton Univ., Princeton, NJ, 1957, pp. 4, 5, 7, 8.

JEVONS (1874)

The inverse application of the rules of probability entirely
depends upon a proposition which may be thus stated, nearly in
the words of Laplace. *If an event can be produced by any one
of a certain number of different causes, all equally probable
à priori, the probabilities of the existence of these causes as
inferred from the event, are proportional to the probabilities
of the event as derived from these causes...*

We may thus state the result in general language. *If it is
certain that one or other of the supposed causes exists, the
probability that any one does exist is the probability that if
it exists the event happens, divided by the sum of all the
similar probabilities.*

* * * *

We form an hypothesis as to the logical conditions under which
the given instances might occur; we calculate inversely the
probability of that hypothesis, and compounding this with the
probability that a new instance would proceed from the same
conditions, we gain the absolute probability of occurrence of
the new instance in virtue of this hypothesis. But as several,
or many, or even an infinite number of mutually inconsistent
hypotheses may be possible, we must repeat the calculation for
each such conceivable hypothesis, and then the complete probability of the future instance will be the sum of the separate
probabilities. The complication of this process is often very
much reduced in practice, owing to the fact that one hypothesis
may be almost certainly true, and other hypotheses, though
conceivable, may be so improbable as to be neglected without
appreciable error.

When we possess no knowledge whatever of the conditions from
which the events proceed, we may be unable to form any probable
hypotheses as to their mode of origin. We have now to fall back
upon the general solution of the problem effected by Laplace,
which consists in admitting on an equal footing every conceivable
ratio of favourable and unfavourable chances for the production
of the event, and then accepting the aggregate result as the
best which can be obtained. This solution is only to be accepted in the absence of all better means, but like other results of the calculus of probability, it comes to our aid where
knowledge is at an end and ignorance begins, and it prevents us
from over-estimating the knowledge we possess.

> —W. Stanley Jevons (1835-1882)
> *The Principles of Science, A Treatise on Logic and
> the Scientific Method* (1874), Dover Pubs., Inc.,
> NY, 1958, pp. 242-243, 268-269.

PEIRCE (1878)

Now, as the whole utility of probability is to insure us in the long run, and as that assurance depends, not merely on the value of the chance, but also on the accuracy of the evaluation, it follows that we ought not to have the same feeling of belief in reference to all events of which the chance is even.   In short, to express the proper state of our belief, not *one* number but *two* are requisite, the first depending on the inferred probability, the second on the amount of knowledge on which that probability is based.

—Charles Sanders Peirce (1839-1914)
"The Probability of Induction," *Popular Science Monthly, 12,* April 1878, p. 710.

PEARSON (1900)

Let x_1, x_2...x_n be a system of deviations from the means of n variables with standard deviations σ_1, σ_2...σ_n and with correlations r_{12}, r_{13}, r_{23}...$r_{n-1, n}$.

...R is the determinant

$$
\begin{vmatrix}
1 & r_{12} & r_{13} & \cdots & r_{1n} \\
r_{21} & 1 & r_{23} & \cdots & r_{2n} \\
r_{31} & r_{32} & 1 & \cdots & r_{3n} \\
\cdot & \cdot & \cdot & \cdots & \cdot \\
\cdot & \cdot & \cdot & \cdots & \cdot \\
r_{n1} & r_{n2} & r_{n3} & \cdots & 1
\end{vmatrix}
$$

and R_{pp}, R_{pq} the minors obtained by striking out the pth row pth column, and the pth row and qth column. S_1 is the sum for every value of p, and S_2 for every pair of values of p and q.

Now let

$$\chi^2 = S_1\left(\frac{R_{pp}}{R}\frac{x_p^2}{\sigma_p^2}\right) + 2 S_2\left(\frac{R_{pq}}{R}\frac{x_p x_q}{\sigma_p \sigma_q}\right)$$

*   *   *   *

PEARSON (1900, cont.)

$$P = \frac{\int_{X}^{\infty} e^{-\frac{1}{2} \chi^2} \chi^{n-1} \, d\chi}{\int_{O}^{\infty} e^{-\frac{1}{2} \chi^2} \chi^{n-1} \, d\chi}$$

This is the measure of the probability of a complex system of n errors occurring with a frequency as great or greater than that of the observed system. ...evaluation...gives us what appears to be a fairly reasonable criterion of the probability of such an error occurring on a random selection being made.

* * * *

We thus seem in a position to determine whether a given form of frequency curve will effectively describe the samples drawn from a given population to a certain degree of fineness of grouping.

If it serves to this degree, it will serve for all rougher groupings, *but it does not follow* that it will suffice for still finer groupings. Nor again does it appear to follow that if the number in the sample be largely increased the same curve will still be a good fit. Roughly the χ^2's of two samples appear to vary for the same grouping as their total contents. Hence if a curve be a good fit for a large sample it will be good for a small one, but the converse is not true, and a larger sample may show that our theoretical frequency gives only an approximate law for samples of a certain size. In practice we must attempt to obtain a good fitting frequency for such groupings as are customary or utile. To ascertain the ultimate law of distribution of a population for any groupings, however small, seems a counsel of perfection.

—Karl Pearson (1857-1936)
"On the Criterion that a given System of Deviations from the Probable in the Case of a Correlated System of Variables is such that it can be reasonably supposed to have arisen from Random Sampling," *Philosophical Magazine and Journal of Science*, London, Taylor and Francis, July-Dec., 1900, pp. 157, 158, 166-167.

DUKENFIELD (1940)

Pick a card, any card.

—Claude William Dukenfield (W.C. Fields, 1880-1946)
Train whistle scene, *My Little Chickadee*,
Universal Pictures, New York, 1940.

EDWARDS (1972)

Since the support function as a whole represents the information, in a non-technical sense, at our disposal, it is clear
that no single number can convey the *amount* of information in
this sense. For support functions display too wide a variety
of form to permit a linear ordering according to some criterion
of general informativeness, even though we may agree that a
constant support conveys no information. The best we can do is
to look on the observed information as indicative of local informativeness near the maximum of the support.

—A.W.F. Edwards (1935-   )
*Likelihood, An Account of the Statistical Concept
of Likelihood and Its Application to Scientific
Inference*, Cambridge Univ. Press, London, 1972,
pp. 152-153.

REFERENCES

Afifi, A.A. and Azen, S.P., *Statistical Analysis, A Computer Oriented Approach*, Academic Press, NY, 1972, pp. 49-50.

Bradley, James V., *Distribution-Free Tests*, WADD Tech. Rept. 60-661, Behavioral Science Lab, Wright-Patterson Air Force Base, Ohio, August 1960.

Christensen, R.A., *Foundations of Inductive Reasoning*, Berkeley, CA, 1964.

_____, "Crossvalidation: Minimizing the Entropy of the Future," *Information Processing Letters*, *4*, Dec. 1975, pp. 73-76. [*Chapter 10 of Volume III.*]

Cochran, W.G., "The Chi-Squared Test of Goodness-of-Fit," *Ann. Math. Statist.*, *23*, 1952, p. 315.

_____, "Some Methods For Strengthening the Common Chi-Squared Tests," *Biometrics*, *10*, 1954, p. 417.

De Moivre, Abraham, "A Method of Approximating the Sum of the Terms of the Binomial $(a+b)^n$ Expanded Into a Series, From Whence are Deduced Some Practical Rules to Estimate the Degree of Assent Which is to be Given to Experiments," Nov. 12, 1733, reproduced in *The Doctrine of Chances* (London, 1756), Chelsea Publ. Co., NY, 1967, pp. 243-259.

Durbin, J., *Distribution Theory for Tests Based on the Sample Distribution Function*,. Soc. for Industrial and Applied Math., Philadelphia, PA, 1973.

Eadie, W.T., *et al.*, *Statistical Methods in Experimental Physics*, North-Holland Pub. Co., Amsterdam, 1971, p. 257.

Edwards, A.W.F., *Likelihood, An Account of the Statistical Concept of Likelihood and Its Applications to Scientific Inference*, Camb. Univ. Press, London, 1972.

Gauss, Karl F., *Theory of the Motion of the Heavenly Bodies Moving about the Sun in Conic Sections* (1809), tr. by C.H. Davis, Dover Pubs., Inc., NY, 1963.

_____, "Theory of the Combination of Observations Which Leads to the Smallest Errors" (Göttingen, 15 Feb. 1921), *Gauss's Work (1803-1826) on the Theory of Least Squares*, tr. by H. Trotter, Tech. Rept. No. 5, DOA No. 5B99-01-004, Statistical Techniques Research Group, Dept. of Mathematical Statistics, Dept. of Math., Princeton Univ., Princeton, NJ, 1957.

Gumbel, E.J., "On the Reliability of the Classical Chi-Squared Test," *Annals of Math. Statistics*, *14*, 1943, pp. 253-263.

Jevons, W. Stanley, *The Principles of Science, A Treatise on Logic and the Scientific Method* (1847), Dover Pubs., Inc., NY, 1908.

Johnson, Norman L. and Samuel Kotz, *Discrete Distributions*, J. Wiley & Sons, Inc., NY, 1969.

_____, *Continuous Univariate Distributions* (2 vols.), J. Wiley & Sons, Inc., NY, 1970.

_____, *Continuous Multivariate Distributions*, J. Wiley & Sons, Inc., NY, 1972.

Kramp, Christian, *Analyse des Refractions Astronomiques et Terrestres*, Strasbourg, 1799.

Langley, Russell, *Practical Statistics*, Dover Pubs., Inc., NY, 1970.

McConnell, Robert K., "Minimax Description Analysis of Faulted Data," paper presented to the Canadian Exploration Geophysical Society - American Geophysical Union, Mining Geophysics Symposium, Toronto, Canada, May 22-23, 1980.

Noel, Paul G., *Intro. to Mathematical Statistics*, J. Wiley & Sons, Inc., NY, 1954, pp. 163, 170.

Pearson, Karl, "On the Criterion that a given System of Deviations from the Probable in the Case of a Correlated System of variables is such that it can be reasonably supposed to have arisen from Random Sampling," *Phil. Mag. and Jour. of Science*, Series 5, *50*, London, Taylor and Francis, 1900, pp. 157-175.

Peirce, Charles Sanders, "The Probability of Induction," *Popular Science Monthly*, *12*, April 1878, pp. 705-718.

Yates, Frank, "Contingency Tables Involving Small Numbers and the χ^2 Test," *J.R. Stat. Soc.*, *Suppl. 1*, 1934, pp. 217-235.

CHAPTER 10

PARAMETRIC AND NONPARAMETRIC PROBABILITY ESTIMATORS

CHAPTER 10

PARAMETRIC AND NONPARAMETRIC PROBABILITY ESTIMATORS

A.   Parametric and Nonparametric Estimators

Methods for estimating numerical values for probabilities can be roughly divided into two categories: parametric and nonparametric. Parametric methods presume a specific functional form for the underlying distribution and adopt a specific rule for using observational data to assign values to parameters in the distribution. Probability estimates are then computed from the presumed distribution using these parameter estimates.

Nonparametric methods do not presume a specific functional form for the underlying distribution.   They are based solely on adoption of a rule for assigning estimated values to the probabilities.   The rule may be merely one of assuming a priori certain values (or of using values derived from other probabilities assumed a priori). Alternatively, the rule may involve the use of empirical observations.

In its specification of a functional form, each parametric method must ultimately be based upon a nonparametric method as the source of the functional form.   Thus, the nonparametric methods are, in this sense, the more fundamental.   The following important and difficult questions arise with respect to the selection of parametric methods:

o   What is the criterion for selecting a particular
    candidate functional form as possibly representing
    the underlying distribution?   If it involves comparing
    sample distributions to various assumed functional
    forms, what measure of similarity is used?   Why?
    Is a "deeper" distribution assumed in justifying the
    similarity measure used?   If so, how is it obtained?

o   What are the criteria for selecting a parameter
    estimation rule?   Is probability involved in these
    criteria (e.g., rate of convergence in probability)?
    If so, how do we estimate this probability?

In this chapter we briefly review examples of both parametric and nonparametric probability estimation:

    Parametric

        equal moments
        minimum T
        maximum likelihood
        maximum support
        minimum entropy decrement

    Nonparametric

        equiprobability
        sample average
        linear shrinkage
        maximum entropy

In accord with conventional practice in statistics, we classify the parametric methods by the assumed rule rather than by the assumed underlying distribution.

Here we will only review some of the basic ideas* so that we can get a picture of the role played by information theory and entropy. The key points in this respect include the following:

o The maximum likelihood ratio test is equivalent to a minimum entropy decrement criterion. As a test-statistic, it can be used for both parametric and nonparametric estimation.

o The maximum entropy principle leads to nonparametric probability estimates with shrinkage properties similar to Stein-James-Lindley estimators.

o The minimum entropy principle provides a criterion for partitioning the independent variable space.

We discuss the first two of these in this chapter. The third point is taken up in subsequent chapters.

------

* For a comprehensive and detailed exposition, a number of excellent compendia of distributions and test-statistics are available. Examples include:

• N.L. Johnson and S. Kotz, *Distributions in Statistics* (4 volumes), J. Wiley & Sons, Inc., NY, 1969-1972.
• James V. Bradley, *Distribution-Free Statistical Tests*, WADD Tech. Rept. 60-661, Behavioral Science Lab., Aerospace Medical Div., Wright-Patterson Air Force Base, Ohio, August 1960.
• W.T. Eadie, D. Drijard, F.E. James, M. Roos and B. Sadoulet, *Statistical Methods in Experimental Physics*, North-Holland Pub. Co., Amsterdam, 1971.

B.   Equal Moments Estimators

The oldest modern method of obtaining point estimates of parameters
is equal moments estimation. It involves making the assumption that
it is a reasonable estimate to set the distribution moments equal
to the sample moments. This gives a set of equations which can be
solved for estimates of the unknown parameters.

Let x be the dependent variable and assume an underlying distribu-
tion function $f(x; \theta_1, \ldots, \theta_m)$ with parameters $\theta_1, \ldots, \theta_m$. The
j^{th} moment of the distribution is defined as*

$$M_j^d(\theta_1, \ldots, \theta_m) = \int_{-\infty}^{\infty} x^j \, f(x; \theta_1, \ldots, \theta_m) \, dx \,.$$

(In all cases, when we use a discrete rather than a continuous distribution, we make the obvious change from an integration to summation using $\Delta x = 1$.)

Now assume that we have a training sample with n observations of the
dependent variable of values $x_1, x_2, \ldots, x_n$. The jth moment of this
sample with respect to the assumed distribution is defined as

$$M_j^s = \frac{1}{n} \sum_{k=1}^{n} x_k^j \,.$$

The equal moments estimators of the parameters are those values
$\hat{\theta}_1, \ldots, \hat{\theta}_m$ for which the first m distribution and sample moments
are equal. This yields a set of m equations in m unknowns:

---

* The jth moment can, alternatively, be defined as the jth derivative of the
moment generating function with respect to a dummy parameter u, evaluated at
u=0, where the moment generating function is defined as $\int e^{ux} f(x) \, dx$.

$$M_1^d(\hat{\theta}_1, \ldots, \hat{\theta}_m) = M_1^s$$

$$M_2^d(\hat{\theta}_1, \ldots, \hat{\theta}_m) = M_2^s$$

$$\vdots$$

$$M_m^d(\hat{\theta}_1, \ldots, \hat{\theta}_m) = M_m^s$$

Once the parameter estimates are known, we can estimate the outcome probabilities for any desired classification of the dependent variable values. For instance, the estimate of the probability for the interval $x_{k-1} \leq x \leq x_k$ is given by

$$P_k = \int_{x_{k-1}}^{x_k} f(x; \theta_1, \ldots, \theta_m) \, dx .$$

As an example, assume that the dependent variable is the discrete set of integers 1, 2,...,K and that the underlying distribution is the equiprobability distribution given by

$$f(x) = \frac{1}{K} \quad \text{for all } x = 1, 2, \ldots, K.$$

This distribution has no parameters (K is fixed), so $m = 0$ equations are required. Thus, equiprobability outcome probabilities,

$$P_k = \frac{1}{K} \quad \text{for all } k = 1, 2, \ldots, K,$$

are implied for the uniform distribution.

As a second example, assume that the dependent variable is the discrete set of two integers 0, 1, and that the underlying distribution is a function of one parameter:

$$f(x) = \begin{cases} 1-p & \text{for } x = 0 \\ p & \text{for } x = 1. \end{cases}$$

We need one equation to determine the one parameter. The first moment of the distribution is given by

$$M_1^d = \sum_{x=0}^{1} x \ f(x;p)$$

$$= p.$$

The first moment of the sample is

$$M_1^s = \frac{1}{n} \sum_{k=1}^{n} x_k.$$

The equal moments estimator $\hat{p}$ for the parameter p is the value for which $M_i^d = M_1^s$,

$$\hat{p} = \frac{1}{n} \sum_{k=1}^{n} x_k.$$

Thus, the equal moments probability estimates are

$$P_o = 1 - \frac{1}{n} \sum_{k=1}^{n} x_k, \text{ and}$$

$$P_1 = \frac{1}{n} \sum_{k=1}^{n} x_k.$$

Note that these probability estimates are just the logical consequences of the assumption of the form of the distribution function and the equal moments assumption $M_1^d = M_1^s$.

As a third example, assume the dependent variable x is continuous and the underlying distribution is normal,

$$f(x;\mu,\sigma) = \frac{1}{\sigma\sqrt{2\pi}} \; e^{-\frac{1}{2}\left(\frac{x-\mu}{\sigma}\right)^2}.$$

Now we have two parameters, μ and σ, for which we need two equations.   The first two distribution moments are

$$M_1^d = \int_{-\infty}^{\infty} x \; f(x;\mu,\sigma) \; dx$$

$$= \mu, \text{ and}$$

$$M_2^d = \int_{-\infty}^{\infty} x^2 \; f(x;\mu,\sigma) \; dx$$

$$= \sigma^2 + \mu^2.$$

The first two sample moments are

$$M_1^s = \frac{1}{n} \sum_{k=1}^{n} x_k, \text{ and}$$

$$M_2^s = \frac{1}{n} \sum_{k=1}^{n} x_k^2.$$

Equating $M_1^d = M_1^s$, our equal moments estimate $\hat{\mu}$ of the parameter μ is

$$\hat{\mu} = \frac{1}{n} \sum_{k=1}^{n} x_k.$$

Equating $M_2^d = M_1^s$ and rearranging terms, our equal moments estimate $\hat{\sigma}$ of the parameter σ is given by

$$\hat{\sigma}^2 = \frac{1}{n} \sum_{k=1}^{n} (x_k - \hat{\mu})^2 .$$

The equal moments probability estimate for the interval from $x_k - 1$ to x_k is

$$P_k = \frac{1}{\hat{\sigma}\sqrt{2\pi}} \int_{x_{k-1}}^{x_k} e^{-\frac{1}{2}(\frac{x-\hat{\mu}}{\hat{\sigma}})^2} dx .$$

These are just a few of the many hundreds of distribution functions for which equal moments parameter estimates have been derived.

Equal moments estimation has a number of disadvantages:

   o It is arbitrary in that it equates only as many
     moments as the function happens to have parameters.

   o Its estimates are not necessarily unbiased. For
     example, $\hat{\sigma}^2$ derived above is not an unbiased estimate
     of σ^2 for the normal distribution.*

   o Its estimates tend to be fairly inefficient. They
     tend to have a large variance, so large sample
     sizes are needed to achieve confidence of being
     near the true value.

---

* An unbiased estimate of σ^2 can be obtained by multiplying the equal moments estimate $\hat{\sigma}^2$ by the factor $n/(n-1)$.

C.   Minimum T Estimators

A more popular parametric method of probability estimation is minimization of a chi-squared type distribution function.

Let x be the dependent variable and assume a functional form $f(x; \theta_1, \ldots, \theta_m)$ for the underlying distribution with parameters $\theta_1, \ldots, \theta_m$. Assume, further, that we have segmented x into K outcome classes, so that the probability for each class,

$$P_k(\theta_1, \ldots, \theta_m) = \int_{x_{k-1}}^{x_k} f(x; \theta_1, \ldots, \theta_m) \, dx,$$

could be computed if we knew the values of the parameters $\theta_1, \ldots, \theta_m$. Finally, assume that we have a training sample of $n = \sum_k n_k$ observations of the dependent variable, where n_k is the number observed in the kth outcome class.

The minimum chi-squared estimates of these parameters are generally defined as those values $\hat{\theta}_1, \ldots, \hat{\theta}_m$ for which the function

$$T = \sum_{k=1}^{K} \frac{[n_k - n \, P_k(\theta_1, \ldots, \theta_m)]^2}{n \, P_k(\theta_1, \ldots, \theta_m)}$$

is a minimum. Strictly speaking, T is only approximately chi-squared distributed, so this is more accurately referred to as *minimum T estimation*.

Minimum T estimators have been developed for a number of different distributions.

D.   Maximum Likelihood Estimators

The most popular parametric method of probability estimation is
maximum likelihood estimation.

Let x be our dependent variable and assume an underlying distribution $f(x; \theta_1, \ldots, \theta_m)$ with parameters $\theta_1, \ldots, \theta_m$.

Assume, also, that we have a training sample of n independent observations of x.   Label them $x_1, x_2, \ldots, x_n$.   If we knew the values
of the parameters $\theta_1, \ldots, \theta_m$, we could compute the value of the distribution function at each sample point: $f(x_1; \theta_1, \ldots, \theta_m), \ldots,$
$f(x_n; \theta_1, \ldots, \theta_m)$.   Then we could multiply all these values together
and obtain a grand product.   This grand product is called the likelihood function of the parameters $\theta_1, \ldots, \theta_m$ for the sample points
$x_1, \ldots, x_n$.

$$L(\theta_1, \ldots, \theta_m) = f(x_1; \theta_1, \ldots, \theta_m) \cdot f(x_2; \theta_1, \ldots, \theta_m) \cdots f(x_n; \theta_1, \ldots, \theta_m)$$

The maximum likelihood estimates of the parameters are those values
$\hat{\theta}_1, \ldots, \hat{\theta}_m$ for which this likelihood function L is a maximum.

Maximum likelihood estimates can be shown to have the following
properties:

   o   consistency,

   o   asymptotic normality, and

   o   asymptotic efficiency.

However, like equal moment estimates, maximum likelihood estimates
are often biased.   They have the advantage of being generally more
efficient than equal moments estimators.

Maximum likelihood estimators have been derived for the parameters
for a large number of different distributions.

For example, assume that x is continuous and that the underlying distribution is normal,

$$f(x;\mu,\sigma) = \frac{1}{\sigma\sqrt{2\pi}} \; e^{-\frac{1}{2}(\frac{x-\mu}{\sigma})^2} .$$

Then, for a set of observations $\{x_1,\dots,x_n\}$ the likelihood function is

$$L(\mu,\sigma) = \frac{\exp\{-\frac{1}{2}(\frac{x_1-\mu}{\sigma})^2\}}{\sigma\sqrt{2\pi}} \; \frac{\exp\{-\frac{1}{2}(\frac{x_2-\mu}{\sigma})^2\}}{\sigma\sqrt{2\pi}} \; \dots \; \frac{\exp\{-\frac{1}{2}(\frac{x_n-\mu}{\sigma})^2\}}{\sigma\sqrt{2\pi}} .$$

With these definitions, it can be shown that the maximum likelihood estimator $\hat{\mu}$ for μ is

$$\hat{\mu} = \frac{1}{n} \sum_i x_i ,$$

whether or not σ is known. If μ is unknown, the maximum likelihood estimator for σ is

$$\hat{\sigma}_{\hat{\mu}} = \sqrt{\frac{1}{n} \sum_i (x_i - \hat{\mu})^2} .$$

If μ is known (an unusual case), then the maximum likelihood estimator for σ is given by

$$\hat{\sigma}_{\mu} = \sqrt{\frac{1}{n} \sum_i (x_i - \mu)^2}$$

$$= \sqrt{\hat{\sigma}_{\hat{\mu}}^2 + (\hat{\mu} - \mu)^2} .$$

The outcome probability estimates are obtained as

$$P_i = \int_{x_{i-1}}^{x_i} f(x;\hat{\mu},\hat{\sigma}) \ dx.$$

The sample average $\hat{\mu}$ is an unbiased estimator of μ, but $\hat{\sigma}_{\hat{\mu}}$ is a biased estimator of σ.

The likelihood function contains all the information available in the training sample about the parameters. Consider, for example, a very small region Δx about each data point. Then

$$f(x_i;\theta_1,\ldots,\theta_m) \ \Delta x$$

is the probability, based on the assumed distribution, of finding the data point in the interval x_i to $x_i + \Delta x$, to an arbitrary accuracy for arbitrarily small Δx.

The information supplied by the ith data point to an observer whose only prior knowledge is the distribution function is

$$I(O(i)) = -\log f(x_i;\theta_1,\ldots,\theta_m) \ \Delta x.$$

Since the data points are assumed independent, the amount of information in their combined occurrence is the sum of their individual amounts of information,

$$I(O(1) \wedge O(2) \wedge \ldots \wedge O(m)) = -\sum_{i=1}^{n} \log f(x_i;\theta_1,\ldots,\theta_m) \ \Delta x$$

$$= -\log \prod_{i=1}^{n} f(x_i;\theta_1,\ldots,\theta_m) - n \log \Delta x$$

$$= -\log L(\theta_1,\ldots,\theta_m) - n \log \Delta x.$$

The last term, n log Δx, is an arbitrary constant,* and thus contains no sample information about the parameters. Thus all of the information is contained in the likelihood function $L(\theta_1, \ldots, \theta_m)$. However, as has been pointed out (Eadie, *et al.*, p. 156), this does not mean that estimating the $\theta_1, \ldots, \theta_m$ by values which maximize the likelihood function is necessarily the best use of this information. For example, this procedure may be inferior to procedures which produce equally consistent but unbiased estimates.

E. Equiprobability Estimators

Equiprobability estimation is the most elementary nonparametric method of probability estimation.

Suppose that we have decomposed the outcome space into a fine-structure of distinguishable ways of occurring such that we have no reason for believing any one to be more likely than any other. If there are w_k ways in which the kth outcome can occur, then the equiprobability estimator is given by

$$P_k = w_k/w,$$

where

$$w = \sum_k w_k.$$

Note that we have only assumed that we have no reason, based on our available information, to believe one fine-structure possibility to be more likely than another. It still may be a fact that their probabilities are unequal. So use of the equiprobability estimator entails additional content beyond the "lack of information" assumption. This additional content is provided by a substantive rule such as the maximum entropy principle.

---

* When considering sample distributions, this term is no longer arbitrary, and enters into the computation of specification entropy.

F.   Sample Average Estimators

Sample average estimation is the most elementary empirically based nonparametric method.

Assume that our dependent variable space is partitioned into K distinct outcomes, and that we have a training sample of n independent observations with the results $\{n_1, n_2, \ldots, n_K\}$.

The sample average estimator of the probabilities for the various outcomes are

$$P_k = n_k/n,$$

where

$$n = \sum_{k=1}^{K} n_k$$

is the total number of events in the sample.

In 1809, Gauss showed that, for normally distributed variables, the sample average is an unbiased maximum likelihood estimator of the mean.   Since many* common distributions are asymptotically normal in the limit of large samples, this has given sample average estimation special status even in nonparametric estimation where we do not assume any particular distribution.

In the 1920s, Fisher showed that the sample average contains all the information which the sample has about the mean.   In the 1940s it was shown that the sample average is the least squared error unbiased estimator of the mean.   In 1951, Blyth, Lehmann and Hodges showed that no other estimator of the mean has so low an expected

---

* Included, for example, are the binomial, multinomial, Poisson, chi-squared, Student's-t and F-distributions.

value of the mean squared error. These results solidified even further the status of sample average estimation.

G. Linear Shrinkage Estimators

In 1955, Charles Stein showed that, even though the sample average estimator P_k is the least squared error estimator for the kth outcome class for events of one type, this is not true if there are many types. In 1960, in collaboration with James, he extended this proof to any situation in which there are more than two types of events. They further produced an example of an estimator, involving shrinkage to zero, which is uniformly better than sample averages for normally distributed populations with three or more types of events. Lindley, commenting upon this development in 1962, suggested shrinkage to the grand average rather than to zero as a better estimator for four or more event types.

To illustrate these concepts, suppose that

$$n_{ki} = \text{number of events of } i^{th} \text{ type observed with } k^{th} \text{ outcome,}$$

$$N_i = \sum_{k=1}^{K} n_{ki},$$

$$n_k = \sum_{i=1}^{L} n_{ki},$$

$$K = \text{number of outcome classes, and}$$

$$L = \text{number of types of events.}$$

Now define the kth outcome average for the ith type as

$$a_{ki} = n_{ki}/N_i,$$

and the grand average for the kth outcome as

$$\bar{a}_k = \frac{1}{n_k} \sum_{i=1}^{L} n_{ki}.$$

The estimator suggested by Stein, James and Lindley for the probability $P(O_k$ given $C_i)$ of the kth outcome given the ith event type is

$$P(O_k \text{ given } C_i) = \bar{a}_k + \alpha_{ki}(a_{ki} - \bar{a}_k),$$

where α_{ki} is a "shrinking" factor. When $\alpha_{ki} = 1$ we use the particularized averages for each event type as the outcome probability estimators for that event type:

$$P(O_k \text{ given } C_i) = a_{ki} \text{ for } \alpha_{ki} = 1.$$

When $\alpha_{ki} = 0$ we shrink the individual estimates to the overall average

$$P(O_k \text{ given } C_i) = \bar{a}_k \text{ for } \alpha_{ki} = 0.$$

The shrinking factor for the Stein-James-Lindley estimator is given by

$$\alpha_{ki} = 1 - \frac{(L-3) \ \sigma_{ki}^2}{\sum_{j=1}^{L} (\bar{a}_k - a_{kj})^2},$$

where σ_{ki}^2 is the variance in the kth outcome mean for the ith event type. For example, a Poisson distribution gives

$$\sigma_{ki}^2 = \bar{a}_k / N_i ,$$

while a binomial distribution gives

$$\sigma^2_{ki} = \bar{a}_k(1-\bar{a}_k)/N_i .$$

If α_{ki} as computed above would be negative, the value of zero is used instead to avoid shrinking in the wrong direction.

No matter what the true means are, the shrunk estimators perform better than the unshrunk estimators in the sense that they have lower expected squared error over all the event types. The improvement is greater the larger the number of event types, particularly when the number exceeds 5 or 6. (The expected squared error is proportional to 3/L.) The improvement is especially great when the true means for the individual types are near each other. There is at least marginal improvement for all values of the true means.

H. Maximum Entropy Estimators

The maximum entropy criterion gives nonparametric estimators that combine aspects of the equiprobability and sample average estimators.

The maximum entropy estimators are values for the probabilities $\{P(O_k \text{ given } C_i)\}$ which maximize the entropy

$$S_i = -\sum_k P(O_k \text{ given } C_i) \log P(O_k \text{ given } C_i),$$

subject to given constraints. When the available information consists exclusively of a sample of $N = \sum_i N_i$ observed events together with a background of w virtual events, this yields

$$P(O_k \text{ given } C_i) = \frac{n_{ki} + w_k}{N_i + w}.$$

Maximum entropy estimators have the shrinkage feature of Stein-James-Lindley estimators inasmuch as the ratios w_k/w represent background information about the grand average for all event types $(i=1,\ldots,L)$.

An important point about maximum entropy estimators is that they depend upon the values of w and N_i for the event type, whereas the equiprobability and sample average estimators do not. Dependence upon N_i is no problem because we know the observed sample size. However, dependence upon w is something new.

For example, having decomposed the outcome space sufficiently to represent uniformly our state of ignorance, why cannot we decompose it further? This would increase w, yet comply entirely with the conditions we require for using the equiprobability estimators. So there must be an additional condition required by the maximum entropy estimator. This additional condition is the requirement that w have a value which minimizes the informational disagreement between the estimated probabilities and the observed data. We shall take up this subject again in Chapter 20 when we discuss entropy minimax crossvalidation.

I. Criteria for Selecting Parameter Estimation Method

A parametric probability estimation method is generally considered desirable to the extent that it possesses the following characteristics:

o Consistency: The estimator converges (in probability) to the true value of the parameter as the number of observations increases.

o Unbiased: The estimator's expectation value is the true parameter value. (A consistent estimator is asymptotically unbiased, but may still be biased for finite sample sizes.)

o Minimum Information Loss: The estimator contains more information about the parameter than any other estimator.

o Minimum Variance (maximum efficiency): The estimator has a small variance (so we can have confidence that, neglecting the possibility of bias, it is near the true value of the parameter).

o Asymptotic Normality: The distribution of the estimator approaches the normal distribution for very large samples.

o Uncorrelated: The estimator of one parameter of the distribution is uncorrelated with the estimate of another parameter.

o Robustness: The estimator is relatively insensitive to the correctness of the choice of the distribution.

o Minimum Disutility: The estimator minimizes the expected disutility resulting from an error, considering the purposes for which the estimate is being used and also the utility of being correct.

o Communicability: The properties of the estimate are sufficiently well understood by the users to make it useful.

o Computational Ease: The human and computer resources required by the estimator are not excessive.

A few of the many statistics proposed to evaluate goodness-of-fit on an independent sample of test data include:

o Pearson's T sum of normalized squared differences test-statistic

o Wald-Wolfowitz's N_r run count test-statistic

o Neyman-Barton's P_s^2 smooth component test-statistic

o David's N_o empty cell count test-statistic

o Smirnov-Cramer-van Mises' W^2 average squared difference of cumulative functions test-statistic

o Kolmogorov's D_n maximum absolute difference of cumulative functions test-statistic

o Freeman-Tukey's Z^2 root deviate test-statistic

o Fisher-Kullback's G^2 log-likelihood ratio test-statistic

The selection of the test-statistic itself, as well as the selection of the significance level cutoff, is not an objective matter, but rather depends upon the decision-making purposes for which goodness-of-fit testing is used. Different test-statistics emphasize errors in different regions of the underlying variable. Depending upon the relative seriousness of errors of different types, one can design

test-statistics to emphasize virtually any region of the underlying variable one wants. Furthermore, there is no such thing as
an inherently "unweighted" means of describing errors. A simple
transformation, such as taking the logarithm or normalizing the
magnitude, for example, changes the emphasis. No matter what one
uses as a test-statistic, one cannot avoid the fact that, in so doing, one is making a choice as to the relative importance of different errors. When the decision-making rationale is unstated,
we are left wondering whether a selection has been made from the
many available test-statistics just so as to produce a desired
result, or whether it is really unknown why the particular test-
statistic was chosen. The question is not whether one has weighted
equal errors "equally" (for this has no absolute meaning), but
whether one has weighted them properly for the intended application.

J. Pearson's T Sum of Normalized Squared Differences Test-Statistic

The most commonly used test-statistic for the goodness-of-fit of
the probabilities $\{P_k\}$ to the frequencies $\{n_k/n\}$ is

$$T = \sum_{k=1}^{K} \frac{(n_k - nP_k)^2}{nP_k} ,$$

where

    K = number of classes,

    P_k = probability of kth class based on assumed distribution
          with assumed parameter values,

    n_k = observed number of events in k^{th} class, and

    $n = \sum_{k} n_k .$

This statistic, which is asymptotically χ^2 distributed for many
important underlying distributions, was discussed in Chapter 9.

The T test-statistic is appropriate for circumstances where equal ratios of error squared to magnitude of the underlying variable have equal disutility, and where the sign of the error is unimportant.

K.   Wald-Wolfowitz's N_r Run Count Test-Statistic

The T-statistic is insensitive to the signs of the errors.   Another statistic, the N_r-statistic, can be used to test whether unusual sequences of signs of the errors have been encountered.   Suppose that in the data order $i = 1,2,3,\ldots,n$ these signs have a sequence such as:

$$+ + + - + - + + - - + + \, .$$

Let

N_+ = number of +'s,

N_- = number of -'s, and

N_r = number of runs,

where a run is a same-sign sequence bounded by opposite signs (or end of sequence).

In the above example

N_+ = 8

N_- = 4

N_r = 7

The critical region is defined as improbably low values of N_r, i.e., $N_r \leq N_{r|min}$.   The expectation and variance of N_r are given by:

$$E[N_r] = 1 + \frac{2\,N_+ N_-}{N_+ + N_-}, \text{ and}$$

$$V[N_r] = \frac{2N_+ N_- [2N_+ N_- - N_+ - N_-]}{[N_+ + N_-]^2 [N_+ + N_- - 1]}.$$

The N_r test-statistic is appropriate for circumstances in which there is disutility associated with non-randomness of the signs of the errors.

L.   Neyman-Barton's P_s^2 Smooth Count Test-Statistic

The Neyman-Barton smooth test is, in effect, a "maximum power" component of Pearson's T.   The smooth test-statistic is defined as

$$P_s^2 = \frac{1}{n} \sum_{r=1}^{s} \left[\sum_{i=1}^{n} \ell_r(Y_i) \right]^2 ,$$

where

$\ell_r(Y_i)$ = Legendre polynomial of order r, orthonormal on $(0,1)$,

$$Y_i = F_o(X_i) = \int_{X_{min}}^{X_i} f(X_i')\,dX_i' , \text{ and}$$

$f(X_i')\,dX_i'$ = probability i^{th} observation between X_i' and $X_i' + dX_i'$.

If $s = K-1$, this is exactly Pearson's T test, i.e., $P_{K-1}^2 = T$.
If $s < K-1$, then P_s^2 is a maximum power component of T.

If Y_i is distributed as

$$f(Y_i) = a(0_1, \ldots, 0_s) \exp \left\{ 1 + \sum_{r=1}^{s} \theta_r \ell_r (Y_i) \right\},$$

then P_s^2 is asymptotically distributed as a non-central $\chi^2(s, \kappa)$ with non-centrality parameter

$$\kappa = n \sum_{r=1}^{s} \theta_r^2 .$$

If $n \geq 20$, then the distributions of P_1^2 and P_2^2 are closely approximated by $\chi^2(1)$ and $\chi^2(2)$.

M. David's N_o Empty Cell Count Test-Statistic

We divide the range of x into equiprobability bins, assuming the distribution f(x). This is done by computing the cumulative function of this distribution

$$F(x) = \int_{-\infty}^{x} f(x') \, dx' ,$$

and dividing F(x) into m equal segments.

Then we look at the n data points and define the empty cell test-statistic as

$$N_o = \text{number of empty bins.}$$

If m and $n \to \infty$ such that $\rho = n/m > 0$, N_o is asymptotically normal with

$$\mu = me^{-\rho}, \text{ and}$$

$$\sigma = m[e^{-\rho} - e^{-2\rho}(1+\rho)].$$

The tendency to normality is fastest for $\rho = 1.255$. So if this value is chosen, standard tables can be used with smaller sample sizes for critical values to define unusually high or low values for N_o.

N. Smirnov-Cramer-von Mises' W^2 Average Squared Deviation
     Test-Statistic

We order the observational data along the variable x, so that $x_i \leq x_{i+1}$ for all i. Then we define the cumulative function of this ordered data as follows:

$$S_n(x) = \begin{cases} 0 & x < x_1 \\ i/n & x_i \leq x < x_{i+1}, \quad i=1,\ldots,n-1 \\ 1 & x_n \leq x. \end{cases}$$

The W^2 test-statistic can be defined as

$$W^2 = \int_{-\infty}^{\infty} [S_n(x) - F(x)]^2 \, f(x) \, dx,$$

where $F(x)$ is the cumulative function of the assumed distribution $f(x)$.

W^2 is an expected squared difference measure of the "distance" between the cumulative function $F(x)$ for the hypothesized distribution $f(x)$ and the cumulative function for the ordered data.

The distribution of W^2 is independent of the distribution of x, even for finite n. The asymptotic characteristic function of nW^2 is

$$\lim_{n \to \infty} E[e^{itnW^2}] = \sqrt{\frac{\sqrt{2it}}{\sin\sqrt{2it}}}.$$

The critical values are given by inversion of this formula.

Significance Level α	Critical Value of nW^2
0.10	0.347
0.15	0.461
0.01	0.743
0.001	1.168

To the accuracy of this table, the asymptotic limit is reached for $n \geq 3$.

The W^2 test-statistic is appropriate for circumstances where equal values of error squared in the cumulative probability distribution over the underlying variable have equal disutility.

O.   Kolmogorov's D_n Maximum Absolute Difference Test-Statistic

Kolmogorov's D_n test-statistic measures the greatest absolute difference between the data and the hypothetical distribution $f(x)$:

$$D_n = \max_{\{x\}} |S_n(x) - F(x)|,$$

where $F(x)$ is the cumulative function of the distribution $f(x)$, and $S_n(x)$ is the cumulative function of the ordered data.

The limiting distribution of $\sqrt{n}\, D_n$ has a left tail ($\sqrt{n}D_n > z$) area given by

$$\lim_{n \to \infty} P(\sqrt{n}\, D_n > z) = 2 \sum_{r=1}^{\infty} (-1)^{r-1} e^{-2r^2 z^2}.$$

This limit is a good approximation to the left tail probability $P(\sqrt{n}\, D_n > z)$ for $n \geq 80$.   The following table gives the size α of the left tail area beyond various critical values of $\sqrt{n}\, D_n$.

Significance Level α	Critical Value of $\sqrt{n}\ D_n$
0.20	1.07
0.10	1.22
0.05	1.36
0.01	1.63

The D_n test-statistic is appropriate in circumstances where equal values of the maximum absolute error in the cumulative probability distribution over the underlying variable have equal disutility.

P.   Freeman-Tukey's Z^2 Root Deviate Test-Statistic

The Freeman-Tukey deviate test-statistic is defined as

$$Z^2 = \sum_{k=1}^{K} z_k^2 \,,$$

where

$$z_k = \sqrt{n_k} + \sqrt{n_k + 1} - \sqrt{4nP_k + 1}\,.$$

The individual deviate z_n is a variance-stabilizing transformation of the cell occupation number n_k. If n_k is Poisson distributed, for example, with mean nP_k, then $\sqrt{n_k} + \sqrt{n_k + 1}$ is approximately normally distributed, with approximate mean $\sqrt{4n_k + 1}$ and unity variance. Thus z_k is approximately normally distributed with zero mean and unity variance. The sum of deviates squared Z^2 is asymptotically chi-squared distributed, with the number of degrees of freedom ν equal to the number of means $\{nP_k\}$ less the number of parameters fixed.

Q.   Fisher-Kullback's G^2 Log-Likelihood Ratio Test-Statistic

Let X_k be an exact (unknown) estimate of the kth outcome frequency, and let P_k be our estimate. Let n_k be the number of test sample events with the kth outcome, $n = \sum_k n_k$. Then the likelihoods of the test data, assuming each of these probabilities, are given by:

$$L_o(\{n_k\}|\{P_k\}) = n! \prod_{k=1}^{K} \frac{P_k^{n_k}}{n_k!}$$

$$L(\{n_k\}|\{X\}) = n! \prod_{k=1}^{K} \frac{X_k^{n_k}}{n_k!} .$$

The maximum likelihood estimator $\hat{X}_k$ of X_k for the test data is found by maximizing L with respect to $\{X_k\}$ subject to the constraint $\sum_k X_k = 1$. This is easily shown to yield the result

$$\hat{X}_k = n_k/n \quad \text{for all } k = 1,\ldots,K$$

on the test sample data. (In making this computation, we ignore the details of the training sample, since we wish to evaluate P_k regardless of how it was obtained.)

The maximum likelihood ratio test-statistic for determining whether or not a particular value of P_k is a good estimate is found by dividing L_o by L, using $\hat{X}_k$ as the estimator of X_k in L:

$$\lambda = L_o/L$$

$$= n^n \prod_{k=1}^{K} \left(\frac{P_k}{n_k}\right)^{n_k} .$$

A log-likelihood ratio test-statistic can now be defined as

$$G^2 = -2 \log \lambda .$$

Note that

$$G^2 = 2n \, \Delta S_e \, ,$$

where

$$\Delta S_e = \sum_{k=1}^{K} \hat{X}_k \, \log \hat{X}_k / P_k$$

is the entropy decrement, expressing the informational disagreement between the prior (training data) estimators P_k and the posterior (test data) estimators $\hat{X}_k$. Thus G^2 is referred to as a discrimination information test-statistic.

The test-statistic G^2 is asymptotically χ^2 distributed with $\nu = K-1-\ell$ degrees of freedom, where ℓ = number of parameters estimated. Fig. 36 illustrates a typical shape for its frequency distribution.

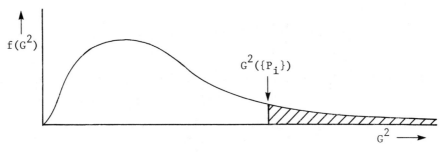

Fig. 36.   Likelihood Ratio Distribution

In likelihood ratio testing we first compute G^2 for a particular set of estimators $\{P_k\}$. We then compute the area under the tail of the $f(G^2)$ curve beyond this value. This shows us how unusual it would be to obtain this high, or higher, a likelihood ratio by chance alone, given the assumed underlying distribution.

Although the G^2-statistic and the T-statistic are both asymptotically χ^2, they can yield quite different assessments for finite sample sizes.

We have given different names to the samples upon which the estimates P_i and $\hat{X}_i$ are based because, in general, the maximum likelihood ratio test can be used to evaluate any estimate P_i, even one
from another method (such as equal moments). Of course, if the test
sample itself is used for "training" the parameters, and the maximum likelihood method is used to get the P_i, then $\lambda = 1$ and the area
under the tail beyond the observed value of G^2 depends only on the
sample size n. This area is then a measure of the tightness of the
maximum likelihood estimator as a function of sample size.

In general

$$\lambda = \frac{L(\hat{\theta}')}{L(\hat{\theta})} ,$$

where

$$L(\hat{\theta}) = \sum_{i=1}^{n} f(x_i; \hat{\theta}_1, \ldots, \hat{\theta}_m) ,$$

$$L(\hat{\theta}') = \sum_{i=1}^{n} f(x_i; \hat{\theta}'_1, \ldots, \hat{\theta}'_m) ,$$

$f(x; \hat{\theta}_1, \ldots, \hat{\theta}_m)$ = frequency function for variable x, with parameters $\theta_1, \ldots, \theta_m$,

$\hat{\theta}_j$ = maximum likelihood estimator of θ_j with no restrictions,

$\hat{\theta}'_j$ = maximum likelihood estimator of θ_j restricted by hypothesis H_o,

n = sample size, and

m = number of parameters (K-1 in previous notation).

For large sample sizes, G^2 is approximately chi-squared distributed
with $\nu = m - \ell$ degrees of freedom, where ℓ is the number of parameters
fixed by the hypothesis H_o.

R.   Support and Evidence

The concept of likelihood has been used by researchers in statistics to define alternative notions of "information". Kullback, for
example, refers to the log-likelihood ratio as the "information" in
the sample.   Jeffreys, Hacking, and Edwards, on the other hand,
refer to this as the relative "support".   They define a "support
function" as the logarithm of the likelihood:

$$U(\underset{\sim}{\theta}) = \log L(\underset{\sim}{\theta}).$$

It is a measure of the extent to which the sample supports the estimated parameter values, given the assumed distribution.

"Support" is defined as the logarithm of the likelihood ratio, $\log \lambda$,
so it differs from the log-likelihood ratio test-statistic only by
a factor of -2.*   It has the following properties:**

- o Transitivity: If hypothesis H_1 is supported better than H_2,
     and H_2 better than H_3, then H_1 is supported better than H_3.

- o Additivity: The amounts of support from two independent sets
     of data are additive.

- o Invariance under data transformation: Support is unchanged
     by any one-to-one transformation of the data.

- o Invariance under parameter transformation: Support is inde-
     pendent of the particular parameter form adopted.

- o Relevance: In the absence of data on which to compare two
     hypotheses, their supports are equal.   If an hypothesis is in
     fact true, then in the limit of infinite data, it will have
     the highest support.

- o Consistency: Support is independent of the particulars of
     the application, i.e., it is value-free.

- o Compatibility: Support is compatible with Bayes' theorem
     in cases where prior probabilities are known.

---

*   The listed properties thus hold also for the discriminant information and
the entropy decrement test-statistics.

** See Edwards, pp. 28-29.

Suppose we expand the support function in a Taylor series, about an estimate $\hat{\theta}$

$$U(\theta) = U(\hat{\theta}) + (\theta-\hat{\theta}) \left. \frac{\partial U}{\partial \theta} \right|_{\theta=\hat{\theta}} + \tfrac{1}{2}(\theta-\hat{\theta})^2 \left. \frac{\partial^2 U}{\partial \theta^2} \right|_{\theta=\hat{\theta}} + \ldots$$

If we use maximum likelihood estimators to obtain $\hat{\theta}$, then

$$\left. \frac{\partial U}{\partial \theta} \right|_{\theta=\hat{\theta}} = 0.$$

In this case, we have approximately

$$U(\theta) = U(\hat{\theta}) + \tfrac{1}{2}(\theta-\hat{\theta})^2 \left. \frac{\partial^2 U}{\partial \theta^2} \right|_{\theta=\hat{\theta}}.$$

Fisher and Edwards refer to $- \left. \frac{\partial^2 U}{\partial \theta^2} \right|_{\theta=\hat{\theta}}$ as the amount of "information," $E(\hat{\theta})$, which the sample contains in support of the parameter estimate $\hat{\theta}$,

$$E(\hat{\theta}) = - \left. \frac{\partial^2 U}{\partial \theta^2} \right|_{\theta=\hat{\theta}}.$$

I would suggest "evidence" as an alternative choice of terminology, if only because the term "information" has become so widely identified with the communications-theoretic usage.

For example, assume a normal distribution

$$f(x; \mu, \sigma) = \frac{1}{\sigma\sqrt{2\pi}} \; e^{-(\frac{x-\mu}{\sigma})^2}.$$

Then the support function for a sample of n data points is

$$U(\mu,\sigma) = -n \log \sigma\sqrt{2\pi} - \frac{1}{2\sigma^2} \sum_{i=1}^{n} (x_i-\mu)^2.$$

The amount of evidence to support the sample average estimator of the mean is

$$E(\bar{x}) = \frac{n}{\sigma^2}.$$

The evidence supporting the square of the standard deviation estimator of the variance is

$$E(s^2) = \frac{n}{\sigma^6} [s^2 + (\bar{x}-\mu)^2] - \frac{n}{2\sigma^4}.$$

Fisher's evidence is distinctly different from communications theory information. For example, evidence is additive over different sets of data. (Its behavior is proportional to an equivalent number of data observations.) This is not necessarily true of information. Additional data may increase or decrease the amount of information one has about an event, depending upon the extent to which prior expectations are reinforced or contradicted.

Evidence pertains to a specific set of past observations, a specific assumed underlying distribution, and specific parameters of which this distribution is regarded a function. It increases monotonically as we acquire more data.

Information pertains to our current state of knowledge about a future event. It can fluctuate with the sequential addition of new observations to our sample, the size of the fluctuation with each added datum becoming less as the sample size grows.

QUOTES

GAUSS (1809)

It has been customary certainly to regard as an axiom the hypothesis that if any quantity has been determined by several
direct observations, made under the same circumstances and with
equal care, the arithmetical mean of the observed values affords
the most probable value, if not rigorously, yet very nearly at
least, so that it is always most safe to adhere to it.

> —Karl F. Gauss (1777-1855)
> *Theory of the Motion of the Heavenly Bodies Moving
> about the Sun in Conic Sections* (1809), tr. by
> C.H. Davis, Dover Pubs., Inc., NY, 1963, p. 258.

FISHER (1922)

If in any distribution involving unknown parameters θ_1, θ_2, θ_3,
..., the chance of an observation falling in the range dx be
represented by

$$f(x, \theta_1, \theta_2, ...) \, dx,$$

...The method of maximum likelihood consists simply in choosing
that set of values for the parameters which makes...

$$\Sigma(\log f)$$

a maximum for variations of θ_1, θ_2, θ_3, &c.

> —Ronald A. Fisher (1890-1962)
> "On the Mathematical Foundations of Theoretical
> Statistics," *Philosophical Transactions of the
> Royal Society of London*, Ser. A., *222*, May 1922,
> pp. 323-324. [Fisher's notation "S" for summation
> over the sample has been replaced by the modern
> "Σ".]

STEIN (1955)

If one observes the real random variables X_1, ..., X_n independently normally distributed with unknown means ξ_1, ..., ξ_n and
variance 1, it is customary to estimate ξ_i by X_i. If the loss
is the sum of squares of the errors, this estimator is admissible for $n \leq 2$, but inadmissible for $n \geq 3$. Since the usual estimator is best among those which transform correctly under trans-
lation, any admissible estimator for $n \geq 3$ involves an arbitrary
choice.

* * * *

STEIN (1955, cont.)

A simple way to obtain an estimator which is better for most
practical purposes is to represent the parameter space (which
is also essentially the sample space) as an orthogonal direct
sum of two or more subspaces, also of large dimension and apply
spherically symmetric estimators separately in each. If the
p^2's (squared length of the population mean divided by the
dimension) are appreciably different for the selected subspaces,
this estimator will be better than the spherically symmetric
one.

> —Charles M. Stein (1920-   )
> "Inadmissibility of the Usual Estimator for the
> Mean of a Multivariate Normal Distribution," *Proc.
> of the Third Berkeley Symposium on Mathematical
> Statistics and Probability*, Dec. 1954 and July-
> August 1955, Vol. I, Univ. of Calif. Press,
> Berkeley, CA, 1956, pp. 197, 198.

JAMES and STEIN (1960)

It has long been customary to measure the adequacy of an estimator by the smallness of its mean squared error.

* * * *

Let X be a normally distributed p-dimensional coordinate vector
with unknown mean $\xi = EX$ and covariance matrix equal to the
identity matrix...The usual estimator is φ_0, defined by

$$\varphi_0(x) = x, \ldots$$

for $p \geq 3$...the estimator φ_1 defined by

$$\varphi_1(X) = 1 - \frac{p-2}{\|X\|^2} X$$

has smaller risk than φ_0 for all ξ.

> —Willard Donald James (1927-   ) and
> Charles M. Stein (1920-   )
> "Estimation With Quadratic Loss," *Proc. of the
> Fourth Berkeley Symposium on Mathematical
> Statistics and Probability*, June 20-July 30,
> 1960, Vol. I, Univ of Calif. Press, Berkeley, CA,
> 1961, pp. 361, 362-363.

LINDLEY (1962)

$$x. + \left(1 - \frac{a_n}{\Sigma(x_i - x.)^2}\right)(x_i - x.).$$

...In other words I suggest using Stein's method for the deviations $x_i - x.$ rather than for the $x_i.$. This avoids the difficulty
of the arbitrary origin...

—Dennis Victor Lindley (1923-   )
"Discussion on Professor Stein's Paper," *Journal
of the Royal Statistical Society*, Ser. A., *125*,
London, 1962, p. 286.

IRELAND, KU and KULLBACK (1969)

We shall use the principle of minimum discrimination information
estimation to obtain RBAN [Regular Best Asymptotically Normal]
estimates of the cell frequencies of an observed $r \times r$ contingency table under hypotheses of either symmetry or marginal
homogeneity...

...For any two contingency tables $p(ij)$, $\Sigma\Sigma p(ij) = 1$ and
$\pi(ij) > 0$, $\Sigma\Sigma\pi(ij) = 1$, having the same classifications and
dimensions we define a distance-like measure from the p-table
to the π-table by the discrimination information [7]

$$I(p{:}\pi) = \sum_i \sum_j p(ij) \, \ell n \frac{p(ij)}{\pi(ij)} .$$

The π-table may be fixed by hypothesis, observed, or estimated
from observations. Let the p-tables be members of a class
subject to restraints consistent with the nature of the problem,
then minimization of $I(p{:}\pi)$ over the class of p-tables leads to
results that are useful in estimation and hypothesis testing.
The minimizing table is denoted by p^*.   ...the procedure we pro-
propose provides estimates of the cell entries under the
hypothesis of marginal homogeneity as well as a test statistic...

—C. Terrance Ireland (1938-   ), H.H. Ku (1918-   )
and S. Kullback (1907-   )
"Symmetry and Marginal Homogeneity of an $r \times r$ Contingency Table," *J. of the American Statistical
Association*, *64*, March 1969, pp. 1325-1326.

[7] Kullback, S., *Information Theory and Statistics*, New York:
Wiley, 1959; New York: Dover Pubs., Inc., 1968.

REFERENCES

Afifi, A.A. and S.P. Azen, *Statistical Analysis, A Computer Oriented Approach*, Academic Press, NY, 1972, pp. 50-51.

Anderson, T.W. and D.A. Darling, "Asymptotic Theory of Certain Goodness-of-Fit Criteria Based on Stochastic Processes," *Ann. Math. Statist*, *23*, 1952, p. 193.

Bartlett, M.S., "Properties of Sufficiency and Statistical Tests," *Proc. Royal Soc., London,* Series A, *160*, 1937, pp. 273 ff.

Berkson, J., "Minimum Discriminant Information, the 'No Interaction' Problem and the Logistic Function," *Biometrics*, *28*, 1972, pp. 443-468.

Bishop, Yvonne M.M., Stephen E. Fienberg and Paul W. Holland, *Discrete Multivariate Analysis: Theory and Practice*, The MIT Press, Cambridge, MA, 1975.

Blyth, C., "On Minimax Statistical Decision Procedures and Their Admissability," *Ann. Math. Statist.*, *22*, 1951, pp. 22-42.

Bradley, James V., *Distribution-Free Statistical Tests*, WADD Tech. Rept. 60-661, Behavioral Sci. Lab., Wright-Patterson Air Force Base, Ohio, August 1960.

Carter, Grace M. and John E. Rolph, "Empirical Bayes Methods Applied to Estimating Five Alarm Probabilities," *J. of the Amer. Statist. Assoc.*, *69*, Dec. 1974, pp. 880-885.

Christensen, R.A., "Entropy Minimax, A Non-Bayesian Approach to Probability Estimation From Empirical Data," *Proc. of the 1973 Intl. Conf. on Cybernetics and Society*, IEEE Systems, Man and Cybernetics Society, Nov. 5-7, 1973, Boston, MA, 73 CHO 799-7-SMC, pp. 321-325. [*Chapter 5 of Volume III.*]

Cochran, W.G., "The Chi-Squared Test of Goodness of Fit," *Ann. Math. Statist.*, *23*, 1952, p. 315.

_____, "Some Methods for Strengthening the Common Chi-Squared Tests," *Biometrics*, *10*, 1954, p. 417.

David, Florence N., "A χ^2 'Smooth' Test for Goodness of Fit," *Biometrika*, *34*, 1947, pp. 299-310.

_____, "Two Combinatorial Tests of Whether a Sample Has Come from a Given Population," *Biometrika*, *37*, 1950, pp. 97-110.

Eadie, W.T., D. Drijard, F.E. James, M. Roos and B. Sadoulet, *Statistical Methods in Experimental Physics*, North-Holland Pub. Co., Amsterdam, 1971.

Edwards, A.W.F., *Likelihood, An Account of the Statistical Concept of Likelihood and Its Application to Scientific Inference*, Cambridge Univ. Press, London, 1972.

Efron, Bradley, "Biased Versus Unbiased Estimation," *Advances in Mathematics*, *16*, 1975, pp. 259-277.

Efron, Bradley and Carl Morris, "Stein's Estimation Rule and Its Competitors-- An Empirical Bayes Approach," *J. of the Am. Statist. Assoc.*, *68*, March 1973, pp. 117-319.

_____, "Data Analysis Using Stein's Estimator and Its Generalizations," *J. of the Am. Statist. Assoc.*, *70*, June 1975, pp. 311-319.

Fisher, R.A., "On the Mathematical Foundations of Theoretical Statistics," *Philos. Transactions of the Royal Society of London,* Ser. A, *222,* 1922, pp. 309-368.

_____, Theory of Statistical Estimation," *Proc. Camb. Phil. Soc., 22,* 1925, pp. 700-725.

_____, *Statistical Methods and Scientific Inference,* Oliver and Boyd, Edinburgh, 1956.

Fourgeand, C. and A. Fuchs, *Statistique,* Dunod, Paris, 1967.

Gauss, Karl F., *Theory of the Motion of the Heavenly Bodies Moving about the Sun in Conic Sections* (1809), tr. by C.M. Davis, Dover Pubs., Inc., NY, 1963.

Hacking, I., *Logic of Scientific Inference,* Cambridge Univ. Press, London, 1965.

Hodges, J.L. and E.L. Lehmann, "Some Applications of the Cramer-Rao Inequality," *Proc. of the Second Berkeley Symposium on Mathematical Statistics and Probability,* Univ. of Calif. Press, Berkeley, CA, 1951, pp. 13-22.

Hoel, Paul G., *Intro. to Mathematical Statistics,* J. Wiley & Sons, NY, 1954, pp. 189ff.

Ireland, C.T., H.H. Ku and S. Kullback, "Symmetry and Marginal Homogeneity of an $r \times r$ Contingency Table," *J. of the Am. Statist. Assoc., 64,* March 1969, pp. 1323-1341.

James, W. and C. Stein, "Estimation with Quadratic Loss," *Proc. of the Fourth Berkeley Symposium on Mathematical Statistics and Probability, 1,* Univ. of Calif. Press, Berkeley, CA, 1961, pp. 361-379.

Jeffreys, H., "Further Significance Tests," *Proc. Camb. Phil. Soc., 32,* 1961, pp. 416-445.

Johnson, N.L. and S. Kotz, *Distributions in Statistics,* 4 Vols., J. Wiley & Sons NY, 1969-1970.

Kempthorne, O., "Some Aspects of Experimental Inference," *J. Am. Statist. Assoc., 61,* 1966, pp. 11-34.

Kullback, S., *Information Theory and Statistics,* J. Wiley & Sons, NY, 1959.

Kullback, S., M. Kupperman and H.H. Ku, "Tests for Contingency Tables in Markov Chains," *Technometrics, 4,* 1962, pp. 573-608.

Lindley, D.V., "Discussion on Professor Stein's Paper," *J. of the Royal Stat. Soc.,* Ser. A, *125,* 1962, pp. 265-296.

Marshall, A.W., "The Small Sample Distribution of nW_n^2," *Ann. Math. Statist. 29,* 1958, p. 307.

Patrick, Edward A., "Maximum Likelihood Estimation," *Fundamentals of Pattern Recognition,* Prentice-Hall, Inc., Englewood Cliffs, NJ, 1972.

Pearson, Karl, "On the Criterion that a given System of Deviations from the Probable in the Case of a Correlated System of Variables is such that it can be reasonably supposed to have arisen from Random Sampling," *Phil. Mag. and Jour. of Science,* Series 5, *50,* Taylor and Francis, 1900, pp. 157-172.

Pitman, E.J.G., "Location and Scale Parameters," *Biometrika, 30,* 1939, pp. 391-421.

Smirnov, N.V., "Sur la distribution de W^2, *C.R. Acad. Sci. Paris, 202,* 1936, p. 449.

Stein, Charles M., "Inadmissibility of the Usual Estimator for the Mean of a Multivariate Normal Distribution," *Proc. of the Third Berkeley Symposium on Mathematical Statistics and Probability, 1,* Univ. of Calif. Press, Berkeley, CA, 1956, pp. 197-206.

Stevens, W.L., "Distribution of Groups in a Sequence of Alternatives," *Ann. Eugenics, 9,* 1939, pp. 10-17.

Wald, A., "Contributions to the Theory of Statistical Estimation and Testing Hypotheses," *Ann. Math. Statist., 10,* 1939, pp. 299-326.

_____, *Theory and Application of Sequential Probability Ratio Tests,* J. Wiley & Sons, Inc., NY, 1947.

Wald, A. and J. Wolfowitz, "On a Test Whether Two Samples are from the Same Population," *Ann. Math. Statist., 11,* 1940, pp. 147-162.

Wilks, S.S., *Mathematical Statistics,* J. Wiley & Sons, Inc., NY, 1962.

PART III

INDEPENDENT VARIABLES: MINIMUM ENTROPY PARTITIONS

CHAPTER 11

INDEPENDENT VARIABLES AND FEATURE SPACE

CHAPTER 11

INDEPENDENT VARIABLES AND FEATURE SPACE

A.   Independent Variables

Commenting in 1972 upon the Stein-James-Lindley estimator, Harding said: "If μ_1 refers to butterflies in Brazil, μ_2 to ball-bearings in Birmingham and μ_3 to Brussels sprouts in Belgium, then an admissible estimator will cause the estimates of these three quite unrelated things to be related to each other. The largest being on the whole pulled down and the smallest pushed up." Although he should have added an extra toe on my foot to bring the list up to four, this comment reveals the disquietude of mathematicians and logicians when faced with what is essentially a scientific problem. When we wish to make reliable predictions about the physical world, we no longer have the freedom to roam through arbitrary classifications. In this part (III), we take up the question of what is similar to what.

Although we pretended otherwise in Part II, we usually observe a large number of things about an event in addition to its outcome. Tables 31, 32, 33, and 34 list a few of the many data generally available for four illustrative examples related to those discussed in Chapter 6. As these may be selected quite arbitrarily by the analyst, they are referred to as *independent* variables. They may be qualitative or quantitative, depending upon whether they are regarded as referencing a classification or a numerical parameter value.

Table 31. Example of Independent Variables
to Help Predict Total Precipitation
Next Year in Sacramento-Tahoe Area

1. Sea surface temperature deviation from seasonal average, July-Sept., two
years ago, Marsden Zones 19-22, 318-321.
2. Change in two-year total Jeffrey Pine tree ring index at Truckee, CA
(this year plus last year minus sum of two previous years).
3. Change in Jan.-March sea surface temperature deviation from seasonal
average, Marsden Zones 308-311 (this year minus last year).
4. Change in yearly average temperature at Darwin, Australia airport (this
year minus last year).
5. Change in Jan.-March sea surface temperature deviation from seasonal
average, Marsden Zones 9-12 (this year minus last year).
6. Change in yearly total precipitation at Auburn, CA (this year minus last
year).
7. Sea surface temperature deviation from seasonal average, Oct.-Dec., total
most recent four years, Marsden Zones 160-163.
8. Total precipitation two years ago (value of dependent variable two years
ago).
9. Change in Oct.-Dec. sea surface temperature deviation from seasonal average, Marsden Zones 9-12 (this year minus last year).
10. Calendar year this year.

Table 32. Examples of Independent Variables to Help
Predict Change in Gold Bullion Price Tomorrow

1. Today's day of week (Monday, Tuesday,...).
2. Whether or not tomorrow is a holiday.
3. High-low price spread.
4. Arrow (close minus midpoint).
5. Whether or not limit hit in gold futures.
6. X on Y moving average.
7. Number days since last peak.
8. Number days since last trough.
9. Change in volume.
10. Change in gold futures open interest.

Table 33. Examples of Independent Variables to Help
Predict Failure Rates for Nuclear Fuel Assemblies

1. Axial convexity of clad creep at time of maximum product of
power and iodine concentration.
2. Fission gas density in plenum and gap.
3. Xenon gas molar fractional density.
4. Fission gas burst magnitude.
5. Range (maximum minus minimum) of axial convexity of clad creep.
6. Mechanistic model estimate of failure probability.
7. Maximum triple product: power $\times$ corrosion $\times$ ramp.
8. Dwell time with power shock exceeding -50 w/cm.
9. Accumulated power shock exceeding -50 w/cm.
10. Number power ramps crossing 2 w/cm/hr.

Table 34.   Examples of Independent Variables to Help
Estimate Two-Year Survival Prognosis after
Treatment for Coronary Artery Disease

1. Left ventricular contraction condition.
2. Aortic diastolic pressure.
3. Whether or not cardiomegaly present.
4. Mitral insufficiency on left ventricular contraction.
5. Left main cineangiogram result.
6. Duration of symptoms of coronary artery disease.
7. Whether or not myocardial ischema (ST-T wave changes) present.
8. Right circumflex artery summary.
9. Left anterior descending summary.
10. Left ventricular ejection fraction.

Let us see how we form *samples* in such cases. As before, our data will contain the identity of each event and the value of the dependent (outcome) variable for each event. Additionally, our sample will now contain the values of the independent variables for each event.

Cases #1 and #2 (rainfall and gold price forecasting) involve *time-series* events. Case #3 (nuclear fuel failure) involves *spatially distributed* events within each reactor core. Case #4 (cause of death prognosis) can be regarded as both time and space distributed. Note that, by using this terminology, we do not commit ourselves as to whether the time and/or space distributions have greater or less predictive power than other variables - or indeed, whether or not these distributions are relevant at all to the dependent variable. Only analysis of actual data will reveal the degree of importance of the various independent variables to prediction.

Table 35 illustrates how we might tabulate raw data for conducting a long-range precipitation forecasting analysis.

Table 35.   Illustrative Raw Data Tabulation
for Precipitation Problem

Current Year	Precip. Next Year	SST Dev July-Sept 19-22 318-321	Tree Ring Index P. Jeff Truckee	SST Dev Jan-Mar 308-311	Temp. Avg WMO Darwin	SST Dev Jan-Mar 9-12	Precip. Auburn	SST Dev Oct-Dec 160-163	SST Dev Oct-Dec 9-12
1976	17.44	*	*	*	*	.454	16.0	*	-.178
1975	16.79	.150	*	-.356	27.0	-.524	33.7	.232	.789
1974	42.97	.473	*	-.296	*	.512	50.6	.957	-.615
1973	52.88	.535	*	-.603	*	.018	43.7	.863	-.277
·	·	·	·	·	·	·	·	·	·
·	·	·	·	·	·	·	·	·	·
·	·	·	·	·	·	·	·	·	·
1851	41.65	*	11.54	*	*	.567	*	*	-.021

* Data not available

In the first column, we record the year. (This is a hydrologically defined "water year" for the particular location. Care must be taken that the definitions of the other variables are consistent with this definition.) The other nine columns tabulate various items of data for the analysis. (Note that these are real-world data and suffer the real-world defect of missing entries.)

Table 36 illustrates how these data are used to formulate a sample.

Table 36. Illustrative Sample Extracted from the Data in Table 35 Using Independent Variables Listed in Table 31

Event I.D. (next year's date)	Dep. Var. (precip. next year)	Indep. Vars.									
		1	2	3	4	5	6	7	8	9	10
1977	−	.535	*	−.606	*	−.494	−17.7	4.68	52.9	−.338	1976
1976	−	.735	*	.307	*	−1.036	−16.9	5.76	62.9	−.005	1975
1975	+	.596	*	−1.410	*	−1.850	6.9	6.83	24.0	−1.957	1974
1974	+	.319	*	.519	*	−1.363	19.7	5.83	40.3	1.897	1973
'	'	'	'	'	'	'	'	'	'	'	'
'	'	'	'	'	'	'	'	'	'	'	'
'	'	'	'	'	'	'	'	'	'	'	'
1852	+	*	*	*	*	*	*	*	*	*	1851

In the first column, we enter event identifiers. These label each event by the year for which the prediction is being made, i.e., *next year's* date.

In the second column we enter the value of the dependent variable for the event (i.e., *next year's* precipitation), according to a selected classification scheme that assigns:

+: values greater than average, and

-: values less than average.

---

* Data not available.

In the remaining columns, we enter our selected independent variables for each event.   The first of these is the sea surface temperature deviation from seasonal average, July-Sept., two years ago, in Marsden Zones 19-22 and 318-321.   The last is *this year's* date.   Several of the independent variables contain moving averages and moving differences.   These are obtained by passing the time-series data through *filters* designed to extract averages and differences.

More complex filters can be designed (such as those which extract high frequency or low frequency information).   Moving averages and differences are just examples to illustrate the extraction of independent variables from serial data.

Since we have included this year's date as one of the independent variables, we can now regard the ordering of the rows in Table 36 as arbitrary.*   We would lose no information at all by shuffling the rows.   A random rearrangement is shown in Table 37.

Table 37.   Rearrangement of Table 36.
(Note that no information is lost)

Dev. Var. (precip. next year)	Indep. Vars.									
	1	2	3	4	5	6	7	8	9	10
−	.277	1.03	**	**	**	−2.1	**	31.6	**	1946
−	−.005	−.36	−.178	−.392	.134	3.5	−7.38	50.7	.325	1911
−	−.834	−.07	−.073	−.383	.202	−16.6	−.01	56.7	−.803	1953
−	**	1.08	−1.487	**	−1.094	16.4	**	49.7	−.599	1876
+	.617	−4.97	−.912	.717	−.321	−21.9	1.98	26.8	.138	1968
'	'	'	'	'	'	'	'	'	'	'
'	'	'	'	'	'	'	'	'	'	'
'	'	'	'	'	'	'	'	'	'	'
−	.615	−2.84	−.103	**	.838	−4.7	**	40.7	−.695	1947

*   The event name column has been dropped as it is not used for analysis purposes.   (It is only kept available in case we wish to cross-check the accuracy of the data.)   Thus, we need to explicitly include the date as one of the independent variables if we want to retain the time-series ordering information.
** Data not available.

B.   Feature Space

Our ultimate objective is to be in a position to predict values for the dependent variable for yet-to-be-observed cases. We collect "independent" variables data in the hope that this information may be of use to us in our endeavor. We must take care to define each independent variable such that its value can be ascertained without first learning the value of the dependent variable.   Otherwise, the independent variables would be of little use in helping us achieve our ultimate objective.

As a preliminary step in the process of putting the independent variables to work, we use them to form the axes of a *feature space* in which to plot our sample.

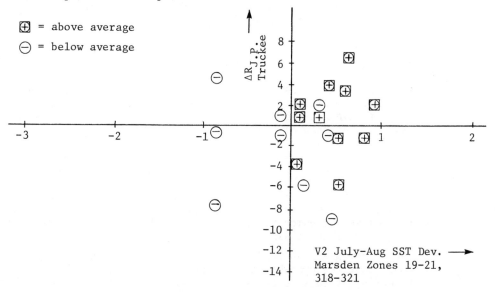

Fig. 37.   Illustrative Two-Dimensional Feature Space
for Precipitation Forecasting Problem

Consider, for example, a two-dimensional feature space in which we take the first independent variable in Table 37 as the horizontal axis and the second as the vertical axis. Fig. 37 shows 20 of the 126 events in Table 37 plotted in this space. Note that the independent variables fix the location of each event in feature space, while the dependent variable is depicted by the symbol representing the outcome of each event.

QUOTES

FREUD (1899)

As a result of the link that has thus been established, next
time this need arises a psychical impulse will at once emerge
which will seek to re-cathect the mnemic image of the perception
and to re-evoke the perception itself, that is to say, to re-
establish the situation of the original satisfaction. An impulse
of this kind is what we call a wish; the reappearance of the
perception is the fulfilment of the wish; and the shortest path
to the fulfilment of the wish is a path leading direct from
the excitation produced by the need to a complete cathexis of the
perception.

> —Sigmund Freud (1856-1939)
> *The Interpretation of Dreams* (1899), ed. and tr.
> by J. Strachey, Avon Books, Discus Edition, NY,
> 1965, pp. 604-605.

KEYNES (1921)

A proposition can be a member of many distinct classes of propositions, the classes being merely constituted by the existence
of particular resemblances between their members or in some such
way. We may know of a given proposition that it is one of a
particular class of propositions, and we may also know, precisely
or within defined limits, what *proportion* of this class are true,
without our being aware whether or not the given proposition is
true. Let us, therefore, call the actual proportion of true
propositions in a class the truth-frequency[2] of the class, and
define the measure of the probability of a proposition relative
to a class, of which it is a member, as being equal to the truth-
frequency of the class.

The fundamental tenet of a frequency theory of probability is,
then, that the probability of a proposition always depends upon
referring it to some class whose truth-frequency is known within
wide or narrow limits.

*   *   *   *

In some cases we may know the exact number which expresses the
truth-frequency of our class; but a less precise knowledge is not
without value, and we may say that one probability is greater
than another, without knowing how much greater, and that it is
large or small or negligible, if we have knowledge of corresponding accuracy about the truth-frequencies of the classes to which
the probabilities refer...

---

2 This is Dr. Whitehead's phrase.

KEYNES (1921, cont.)

On the frequency theory, therefore, there is an important sense in which probabilities can be unknown, and the relative vagueness of the probabilities employed in ordinary reasoning is explained as belonging not to the probabilities themselves but only to our knowledge of them. For the probabilities are relative, not to our knowledge, but to some objective class, possessing a perfectly definite truth-frequency, to which we have chosen to refer them.

— John Maynard Keynes (1890-1962)
A Treatise on Probability (1921), Macmillan & Co.,
Ltd., London, 1957, pp. 101, 102.

WRINCH and JEFFREYS (1921)

...it will never be possible to attach appreciable probability to an inference if it is assumed that all laws of an infinite class, such as all relations involving only analytic functions, are equally probable *a priori*. If inference is possible, the admissible laws must not be all equally probable *a priori*.

— Dorothy Wrinch (1894-1976) and Harold Jeffreys
(1891-   )
"On Certain Fundamental Principles of Scientific
Inquiry," *Phil. Mag.*, *42*, 1921, p. 389.

WHITEHEAD (1929)

In every inductive judgment, there is therefore contained a presupposition of the maintenance of the general order of the immediate environment, so far as concerns actual entities within the scope of the induction. The inductive judgment has regard to the statistical probabilities inherent in this given order.

...The question, as to what will happen to an unspecified entity in an unspecified environment, has no answer. Induction always concerns societies of actual entities which are important for the stability of the immediate environment.

— Alfred North Whitehead (1861-1947)
Process and Reality (1929), The Free Press, NY,
1957, pp. 235-236.

WHORF (1941)

In English, the sentences 'I pull the branch aside' and 'I have an extra toe on my foot' have little similarity...Common, and even scientific, parlance would say that the sentences are unlike because they are talking about things which are intrinsically unlike...

WHORF (1941, cont.)

We find that in Shawnee these two statements are, respectively, *ni-l'θawa-'ko-n-a* and *ni-l'θawa-'ko-θite* (the θ here denotes *th* as in 'thin' and the apostrophe denotes a breath-catch). The sentences are closely similar; in fact, they differ only at the tail end. In Shawnee, moreover, the beginning of a construction is generally the important and emphatic part. ...the first sentence means 'I pull it (something like branch of tree) more open or apart where it forks.' In the other sentence, the suffix -*θite* means 'pertaining to the toes,' and the absence of further suffixes means that the subject manifests the condition in his own person. Therefore the sentence can mean only 'I have an extra toe forking out like a branch from a normal toe."

Shawnee logicians and observers would class the two phenomena as intrinsically similar..."Facts are unlike to speakers whose language background provides for unlike formulation of them."

> —Benjamin Lee Whorf (1897-1941)
> "Language and Logic," *The Technology Review, 43,*
> April 1941, pp. 250-251.

WATANABE (1965)

If we decide to measure the similarity between two objects by the number of logically possible predicates shared by them simultaneously, we come to a disturbing consequence than any pair of two objects (say, a duckling and a swan) are as similar to each other as any other pair of two objects (say, two swans). The situation is essentially the same in the case of continuously-valued vectors or in that of functions. The existence of classes of objects (which is possible only if some pairs of objects are more similar than some others) therefore presupposes the non-uniform importance of predicates or variables. For this reason the problem of clustering and recognition often reduces to a problem of weighting given variables.

> —Satosi Watanabe (1910-   )
> "Karhunen-Loève Expansion and Factor Analysis,
> Theoretical Remarks and Applications," *Transactions
> of the Fourth Prague Conference on Information
> Theory, Statistical Decision Functions, Random
> Processes,* Prague, Aug. 31 - Sept. 11, 1965,
> Academia, Pub. House of the Czechoslovak Academy
> of Sciences, Prague, 1967, p. 636.

REFERENCES

Atchley, William and Edwin H. Bryant (eds.), *Multivariate Statistical Methods: Among-Groups Covariation,* Dowden, Hutchinson & Ross, Inc., Stroudsburg, PA, 1975.

_____, *Multivariate Statistical Methods: Within-Groups Covariation,* Dowden, Hutchinson & Ross, Inc., Stroudsburg, PA, 1975.

Bishop, Yvonne M.M., Stephen E. Fienberg, and Paul W. Holland, *Discrete Multivariate Analysis: Theory and Practice,* The MIT Press, Cambridge, MA, 1975.

Christensen, R.A., "Induction and the Evolution of Language," Physics Dept., Univ. of Calif., Berkeley, CA, July 19, 1963. *[Chapter 7 of Volume II.]*

_____, "A General Approach to Pattern Discovery," Tech. Rept. No. 20, Computer Center, Univ. of Calif., Berkeley, CA, June 29, 1967 (revised Nov. 15, 1967). *[Chapter 1 of Volume III.]*

_____, "A Pattern Discovery Program for Analyzing Qualitative and Quantitative Data," *Behavioral Science, 13,* Sept. 1968, pp. 423-424. *[Chapter 2 of Volume III.]*

_____, "Seminar on Entropy Minimax Method of Pattern Discovery and Probability Determination," presented at Carnegie-Mellon Univ. and MIT, Feb. and March 1971, Tech. Rept. No. 40.3.75, Carnegie-Mellon Univ., Pittsburgh, PA, April 1971. *[Chapter 3 of Volume III.]*

_____, "Entropy Minimax Method of Pattern Discovery and Probability Determination, Arthur D. Little, Inc., Acorn Park, Cambridge, MA, March 7, 1972. *[Chapter 4 of Volume III.]*

_____, "Statistical Model of UO_2 Nuclear Fuel Under Direct Electrical Heating," *Thermal Mechanical Behavior of UO_2 Nuclear Fuel. Vol. 1--Statistical Analysis of Acoustic Emission, Diametral Expansion, Axial Elongation, and Crack Characteristics,* final report to EPRI, Contract No. RP508-2, Entropy Limited, Lincoln, MA, June 1978.

_____, *SPEAR Fuel Performance Reliability Code System: General Description,* EPRI NP-1378, Electric Power Research Institute, Palo Alto, CA, March 1980.

Christensen, R.A., R.E. Eilbert, O. Lindgren and L. Rans, "Successful Hydrologic Forecasting for California Using an Information Theoretic Model," *Journal of Applied Meteorology* (to be published).

Cooley, William and Paul R. Lohnes, *Multivariate Data Analysis,* J. Wiley & Sons, Inc., NY, 1971.

Cox, D.R., *The Analysis of Binary Data,* Methuen, London, 1970.

Fleiss, J.L., *Statistical Methods for Rates and Proportions,* J. Wiley & Sons, Inc., NY, 1973.

Freud, Sigmund, *The Interpretation of Dreams* (1899), ed. and tr. by J. Strachey, Avon Books, Discus Edition, NY, 1965.

Good, I.J., *The Estimation of Probabilities,* The MIT Press, Cambridge, MA, 1965.

Harding, E.F., comments following D.V. Lindley and A.F.M. Smith, "Bayes Estimates for the Linear Model (with discussion)," *J. of the Royal Statistical Society, B34(1),* 1972, pp. 1-41.

Lancaster, H.O., *The Chi-Squared Distribution*, J. Wiley & Sons, NY, 1969.

Maxwell, A.E., *Analyzing Qualitative Data*, Methuen, London, 1961.

Reichert, T.A. and R.A. Christensen, "Validated Predictions of Survival in Coronary Artery Disease," Duke Medical Center, Durham, NC, and Entropy Limited, Lincoln, MA, 1981.

Reichert, T.A. and A.J. Krieger, "Quantitative Certainty in Differential Diagnosis," *Proc. of the Second Intl. Joint Conf. on Pattern Recognition*, 74-CHO-885-4C, Copenhagen, Denmark, Aug. 13-15, 1974, pp. 434-437.

Watanabe, Satosi, "Karhunen-Loève Expansion and Factor Analysis, Theoretical Remarks and Application," *Transactions of the Fourth Prague Conference on Information Theory, Statistical Decision Functions, Random Processes*, Prague, Aug. 31 - Sept. 11, 1965, Academia, Pub. House of the Czechoslovak Academy of Sciences, Prague, 1967, pp. 635-660.

Whitehead, Alfred North, *Process and Reality* (1929), The Free Press, NY, 1957.

Whorf, Benjamin Lee, "Language and Logic," *The Technology Review*, *43*, April 1941, pp. 250-252, 266, 268, 270, 272.

Wrinch, Dorothy and Harold Jeffreys, "On Certain Fundamental Principles of Scientific Inquiry," *Phil. Mag.*, *42*, 1921, pp. 369-390.

CHAPTER 12

MINIMUM ENTROPY PARTITIONS

CHAPTER 12

MINIMUM ENTROPY PARTITIONS

A. Clustering

It is after the baseball season in 1927. We wish to predict the batting average for George Herman (Babe) Ruth in 1928. The Babe was at bat 540 times in 1927 and made a hit 192 times, for an average of 0.356. Nothing could be simpler. Estimate 0.356 for 1928.

But is it so simple?

On what category of sample events are we basing our prediction? Babe-at-bat. What is that? An-assemblage-of-molecules-constitut-ing-a-baseball-player-called-"Babe-Ruth"-interacting-with-other-molecules-constituting-a-"baseball-game"-in-a-configuration-re-ferred-to-as-"at-bat". Consider two of the 540 sample events in this category:

   (1) Babe-at-bat-at-4:10 p.m., April 23, 1927, and

   (2) Babe-at-bat-at-4:35 p.m., September 22, 1927.

How much does this assemblage of molecules called "Babe Ruth" weigh? About 215 pounds. How many of these pounds were the same molecules at 4:35 p.m., September 22 as at 4:10 p.m., April 23? Probably about 180. Almost total turnover (99%) takes about four-teen years, those in the gastrointestinal system being replaced quite rapidly, those in bone and nerve cells much more slowly.

The two events (1) and (2) differ in many ways. Just for starters, about 35 pounds of matter constituting what we conventionally refer to as "Babe Ruth" are different. It is entirely possible that this could be a critical difference with respect to the probability of getting a hit.

Were the stadium and opposing team the same on September 22, 1927, as on April 23? No, they differed in innumerable ways. On the earlier date, he played against Philadelphia at their stadium, while on the later date he played against Detroit at his home grounds, Yankee Stadium in New York. Perhaps these included important differences with respect to the probability of getting a hit.

The event category "Babe-at-bat" is quite arbitrary. It is indeed a convenient category for conversational purposes. However, it is not necessarily a very good category for predictive purposes, even for predicting the average for "Babe-at-bat-in-1928."

There may be others that are more appropriate for purposes of prediction. For example, it may be better to sort "Babe-at-bat-in-1927" into categories such as: "Person-at-bat-with-X-experience-and-Y-physical-attributes-in-Z-condition." This may include some of the "Babe-at-bat-in-1927" events and exclude others. It may also include instances of other people at bat. Other categories we can imagine could even include instances of Babe Ruth attempting to succeed in other activities, or even people in general attempting to do things.

Fig. 38 shows an analysis conducted simply using the class for which the prediction is wanted as the template, translated in time, for event classification.

Fig. 39 illustrates an alternative classification of the sample events. We will still be able to obtain the desired overall 1928 prediction, but now it must be accomplished by assembling the predictions for various conditions.

The role of entropy minimization in this process is to determine which categories of past events are best for predictive purposes. This determination is based both on available information in data on past events, and on what is being predicted about future events.

Fig. 38. Analyses Conducted in Classification of Set
of Events for Which Prediction Wanted

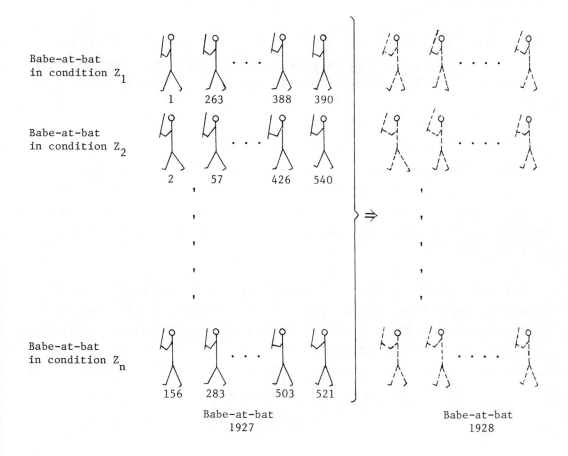

Fig. 39. Analysis Conducted in Minimum Entropy Classification

In the previous chapter, we saw how events may be plotted in feature space based on the information we have about the independent
variables. If we have been fortunate, events with the same dependent variable value will be *clustered* together in feature space,
and the clusters for different dependent variable values will be
clearly separated.

When this is so, we can draw definitive boundaries about each
cluster. The entire set of boundaries is said to form a *partition*
of the feature space. The boundary defining each cluster is called
a *pattern*. Events falling within the boundary are said to *match*
the pattern.

Often, however, data are not definitively clustered. In such
cases, the best we can do is seek diffused clusters described by
probability distributions. This is the task of entropy minimization.

B.   Entropy Measure of Data Clustering

Consider the problem of predicting whether there will be precipitation on a given day at Ft. Dodge, Iowa, assuming the only available independent variable data is the barometric pressure at 10 p.m.
on the previous day. Displayed in Fig. 40 is a sample of 343 days
for 1980.*

The total number of days with precipitation in our sample is 116.
Ignoring for the moment any possible information in the barometric
pressure about precipitation proneness, the probability of rain may
be computed using the equation we derived in Chapter 8:

$$P_{precip} = \frac{116+t+a}{343+t+f+a+b}$$

$$= \frac{116+1+10}{343+2+30}$$

$$= 0.339.$$

---

* Data supplied by Ruth E. Peterson, KVFD, Ft. Dodge, IA.

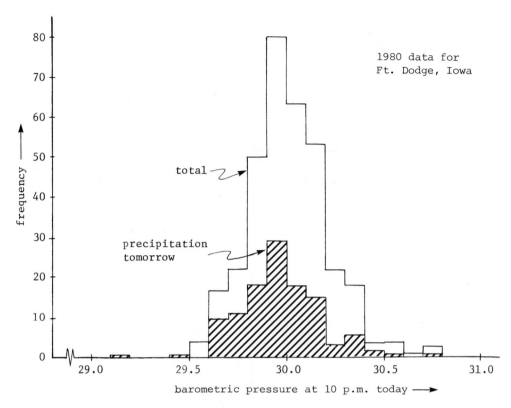

Fig. 40. Frequency of Precipitation Tomorrow versus
Barometric Pressure at 10 p.m. Today.

For illustrative purposes, we have taken values for the weights of
the various categories of background information as t=f=1, a=10, b=20.
Procedures for determining these weights will be given in Chapter 20.

We can now compute the entropy associated with this probability:

$$S(0) = -[P_{rain} \log P_{rain} + P_{nonrain} \log P_{nonrain}]$$

$$= -[P_{rain} \log P_{rain} + (1-P_{rain}) \log (1-P_{rain})]$$

$$= -[0.339 \log 0.339 + 0.661 \log 0.661]$$

$$= 0.640.$$

At this point, we consider making use of the barometric pressure data to improve our prediction. We investigate this prospect by tentatively defining a "threshold" on the barometric pressure. See Fig. 41.

For any given choice of this threshold, we can count the number of below-threshold events which were rainy and the number which were nonrainy; and the number of above-threshold cases which were rainy and the number which were nonrainy.

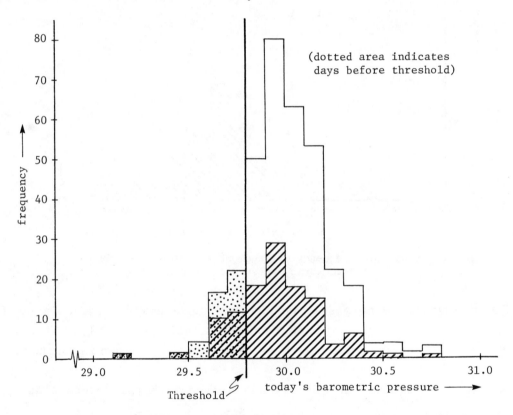

Fig. 41. Illustration of Effect on Outcome Class Counts in Region above and below Threshold (as one sweeps the threshold across the range of values of today's pressure)

We can then compute the rain-tomorrow probability for below-threshold cases, and the rain-tomorrow probability for above-threshold cases. Finally, we can compute the entropy associated with these probabilities. We will refer to this as the entropy of separability associated with the threshold. It is a measure of how well the threshold separates our data according to rain, lower values implying a clearer separation.

As an example for a threshold pressure of 29.8 mmHg we have:

Number of days below (or equal to) threshold		Number of days above threshold	
Precip. tomorrow	= 23	Precip. tomorrow	= 93
No precip. tommorrow	= 22	No precip. tomorrow	= 205
subtotal	45	subtotal	298
P(precip. given below thres.) = $\frac{23+1+10}{45+2+30}$		P(precip. given above thres.) = $\frac{93+1+10}{298+2+30}$	
= 0.442		= 0.315	

Assuming no background information for whether or not an event will be below the threshold:

$$P\left(\begin{array}{c} \text{event} \\ \text{below} \\ \text{threshold} \end{array}\right) = \frac{45+1}{343+2} = 0.133.$$

Then the entropy is:

$$S(O \text{ given Threshold}) = -0.133[0.442 \log 0.442 + (1-0.442) \log (1-0.442)]$$
$$-(1-0.133)[0.315 \log 0.315 + (1-0.315) \log (1-0.315)]$$
$$= 0.630 .$$

C.   Entropy Minimization Threshold Criterion

The entropy is, in general, dependent upon the position of the threshold. Carrying out this entropy computation for all possible threshold pressures, we obtain the functional dependence of entropy upon the position of the threshold. This is illustrated in Fig. 42.

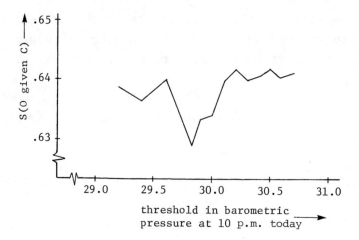

Fig. 42.   Entropy of Rain Tomorrow (given whether today's
           barometric pressure is above or below threshold,
           as a function of the threshold position).

We find that the entropy is lowest for a threshold pressure of about
29.8 mmHg. *For these data,* 29.8 mmHg is, in an important sense, the
best threshold *because* it has lower entropy than any other threshold.

D.   Minimum Entropy Cluster Boundaries

We have seen the process by which a minimum entropy criterion is
used to select a boundary which divides the sample into two portions
in a most outcome-ordered fashion in a one-dimensional feature space.

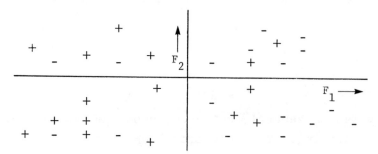

Fig. 43.   Illustrative Sample Displayed in a
           Two-Dimensional Feature Space

We now generalize this idea to selection of more complex boundaries
that pertain to multidimensional feature space. For simplicity, we
adopt the rule of drawing piecewise rectilinear boundaries. Fig. 43

shows a two-dimensional feature space. Fig. 44 shows four patterns defined by piece-wise rectilinear boundaries in this space. We will see shortly that flexibility in feature definition makes this a much more general way of specifying patterns than it may at first seem.

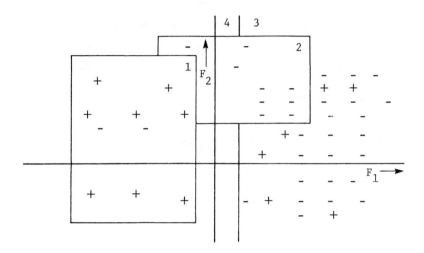

Fig. 44. Examples of Patterns Defined by
Piecewise Rectilinear Boundaries

Recall that entropy is, by definition, the expected value of information. In our example this is, specifically, the information we would gain by observing more tomorrows and learning whether or not it rained after the pressure was above or below the threshold.

The independent variable threshold defines the *condition* for the conditional probability of rain. If we condition our rain-tomorrow probabilities on whether today's pressure is above or below the 29.8 mmHg threshold, we expect to gain less information by further observation than if we had, for example, conditioned them on a 30.2 mmHg threshold. It is by setting the threshold that we extract information from the independent variable. The amount of information we extract by using any specified condition C is given explicitly as the difference between the unconditional entropy $S(O)$ and the conditional entropy $S(O$ given $C)$. This difference is identical to the mutual information exchange between O and C:

$$\Delta S(O \text{ mutual } C) = S(O) - S(O \text{ given } C).$$

The less information we expect to gain by further observation, the more information we have extracted from past observation. Thus, by selection of a threshold for which S(O given C) is a minimum, we maximize the information extracted from the available data.*

Having stated this general result, which carries with it implications for further generalizations to multiple thresholds, multiple independent variables, etc., two questions immediately arise.

First Question

The reasoning behind the entropy minimization principle sounds very similar to that behind the entropy maximization principle. Yet the conclusion is exactly the reverse. Does this make sense?

To answer this question, Table 38 reviews the two lines of reasoning in parallel fashion.

Table 38.   Entropy Minimax Reasoning

Entropy Maximization	Entropy Minimization
What is being varied is the set of numerical values of outcome probabilities for each given condition.	What is being varied is the set of feature space boundaries defining the conditions to which the outcome probabilities apply.
The objective is to minimize bias, i.e., assertion of information in excess of that actually available.	The objective is to maximize information extraction from the independent variable data.
This is achieved by maximizing the entropy of the individual outcome frequencies, subject to given constraints and data.	This is achieved by minimizing the conditional outcome entropy (subject, of course, to minimization of bias under each condition).
Intuitively: This fixes the outcome probabilities at values which contain the least bias by maximizing the information expected to be learned by further observation.	Intuitively: This organizes (partitions) the independent variable data so as to extract from it a maximum of information about the dependent variable by minimizing the information expected to be learned by further observation.

---

* R.A. Christensen, "Induction and the Evolution of Language," Physics Dept., Univ. of Calif., Berkeley, CA, 1963, (*Volume II*, p. 125); *Foundations of Inductive Reasoning*, Berkeley, CA, 1964, Sec. 8.D.

Entropy maximization and minimization perform different but comple-
mentary tasks. The bulk of statistical research has focused upon
the probability determination task, the problem addressed by entropy
maximization. However, event categorization is equally important.
Except for recent work in pattern recognition and multivariate data
analysis, definition of categories in independent variable space has
generally been left to intuition. Entropy minimization gives categorization a foundation in terms of a physical ordering principle.

Second Question

Why, precisely, does the entropy exchange ΔS(O mutual C) correspond
to the quantity of information *about* O extracted *from* C? The answer
may be seen by considering two extreme cases.

1) At one extreme, consider a C which is completely independ-
    ent of O. Then

$$P(O \text{ given } C) = P(O),$$

and it follows that

$$S(O \text{ given } C) = S(O).$$

In this case we gain no information about O by knowing C,
and the entropy exchange takes on its minimum value:

$$\Delta S(O \text{ mutual } C) = 0.$$

This minimum information about O is extracted from C when
O is independent of C.

2) At the other extreme, consider a C which definitely
    determines whether or not O occurs. Then

$$P(O \text{ given } C) = 1 \text{ or } 0,$$

and it follows that

$$S(O \text{ given } C) = 0.$$

In this case the entropy exchange is maximum (for this
particular O):

$$\Delta S(O \text{ mutual } C) = S(O).$$

This maximum information about O is extracted from C when
C completely determines O.

Fig. 45 illustrates the situation. Note that conditional entropy
S(O given C) references one time, while entropy exchange links different times. In the conceptual picture, entropy exchange is a
conserved quantity with an associated frame invariance symmetry.

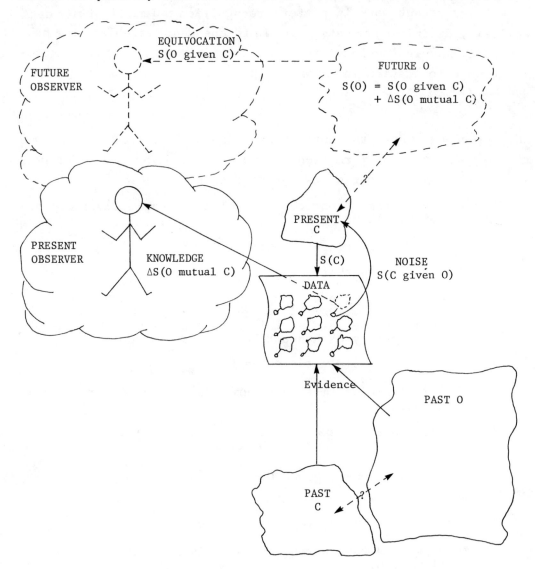

Fig. 45.   Entropy Is Our Conceptual Link to the Future.
           The expected information of future outcome O is
           S(O). To maximize the knowledge ΔS(O mutual C)
           about outcome O separated from the noise in condition
           C, we minimize the conditional entropy S(O given C).

E.   Information Transmission Rate Limitations on Induction

Recall that in Chapter 7 we defined the equivocation as the conditional entropy S(0 given C).

Shannon's two fundamental transmission rate theorems for discrete
channels are:

Discrete Channel   Let a source have entropy S (bits per symbol)
without Noise   and a channel have capacity K (bits per time
                  interval). Then it is possible to encode the
                  output of the source so as to transmit at the
                  average rate K/S-ϵ symbols per time interval
                  over the channels where ϵ is an arbitrarily
                  small positive number. It is not possible to
                  transmit at an average rate greater than K/S.

Discrete Channel   Let a discrete channel have the capacity K and
with Noise   a discrete source the entropy per time interval
                  S'. If S' ≤ K there exists a coding system
                  such that the output of the source can be trans-
                  mitted over the channel with an arbitrarily
                  small error frequency. If S' ≥ K, there exists
                  an encoding system such that the equivocation
                  is less than S' - K + ϵ where ϵ is arbitrarily
                  small. There is no method of encoding which
                  gives an equivocation less than S' - K.

Our objective in adjusting pattern boundaries is to minimize the
equivocation S(0 given C). This encodes the signals in our data
to maximize the rate of information extraction. Think of the time
interval as being, for instance, the duration of a computer analysis. Then channel capacity has units of bits per analysis. These
theorems provide fundamental limitations on the rate at which we
can gain information about the future in an inductive analysis
sequence.

F. Step-Wise Approximation Entropy Minimization

The number of possible partitions of feature space is very large,
so large that it would be impractical to seek the minimum entropy
partition by exhaustively computing the entropy of each possibility.
The difficulty is that this is a minimization problem in a discrete

space. The entropy changes discontinuously with boundary alterations each time we cross a data point. It is constant for alterations which do not cross data points. So the slope climbing techniques of the differential calculus cannot be employed.

One practical approach is to build up a step-wise approximation to
entropy minimization. Using this concept, we first find the minimum entropy single pattern satisfying the dimensional and shape
restriction imposed. Fig. 46 shows an example in a two-dimensional
feature space.

After this first minimum entropy pattern is found, we remove the
data matching it from the sample. Then we proceed to find the
lowest entropy pattern which can be formed for the remaining data.

We repeat this process until the data are exhausted. The final result is a complete partition of the space into patterns, as illustrated in Fig. 46. We call this a *screening* of feature space.

Each pattern is described by a statement in which the logical operators: "not", "and", and "or", are applied to a subset of feature
conditions $\{C_i\}$. Each feature condition is a statement that the
magnitude of a variable lies above or below a given threshold.

Note that when the definitions overlap, the earliest pattern in the
sequence is the one to which the data were matched. So this is the
one which controls for predictive purposes.

Also note that, in general, it is possible to define very complex
(event disjoint) patterns by this process.

G. Stitch-Wise Approximation Entropy Minimization

In some circumstances, lower overall entropy may be achieved by
sacrificing entropy minimization on the first pattern in order to
achieve even lower entropy on the second. The same concept can be

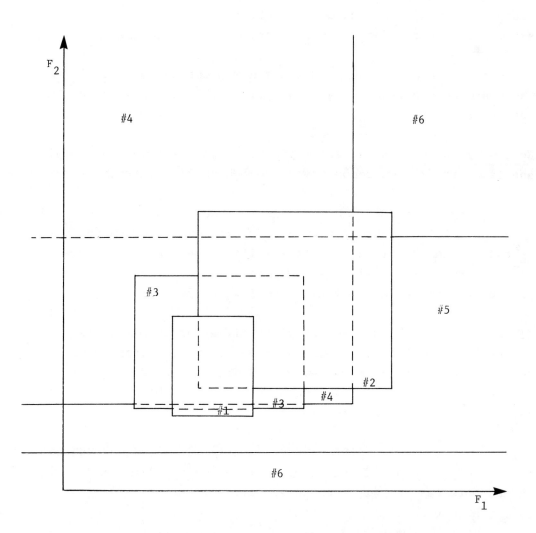

Fig. 46. Illustrative Sequential Generation of Rectilinear
          Patterns in a Two-Dimensional Feature Space

applied to tradeoffs among three or more patterns. Pair-wise
tradeoffs can be taken into consideration by using a "stitch-wise"
search algorithm. Fig. 47 shows its logical structure.

The process begins as we find the minimum entropy pattern in the
whole sample and label it #1. We remove all events matching this
pattern.

Next we find the minimum entropy pattern in the remaining sample
and label it #2. We remove events matching this pattern.

Now we put the events matching pattern #1 back into the sample and
find a new minimum entropy pattern. If the new pattern is the same
as the original #1 (i.e., if it specifies the same sample events)
then we are finished with #1. We remove events matching #1 and
proceed to find a new minimum, labeling it #3.

If, on the other hand, the new #1 is different from the original
#1, we have a possible tradeoff for lower overall entropy. We re-
move the events matching the new #1 and find a new #2. If this is
the same as the original #2, we have fixed #1. We remove events
matching #2 and proceed to find #3.

If, however, the new #2 differs from the original #2, we are thrown
into a loop. To avoid extensive cycling, we may limit the logic
to one more pass to #1, then to #3 directly, or to #2 followed by
#3 (with return to #1 forbidden).

H. Entropy Exchange Maximization

Our objective is to come as close as possible to finding a partition $\{C_i\}$ that minimizes the overall conditional entropy, given by

$$S(O \text{ given } C) = - \sum_{i=1}^{L} P(C_i) \sum_{k=1}^{K} P(O_k \text{given } C_i) \log P(O_k \text{ given } C_i).$$

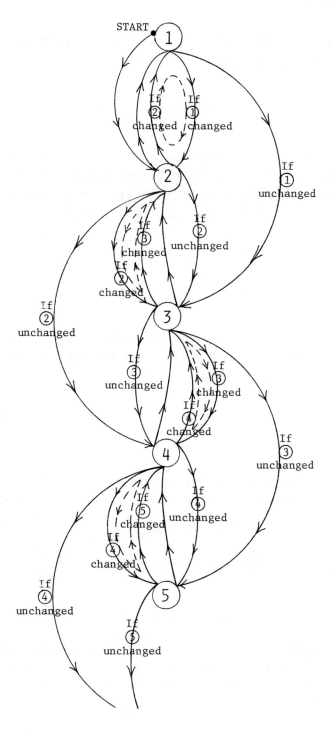

Fig. 47. Decision Logic for Generation of Stitch-Wise
Approximation to Minimum Entropy Partition

The step-wise and stitch-wise approximations minimize on each step
the individual pattern entropy, given by

$$S(0 \text{ given } C_i) = -\sum_{k=1}^{K} P(O_k \text{ given } C_i) \log P(O_k \text{ given } C_i).$$

This algorithm has a preference for patterns at each step which
differ as much as possible from equiprobabilities, with a tradeoff
for large pattern population built in via the weight normalization
(see Chapters 20 and 23). The general behavior is to deplete first
the most populous outcome class, then the second most populous,
continuing until all are of roughly equal population. Then which
outcome class is predominant oscillates as further patterns are
generated until the entire sample is exhausted.

An equivalent way of expressing our objective is to reframe it as
a problem of finding the partition $\{C_i\}$ which maximizes the total
entropy exchange,* given by

$$\Delta S(0 \text{ mutual } C) = \sum_{i=1}^{L} P(C_i) \sum_{k=1}^{K} P(O_k \text{ given } C_i) \log \frac{P(O_k \text{ given } C_i)}{P(O_k)}.$$

This formula suggests an approximation in which we maximize, on
each step, the entropy exchange between the pattern and the overall sample:

$$\Delta S(0 \text{ mutual } C_i) = \sum_{k=1}^{K} P(O_k \text{ given } C_i) \log \frac{P(O_k \text{ given } C_i)}{P(O_k)}.$$

_____

* Note that it would be inconvenient to deal with the noise, S(C given O), in
the expression

$$\Delta S(0 \text{ mutual } C) = S(0) - S(0 \text{ given } C) = S(C) - S(C \text{ given } 0),$$

since S(C) is affected by our feature space partition $\{C_i\}$, whereas S(0) is
fixed by the outcome partition $\{O_k\}$ reflecting the purposes for which we wish
to make the predictions.

This algorithm has a preference for patterns which differ as much as possible from the sample average, again with a tradeoff for large pattern population via the normalization. In this case, there is a preference for unusually high purity patterns.

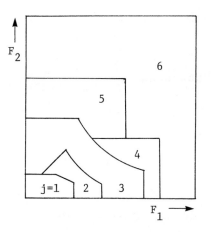

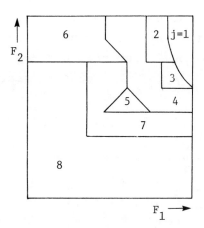

(a) Sequential Approximation to Minimum $S(O$ given $C)$

(b) Sequential Approximation to Maximum $\Delta S(O$ mutual $C)$

Fig. 48. Patterns Formed by Sequentially (a) Minimizing $S(O$ given $C_i)$, and (b) Maximizing $\Delta S(O$ mutual $C_i)$ for a Two-Dimensional Example in which the Rarer Outcomes Tend to Occur at Larger Values of F_1 and F_2

The contrasting behavior of the two algorithms is to sweep the feature space in roughly opposite directions in the projection of outcome distribution into feature space. See Fig. 48. The minimum S algorithm starts with the high population classes. The maximum ΔS algorithm starts with the low population classes.

Note that in Fig. 48 we show slanted and curved as well as rectilinear boundaries. The slanted boundaries are formed by linear combinations of the feature space axes, and the curved boundaries are formed by nonlinear combinations (e.g., products). These are discussed in Chapter 13.

I. Amalgamated Predictions

With two sets of patterns, one based on minimum S and the other on
maximum ΔS, we have two "predictions" for each point in feature
space. Thus, we need a procedure to amalgamate them.

This is accomplished by weighting each prediction by the ratio of
the number of data events on which it is based to the weight normalization.* For each outcome class O_k, we compute the amalgamated
probability at any point in feature space as

$$ P = \frac{(n_o/w_o)P_o + (n_1/w_1)P_1}{(n_o/w_o) + (n_1/w_1)} \, , $$

where the maximum entropy probabilities (see Chapter 8) are

$$ P_{o,1} = \frac{x_{o,1} + a_{o,1} + t}{n_{c,1} + a_{o,1} + b_{o,1} + t + f} \, , $$

and where the counts based on the approximations to a minimum
entropy partition are:

$$ \left. \begin{array}{c} x_o \\ n_o \\ a_o \\ b_o \end{array} \right\} = \left\{ \begin{array}{l} \text{real and virtual event counts for the predicted} \\ \text{point using pattern taken from the } S(O \text{ given } C_i) \\ \text{minimizing pattern set , and} \end{array} \right. $$

$$ \left. \begin{array}{c} x_1 \\ n_1 \\ a_1 \\ b_1 \end{array} \right\} = \left\{ \begin{array}{l} \text{real and virtual event counts for the predicted} \\ \text{point using pattern taken from the } \Delta S(O \text{ mutual } C_i) \\ \text{maximizing pattern set.} \end{array} \right. $$

The overall performance of the amalgamated predictor is superior
to that of a predictor for either pattern set individually. (Note
that each pattern set has its own weight normalization,
$w_{o,1} = a_{o,1} + b_{o,1} + t + f$, determined as described in Chapter 20.)

---

* The weights determine the normalized "unit" of data. For $w_o = w_1$, in the limit
of large same sizes, this becomes simply the frequency $P \approx (x_o + x_1)/(n_o + n_1)$.

QUOTES

MACCACARO (1958)

Therefore we can arrange the tests in a descending hierarchy in accordance with their capacity for identification, according to the value of

$$H = -\left(\frac{c}{N} \log_2 \frac{c}{N} + \frac{N-c}{N} \log_2 \frac{N-c}{N}\right)$$

and construct a key repeating the sequence from this hierarchy.

> —Giulio Alfredo Maccacaro (1924-   )
> "La misura della informazione contenuta nei criteri di classificazione," *Annali di Microbiologia, 8,* May 1958, p. 236, tr. by R. Christensen.

TANIMOTO (1958)

Let $A = \{a_i\}$, $i = 1, 2, \ldots, m$ be a finite set of m attributes (of which some may be the absence of particular attributes) associated with the finite set $B = \{b_j\}$, $j = 1, 2, \ldots, n$ of n objects. Define the $m \times n$ matrix $R = (r_{ij})$ so that $r_{ij} = 1$ if b_j possesses the attribute a_i and $r_{ij} = 0$ if b_j does not possess the attribute a_i...

Let us denote by A_i the ith row vector of R (and B_j the jth column vector of R)...

...define the dual similarity coefficient s_{jh} of a pair of objects b_j and b_h with respect to the set of equally weighted attributes A by

$$s_{jh} = \frac{N(B_j \cap B_h)}{N(B_j \cup B_h)}$$

...we consider the objects b_j, $j = 1, 2, \ldots, n$ as points in a semi-metric space H with the distance $d_{ij} \geq 0$ between points b_i and b_j defined by

$$d_{ij} = -\log_2 s_{ij}.$$

TANIMOTO (1958, cont.)

...We define the hierarchical power $H(b_i)$ of the point b_i by

$$H(b_i) = \sum_j d_{ij}$$

Thus we have introduced an order in H so that the set B is a lattice L. In terms of information theory, $H(b_i)$ is the "total" entropy of the system associated with b_i (Ref. #3). The element b_{i_0} determined by

$$H(b_{i_0}) = \min_i \sum_j d_{ij}$$

will be called the apex of the lattice L. b_{i_0}, in general, is not necessarily unique.

*   *   *   *

The problem in prediction or in diagnosis is the following: given a set of attributes, what object, among the considered objects, is this set of attributes most likely to represent? (e.g., in medical diagnosis: given a set of symptoms, what disease is most likely to be associated with this set of symptoms?) The solution to this problem is accomplished by augmenting the matrix R...so that the last column corresponds to the unknown object x with the absence or presence of its attributes a_i.

—Taffe T. Tanimoto (1917-   )
"An Elementary Mathematical Theory of Classification and Prediction" (1958), *The IBM Taxonomy Application*, M&A-6, ed. by T.T. Tanimoto and R.G. Loomis, Mathematics and Applications Dept., IBM, New York, 1960, pp. 33-36, 38.

#3 C.E. Shannon, "A Mathematical Theory of Communication," *Bell System Tech. Journal*, 27, 379, 623, (1948).

RESCIGNO and MACCACARO (1960)

To classify is to recompose the ensemble by grouping hierarchi-
cally the subsets according to a plan which is formulated in
terms of an ordered sequence of the properties. What makes one
classification different from another is the order criterion of
the sequence. Generally it depends on practical or theoretical
considerations which form no part of the items of knowledge to
be set in order.

$$* \quad * \quad * \quad *$$

$$H = -\sum \frac{a_i}{n} \log \frac{a_i}{n}$$

$$* \quad * \quad * \quad *$$

$$H(\Theta_a, \Theta_{b_1}, \Theta_{b_2}) - H(\Theta_a) = Q(\Theta_a; \Theta_{b_1}, \Theta_{b_2})\dots$$

$$S(\Theta_a; \Theta_{b_1}, \Theta_{b_2}) = Q(\Theta_a; \Theta_{b_1}, \Theta_{b_2}) - \tfrac{1}{2}[Q(\Theta_a; \Theta_{b_1}) + Q(\Theta_a; \Theta_{b_2})], \dots$$

$$* \quad * \quad * \quad *$$

The ratios:

$$R_1(\Theta_a) = \sum_{b_1} S(\Theta_a; \Theta_{b_1}) / \sum_{b_1} S^*(\Theta_a; \Theta_{b_1}),$$

$$R_2(\Theta_a) = \sum_{b_1, b_2} S(\Theta_a; \Theta_{b_1}, \Theta_{b_2}) / \sum_{b_1, b_2} S^*(\Theta_a; \Theta_{b_1}, \Theta_{b_2}),$$

measure the *relative disorder* existing inside the subsets defined by Θ_a in respect of all other operators, taken one by one,
two by two, and so on.

...The operator Θ_a for which the following sum is minimal:

$$R(\Theta_a) = R_1(\Theta_a) + R_2(\Theta_a) + \dots + R_{m-1}(\Theta_a)$$

...divides the set X into subsets of order one which are called
classes of order one. The group of classes of order one is more
ordered than any other group of subsets of order one...

—Aldo Rescigno (1924-   ) and
Giulio A. Maccacaro (1924-   )
"The Information Content of Biological Classifica-
tions," *Symposium on Information Theory*, Royal
Inst., (London, August 29 - September 2, 1960)
C. Cherry (ed.), Buttersworths Scientific Publi-
cations, London, 1961, pp. 437, 439, 442, 443.

ROGERS and TANIMOTO (1960)

...s_{12} is the ratio of the number of attributes in common in cases 1 and 2 to the number of distinct attributes possessed by cases 1 and 2...

One may define a "distance" d_{ij} between case i and case j by

$$d_{ij} = - \log_2 s_{ij}$$

...That case i_o for which the corresponding value of H_{i_o} is the least,

$$H_{i_o} = \min_i H_i = \min_i \sum_j -\log_2 s_{ij}$$

we define as the typical case.

$$* \quad * \quad * \quad *$$

...we define the total entropy $E_n[(d_{ij})]$ of a given set of points determined by the cases whose distances are the elements of the matrix (d_{ij}) by

$$E_n[(d_{ij})] = -\tfrac{1}{2} \sideset{}{'}\sum_{ij} \frac{d_{ij}}{T_n[(d_{ij})]} \log_2 \frac{d_{ij}}{T_n[(d_{ij})]}$$

where the normalization factor is given by

$$T_n[(d_{ij})] = -\tfrac{1}{2}\left(\sideset{}{'}\sum_{ij} d_{ij} \right)$$

where Σ' indicates summation only of the finite terms after repeated rows and columns are deleted. If g is the number of zeros in the symmetric matrix (d_{ij}) which lie strictly above the main diagonal, and h is the number of infinite elements above the main diagonal which are not on the same rows and columns as the g zeros, then the maximum entropy expression $\mathcal{E}_n$...

$$\mathcal{E}_n(g,h) = \log_2 \left[\tfrac{n-g}{2}(n-g-1) - h \right]$$

ROGERS and TANIMOTO (1960, cont.)

...a reasonable measure of inhomogeneity $U_n[(d_{ij})]$ determined by the matrix (d_{ij}) can be given the normalized difference

$$U_n[(d_{ij})] = \frac{\mathcal{E}_n(g,h) - E_n[(d_{ij})]}{\mathcal{E}_n(g,h)}$$

$$= 1 - \frac{E_n[(d_{ij})]}{\mathcal{E}_n(g,h)}$$

This expression is identical with Shannon's definition of redundancy.*

—David J. Rogers (1918-   ) and Taffe T. Tanimoto
(1917-   )
"A Computer Program for Classifying Plants,"
Science, 132, Oct. 21, 1960, pp. 1117, 1118.

HYVÄRINEN (1962)

A method has been described for classifying a collection of items characterized by qualitative attributes. The method uses mainly two concepts, typicality and similarity. The first of these has been derived by applying information-theoretical tools to the data. Typicality is a relation of one item to the whole collection. Similarity is a relation of one item to the two items and depends on the number of common properties shared by them.

...The method uses information-theoretical concepts for defining the typicality of item a_i with respect to the set A. The most typical item a_d is chosen as the model for the first group or class. Next, a measure of similarity between the model and any other item is defined. Those items whose similarity exceeds a preassigned value are admitted to this class. The same procedure is repeated for the remainder set A minus the classified items until the set has been exhausted...

This classification method has been programmed for the computer IBM 1620 and is available for the Users Group [1]. Tanimoto and Loomis [2] have written a classification program for IBM 704 but this method is based on a different mathematical background...

—Lassi P. Hyvärinen (1930-   )
"Classification of Qualitative Data," *BIT (Nordisk Tidskrift for Informationsbehandling), 2* (1962), Swets und Zeitlinger N.V., Amsterdam, 1968, pp. 83-84.

---

* C.E. Shannon, *Bell System Tech. J., 27,* 379, 623 (1948).
[1] L. Hyvärinen: *1620 Taxonomy Program,* IBM 1620 Users Group Library, 1962.
[2] T.T. Tanimoto, R.G. Loomis: *The IBM Taxonomy Application (IBM 704),* IBM Application Library, 1960.

REFERENCES

Anderberg, Michael R., *Cluster Analysis for Applications*, Academic Press, NY, 1973.

Ball, Geoffrey, "Data Analysis in the Social Sciences: What About the Details?" *Proceedings of the Fall Joint Computer Conference*, 1965, pp. 533-559.

Bledsoe, W.W., "Some Results on Multicategory Pattern Recognition," *J. Assoc. Computing Machinery*, *13*, 1966, pp. 304-316.

Bongard, M., "Useful Information," *Pattern Recognition*, Spartan Books, NY, 1970, pp. 94-112.

Boulton, D.M. and C.S. Wallace, "The Information Content of a Multisize Distribution," *Jour. of Theo. Biology*, *23*, 1969, p. 269.

_____, "A Program for Numerical Classification," *The Computer Journal*, *13*, 1970, pp. 63-69.

Christensen, R.A., "Induction and the Evolution of Language," Physics Dept., Univ. of Calif., Berkeley, CA, July 19, 1963. [*Chapter 7 of Volume III.*]

_____, *Foundations of Inductive Reasoning*, Berkeley, CA, 1964.

_____, "Inductive Reasoning and the Evolution of Language," Physics Dept., Berkeley, CA, 1964. [*Chapter 8 of Volume II.*]

_____, "A General Approach to Pattern Discovery," Tech. Rept. No. 20, Computer Center, Univ. of Calif., Berkeley, CA, June 29, 1967 (revised Nov. 15, 1967). [*Chapter 1 of Volume III.*]

Field, J.G., "The Use of the Information Statistic in the Numerical Classification of Heterogeneous Systems, *J. Ecol.*, *57*, 1969, pp. 565-569.

Fukunaga, Keinosuke, "Additional Clustering Procedures," *Intro. to Statistical Pattern Recognition*, Academic Press, NY, 1972, pp. 347-349.

_____, *Introduction to Statistical Pattern Recognition*, Academic Press, NY, 1972.

Good, I.J., "Speculations Concerning the First Ultra-Intelligent Machine," *Advances in Computers*, F.L. Alt (ed.), Vol. 6, Academic Press, NY, 1965, pp. 31-88.

Gower, J.C., "The Basis of Numerical Methods of Classification," *The Soil Ecosystem*, J.G. Sheals (ed.), Systematics Assoc., London, 1969, pp. 19-30.

Hall, A.V., "Studies in Recently Developed Group-Forming Procedures in Taxonomy and Ecology," *J. South Afr. Bot.*, *33*, 1967, pp. 85-96.

Harrison, P.J., "A Method of Cluster Analysis and Some Applications," *Applied Statistics*, *17*, 1968, pp. 226-236.

Hartigan, John A., *Clustering Algorithms*, J. Wiley & Sons, Inc., NY, 1975.

Hyvärinen, L.P., "Classification of Qualitative Data," *BIT (Nordisk Tidskrift for Informationsbehandling)*, 2 (1962), Swets und Zeitlinger N.V., Amsterdam, 1968, pp. 83-89.

Hyvärinen, L.P., *Information Theory for Systems Engineers*, Springer-Verlag, NY, 1970.

Lambert, J.M. and W.T. Williams, "Multivariate Methods in Plant Ecology. VI. Comparison of Information-Analysis and Association-Analysis," *J. of Ecology, 54*, 1966, pp. 635-664.

Lance, G.N. and W.T. Williams, "Computer Programs, for Hierarchical Polythetic Classification ('Similarity Analysis')," *The Computer Journal, 9,* 1966, pp. 60-64.

_____, "A General Theory of Classificatory Sorting Strategies. 1. Hierarchical Systems," *The Computer Journal, 9,* pp. 373-380.

_____, "Note on a New Information-Statistic Classificatory Program," *The Computer Journal, 11,* 1968, p. 195.

MacArthur, R.H. and J.W. MacArthur, "On Bird Species Diversity," *Ecology, 42,* 1961, pp. 594-598.

Maccacaro, G.A., "La misura della informazione contenuta nei criteri di classificazione," *Ann. Microbiol. Enzimol., 8,* May 1958, pp. 231-239.

Macnaughton-Smith, P., "Some Statistical and Other Numerical Techniques for Classifying Individuals," *Studies in the Causes of Delinquency and the Treatment of Offenders, 6,* Home Office, London, 1965.

Monin, Andrei S., *Weather Forecasting as a Problem in Physics* (1969), tr. by Paul Superak, MIT Press, Camb., MA, 1972, pp. 173-176.

Orloci, L., "Information Theory Models for Hierarchic and Non-Hierarchic Classifications," *Numerical Taxonomy* (A.J. Cole, ed.), Academic Press, NY, 1969, pp. 148-164.

Ornstein, L., "Computer Learning and the Scientific Method: A Proposed Solution to the Information-Theoretical Problem of Meaning," *J. Mt. Sinai Hosp., 33,* July 1965, pp. 437-494.

Quastler, H., "The Measurement of Specificity," *Information Theory in Biology,* (ed. by H. Quastler), Univ. of Illinois Press, Urbana, IL, 1953.

Rescigno, A. and G.A. Maccacaro, "The Information Content of Biological Classifications," *Symposium on Information Theory,* Royal Inst., London, Aug. 29-Sept. 2, 1960, C. Cherry (ed.), Butterworths Scientific Publications, London, 1961, pp. 437-446.

Rogers, D.J. and H. Fleming, "A Computer Program for Classifying Plants, II, A Numerical Handling on Non-Numerical Data," *Bioscience, 14,* 1964, p. 15.

Rogers, D.J. and T.T. Tanimoto, "A Computer Program for Classifying Plants," *Science, 132,* Oct. 21, 1960, pp. 1115-1118.

Sebestyen, George S., *Decision-Making Processes in Pattern Recognition,* Macmillan & Co., NY, 1962.

Sokal, R.R. and P.H.A. Sneath, *Principles of Numerical Taxonomy,* W.H. Freeman, San Francisco, CA, 1963.

Tanimoto, T.T., "An Elementary Mathematical Theory of Classification and Pre-diction" (1958), *The IBM Taxonomy Application*, M&A-6, ed. by T.T. Tanimoto and R.G. Loomis, Mathematics and Applications Dept., IBM, New York, 1960, pp. 30-39.

Terrot, Bishop, "On the Possibility of Combining Two or More Probabilities of the Same Event, So as to Form One Definite Probability," *Edin. Phil. Trans.*, *XXI*, 1856, pp. 369-376.

Tryon, Robert C. and Daniel E. Bailey, *Cluster Analysis*, McGraw-Hill Book Co., NY, 1970.

Turkin, Hy and S.C. Thompson, *The Official Encyclopedia of Baseball*, A.S. Barnes and Co., NY, 1964.

Wallace, C.S. and D.M. Boulton, "An Information Measure for Classification," *The Computer Journal*, *11*, 1968, p. 185.

Watanabe, S., "Entropic Measure of Simplicity and Cohesion," *Knowing and Guessing, A Quantitative Study on Inference and Information*, J. Wiley & Sons, Inc., NY, 1969, pp. 403-408.

Williams, W.T. and J.M. Lambert, "Multivariate Methods in Plant Ecology. V. Similarity Analyses and Information-Analysis," *J. of Ecology*, *54*, 1966, pp. 427-445.

CHAPTER 13

FEATURE SELECTION

CHAPTER 13

FEATURE SELECTION

A.   Overview of Feature Selection

Fig. 49 outlines the entropy minimax pattern discovery process as implemented in the SWAPDP algorithm.

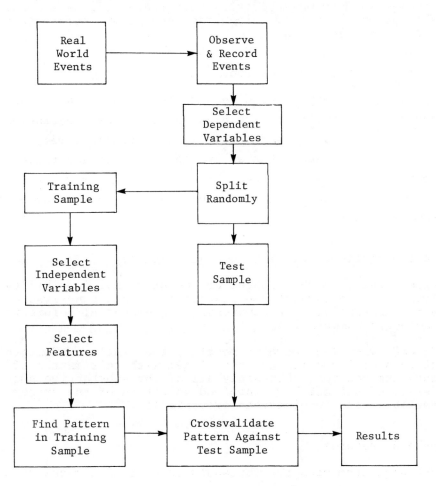

Fig. 49.   SWAPDP Pattern Discovery

Feature selection occurs at two stages of this process:

o Selection of Independent Variables

This is a data collection stage function. Although conceptually distinct from sample selection, it does affect the
definition of the events which qualify for inclusion in the
sample and decision concerning what data to take from observational records. It may also involve decisions to make additional observations.

o Selection of Feature Space Axes

This is an analysis stage function. Most practical problems
involve so many possibilities that there must be a considerable amount of preliminary human weeding out before computer
processing can begin.

B. Necessity for Data Compression

The theorist may ask why we need feature selection at all. Why not
gather and use all data that have even the slightest bearing on the
dependent variable? To be sure, this will result in a very high
dimensional feature space. But we are going to let a computer do
the tedious data processing work. In the meantime, why not get on
with the *fundamental* aspects of the problem and avoid getting bogged
down in these details?

There are two difficulties with this suggestion:

o First, there is the problem of computational feasibility.
Even with the largest and fastest computers conceivable,
very substantial data compression is essential before we
can begin computations.

o Second, many (and perhaps most) of the really fundamental
aspects of induction are associated with this matter of
data compression. Certainly all of the substantive successes of science have involved selection of the proper
variables and combinations of variables.

C. Threshold Features

The SWAPDP algorithm uses the threshold concept as a means of extracting compressed feature information from quantitative variables.
Events are discretely categorized according to whether they are

above or below a given threshold, or between two thresholds.   Complex events initially represented in a data base by many binary bits of information are reduced to a few bits by such categorization.

In this manner, SWAPDP partitions a continuous multidimensional feature space into cells with boundaries perpendicular to axes in feature space as shown in Figs. 50 and 51.

Fig. 50 gives an example showing four such cells in a two-dimensional space.   Fig. 51 shows an example for a three-dimensional space. (Note that the cells are not necessarily convex.   Since they are defined sequentially, one cell may loop around another.)

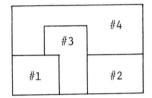

Fig. 50.   Examples of Patterns Formed Using Thresholds
in a Two-Dimensional Feature Space

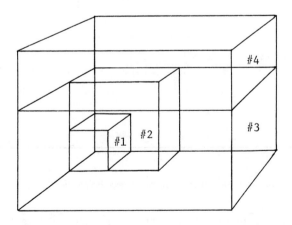

Fig. 51.   Examples of Patterns Formed Using Thresholds
in a Three-Dimensional Feature Space

Entropy minimization is the criterion for selecting thresholds. Fig. 52 illustrates how the entropy varies as the threshold is swept across the range of a variable.

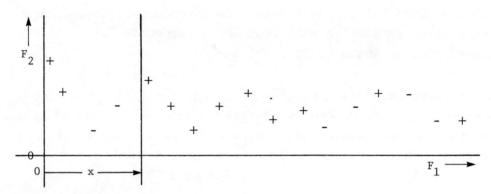

(a) Threshold on Feature F_1 Swept across Data Range

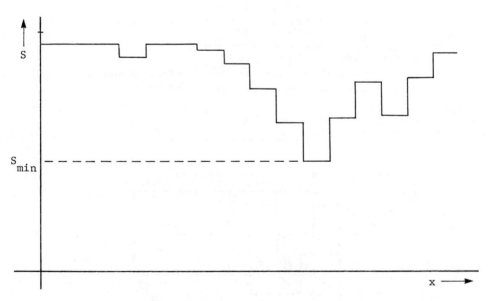

(b) Entropy as Function of Threshold Position

Fig. 52.   Functional Dependence of Entropy on Threshold is Determined by Changes in Left and Right Occupancy Counts as Threshold is Swept across Range

The entropy curve is flat between data points. In multidimensional feature space this forms a "security hyperannulus" providing informational leeway for boundary setting.

The SWAPDP user interacts with the computer in making feature selection decisions in the following ways:

o   The user defines the upper limit on the number of thresholds which may be defined for each quantitative variable.

o   The user specifies how low the entropy must be with the variable partitioned at the threshold, compared to the unpartitioned entropy, in order for the threshold to be accepted as significant.

o   The user specifies how correlated two thresholds can be before one is rejected as supplying insufficiently independent information.

Selection of these parameters is made to hold computer run time down to manageable proportions. The user must, however, recognize that the algorithm only minimizes entropy subject to the imposed restrictions. These restrictions will, in general, result in suboptimal performance compared to unrestricted minimization. As a practical matter, one uses sensitivity analyses and prior experience as guides to making these decisions. Sensitivity analyses involve relaxing restrictions and determining the extent of change in the results. Prior experience refers to previous analyses of similar data sets.

D.   Linear Combination Features

Threshold specification is the basic means by which SWAPDP specifies its patterns. However, this does not mean that the pattern boundaries can have only rectangular faces. New axes can be formed from linear combinations of old axes. This is equivalent to rotating an axis in the original feature space.

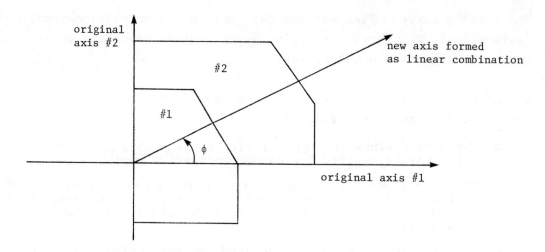

Fig. 53.   Linear Combination Features Are Formed by Rotations
in the Original Feature Space

Fig. 53 above shows two cells formed in the original two-dimensional feature space augmented by a third axis at a rotated position.   The SWAPDP algorithm allows boundaries to be at right angles
to any axis.   In this case the new axis is formed at an angle ϕ with
respect to an original axis.

The principle of entropy minimization is used to select rotation
angles for the formation of these new axes.   Fig. 54 shows the entropy as the rotation angle sweeps from 0 to 180°.

User decisions to be made in the construction of a set of rotated
axes are:

o   How many original axes are to be used to form linear com-
    binations?

o   Which ones?

o   How many new axes may be formed?

o   What is the criterion for rejecting a candidate new axis
    as too closely correlated to an existing axis?

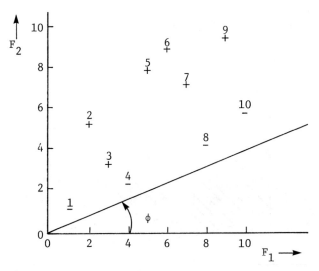

(a)   New Axis $F_3 = F_1 \cos\phi + F_2 \sin\phi$ Formed by
Performing Rotation Through Angle ϕ

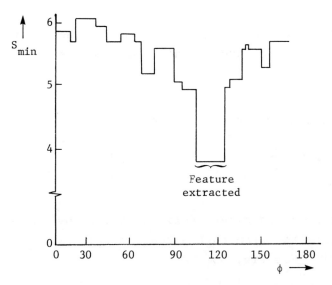

(b)   Minimum Entropy (for Threshold Separation)
as Function of Rotation Angle

Fig. 54.   Dependence of S_{min} on Rotation Angle, Determined by
Varying the Angle and Determining the Corresponding
Minimum Entropy for Threshold Separation

E. Nonlinear Combinations - Curved Boundaries

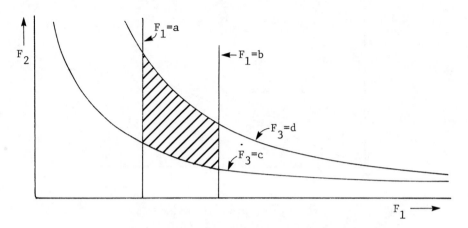

Fig. 55. Curved Boundaries Formed by Nonlinear Combinations

Suppose our "raw" features are F_1 and F_2, and that we, as the SWAPDP users, define a third feature as the product of these two. That is

$$F_3 = F_1 \, F_2.$$

Now imagine that SWAPDP finds a cell defined as:

$$a \leq F_1 \leq b, \text{ and}$$
$$c \leq F_3 \leq d.$$

This cell has two curved sides, defined in part by ourselves and in part by SWAPDP, in addition to the two straight sides defined entirely by SWAPDP. See Fig. 55.

The SWAPDP algorithm does not perform selection of nonlinear combinations similar to its selection of linear combinations. Rather, this process is left to the user. However, once nonlinear combinations have been specified, SWAPDP uses them to form potential cell boundaries by defining the thresholds. Further, it can form linear combinations between these nonlinear features and other features.

The user may be very creative in forming nonlinear combinations.
Consider, for example, Fig. 56.

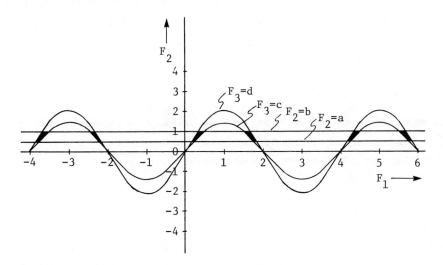

Fig. 56.   Even Disjoint Regions Can be Defined by Nonlinear
Boundaries and Threshold Conditions.

Here a third feature is defined as:

$$F_3 = \frac{F_2}{\sin F_1} \, .$$

A cell boundary is defined as:

$$a \leq F_2 \leq b, \text{ and}$$

$$c \leq F_3 \leq d.$$

This cell happens to consist of a series of disjoint regions in
the original (F_1, F_2) feature space.

Some nonlinear combinations can be transformed to linear combinations quite easily.   Take, for example, the case of products to
powers:

$$F = V_1^{\alpha_1} \ V_2^{\alpha_2} \ V_3^{\alpha_3} \ldots$$

Suppose we write

$$\log F = \alpha_1 \log V_1 + \alpha_2 \log V_2 + \alpha_3 \log V_3.$$

If we enter the variables $\log V_1$, $\log V_2$, $\log V_3$,...into an entropy minimax rotations analysis, we obtain high conditional information sets of coefficients $(\alpha_1, \alpha_2, \alpha_3, ...)$. Since $\log F$ is monotonic in F, the two contain identical information about the dependent variable.

Not all nonlinear combinations are so easily handled by simple rotation. Most must be entered explicitly to enable assessment of their conditional information content.

F.   Mechanistic Model Generated Features

Once we have recognized the possibility of forming linear and/or nonlinear combinations of variables as features for the purpose of pattern boundary definition, the potential contribution of theoretical models and computer codes to the pattern discovery process becomes clear.

Suppose we wish to discover patterns in the behavior of a very complex system for which we have both empirical data and one or more crude mechanistic theories. We can use the input-output response characteristics of these theories to generate informationally enriched features for the pattern building process. See Fig. 57. Though an inaccurate theory may not function well as a stand-alone predictor, it may still prove useful in contributing to the description of the pattern boundaries.

Of course, if we are fortunate enough to possess a mechanistic model which predicts the behavior of the system perfectly, then a single axis corresponding to this model's output contains all possible dependent variable information and we need look no further. However, in real life the perfect theory is seldom available, and we generally must augment the model generated features with empirical observation data.

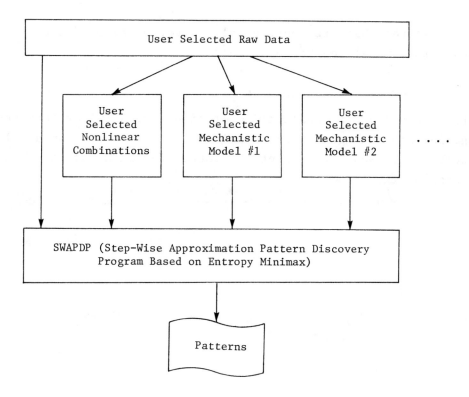

Fig. 57. Feature Input to SWAPDP Can Be Output from
Mechanistic Models as well as Raw Data and
User Selected Combinations of Raw Data

We may, in fact, want to try several alternately proposed models as
feature generators. It is clear that, using this approach, SWAPDP
can perform no worse than the best of the user selected models.
Further, SWAPDP can act as a selector among theoretical models that
is more nearly free of human bias.

If a feature generated by one of the models turns out to be a superior predictor to any other feature input to SWAPDP, or any combination, then SWAPDP will simply select this feature to make its
prediction.

In practice, however, when applied to complex problems, SWAPDP is
able to find combinations of the theoretical model features and raw

data features that are substantially superior as predictors to any theoretical model feature by itself.

G. Time-Series Filter Features

As an example of the application of time-series filter features, Fig. 58 shows the gold bullion prices for the period January 1 - December 31, 1980. One point is plotted for each day, the London afternoon fix.

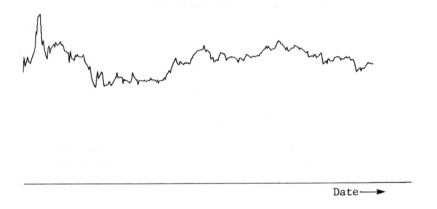

Date⟶

Fig. 58. Time-Series Data on Gold Bullion Price

Imagine that we are interested in predicting the following dependent variable:

Price difference between
tomorrow's fix and   =   { 1 if positive
today's fix   0 if negative or zero.

The time-series shown in Fig. 58 covers 240 trading days. Suppose we consider any given day as "today". Assume that "today" refers to a time after the market has closed, so that all information generated today, as well as in preceding days, is available to our predictor.

Fig. 59. Dependent Variable Selected as Price Change

Fig. 59 shows whether the price change for each trading day from the previous trading day, prior to and including today, was up or down. Our objective is to predict whether it will be up or down tomorrow. We need this information to be able to place a "buy", "sell", or "stay" order with our broker.

As a first step in making a prediction for tomorrow, we have to decide what to use for independent variables. The previous history of price changes is one likely candidate. Other candidates are time-series data about related variables such as the volume of gold traded, monetary exchange rates, and stock market indices.

As a second step, we must decide how to extract features from the time-series we have selected. Take the gold price movement itself, for example. Suppose we decide to restrict our independent variable data for any given "today" to the immediately preceding ten trading days (including today).

With this restriction, let us see how many data points we have in our training data set.

Necessary historical days   =   9
Earliest possible "today"   =   1
Sequence of more "todays"   = 228   230 training events
Latest possible "today"   =   1
Last day in available data   =   1
                              240

We lose nine days at the beginning of our data because they are necessary to make up the ten-day history from which we will build the independent variable data set for each event.

We lose one day at the end of our data because we are developing patterns to predict one day into the future.

This leaves us with 240-9-1 = 230 training events out of our original 240 days of data.

Now consider a typical ten-day history sequence from which we wish to form features. See Fig. 60. (The sequence may be a segment of the dependent variable time-series or of any other time-series we are using.)

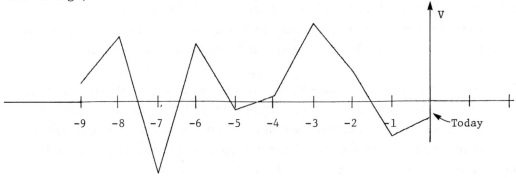

Fig. 60. Independent Variable Selected as a Preceding
Portion of Gold Bullion Price Change Time-Series

We number today as "0", one day before as "-1", two days before as "-2", and so forth.

We may define one feature as the average of today's value and yesterday's value, $F_1 = \frac{1}{2}(V_0 + V_{-1})$.

This two-day-wide computational box surrounding the time-series outputs F_1, the average. See Fig. 61.

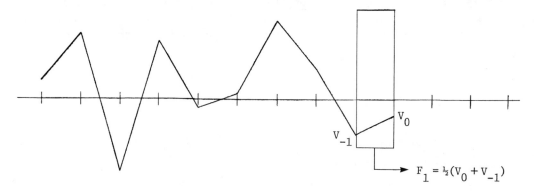

Fig. 61. Two-Day Average Computer Focused on a
Specific Portion of the Time-Series

If we pass the entire time-series through the box, we obtain a series of two-day averages, one for each "today" in the training data. See Fig. 62.

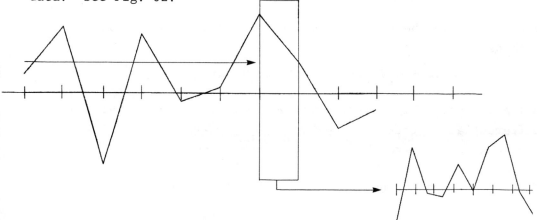

Fig. 62. Filter Extracting a Two-Day Moving
Average from an Input Time-Series

Thus, what we have is a *moving* two-day average. Our computational box acts as a *filter* extracting this average as the time-series passes through it.

It really doesn't make any difference whether we think of the box or of the time-series as moving. In the case of computerized analysis, the computer program is the computational box and the input data which passes through it constitute the time-series, so this may be a good way to view it.

It is, of course, possible to construct many different filters to extract

- today's value: $F = V_0$
- yesterday's value: $F = V_{-1}$
- lag-m value: $F = V_{-m}$
- 2-day moving average, lag-0: $F = \frac{1}{2}(V_0 + V_{-1})$
- n-day moving average, lag-m: $F = \frac{1}{n}(V_{-m} + V_{-m-1} + \ldots + V_{-m-n})$
- 2-day moving difference, lag-0: $F = V_0 - V_{-1}$
- 2-day moving difference, lag-m: $F = V_{-m} - V_{-m-1}$

More complex filters may be defined, such as:

- $F = V_{-m} - V_{-m-1} + V_{-m-2}$
- $F = V_{-m} + V_{-m-1} - V_{-m-2} - V_{-m-3}$

The concept may be extended to general linear combination filters, such as:

- $F = a_0 V_0 + a_1 V_{-1} + a_2 V_{-2} + \ldots + a_9 V_{-9}$

A linear combination filter applied to variables with zero mean is differential or integral, depending upon whether or not its coefficients sum to zero.

Finally, we can define nonlinear combination filters, such as:

- $F = V_0 V_{-1}$
- $F = V_0 / \sin V_{-1}$

Entropy minimax rotations can be used as a guide to the selection of coefficients for linear combination filters. For most nonlinear combinations, separate analyses are required. Some special filters include expansions in terms of Fourier, Karhunen-Loève and entropy minimax basis functions. These are discussed in succeeding sections.

H. Fourier Analysis Features

A commonly employed way of representing serial data is by means of Fourier analysis. We shall only discuss one-dimensional cases here, to illustrate basic concepts of using basis functions to extract features from waveforms. Generalization to multiple dimensions is straightforward.

Fig. 63 shows, as an illustrative waveform, the electrocardiogram signal from a human patient.

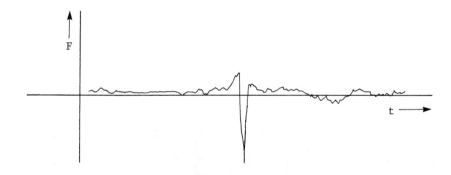

Fig. 63. Illustrative Input Data Waveform

Fourier analysis seeks to represent this irregular waveform as the summation of a set of regular basis waveforms. The basis waveforms, in this case, are defined as sine waves of various amplitudes and frequencies. Thus, the waveform is represented as

$$F(t) = \sum_{k=-\infty}^{\infty} a_k \, e^{ik\pi t/T},$$

where the coefficients are given by

$$a_k = \frac{1}{2T} \int_{-T}^{T} F(t) e^{-ik\pi t/T} dt \, ,$$

and where $\pm$ T defines the limits on the time axis for the waveform.

Suppose we perform a Fourier analysis of a specific segment preceding each training event.

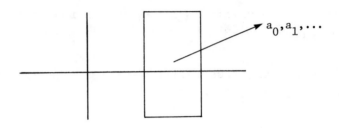

Fig. 64.   Fourier Analysis Outputs

The outputs from this Fourier analyzer (Fig. 64) can be used as features to help us form a feature space (Fig. 65) for entropy minimax analysis.

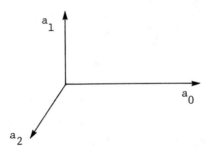

Fig. 65.   Feature Space with Axes from Selected
Fourier Analysis Outputs

For future reference, it may be noted that if we let T→∞, then the Fourier equations become

$$F(t) = \frac{1}{\sqrt{2\pi}} \int_{-\infty}^{\infty} D(\omega) e^{i\omega t} \, d\omega, \text{ and}$$

$$D(\omega) = \frac{1}{\sqrt{2\pi}} \int_{-\infty}^{\infty} F(t) e^{-i\omega t} \, dt,$$

where the spectral density $D(\omega)$ is now a point in an infinite dimensional feature space.

I.   Karhunen-Loève Minimum Unconditional Entropy Features

A basic difficulty with Fourier analysis is that it commonly takes a very large number of basis waveforms to represent an input waveform to adequate accuracy, especially if there are high frequencies (sharp edges) in the original waveform. This can result in very inefficient pattern analysis, as the information in the original waveform is spread over numerous Fourier amplitudes, each individually containing relatively little information.

In order to overcome this problem and produce a more compact and efficient representation of the shapes of waveforms, Karhunen and Loève developed an alternative means of analysis.

Whereas Fourier analysis uses sine functions of various phases and frequencies as the basis functions, K-L analysis uses linear combinations of the training waveforms themselves. For a given set of training waveforms, each K-L basis function is defined by a set of coefficients which specify the contribution of each training waveform to the basis function.

Suppose there are N training waveforms and we wish to specify M basis functions, each defined by N coefficients. In Fourier analysis, it is usual to arrange the components of a waveform in order of increasing frequency. In K-L analysis, the basis functions are arranged in the order of their importance to minimizing the error in representing the shape of waveforms like those in the training sample. This ordering ensures that the first M basis functions represent the average training waveform with less error than could any other set of M basis functions, for any number M.

To the extent that accurate representation of the shape of the waveform is relevant to our analysis, K-L analysis is more efficient than Fourier analysis. It requires fewer coefficients, hence a lower dimensional feature space, to represent shape information.

Consider a training set of N waveforms denoted by

$$\psi_1(t), \ \psi_2(t), \ldots, \ \psi_N(t).$$

We wish to construct M orthonormal basis functions by means of a set of *basis coefficients* $\{b_{mi}\}$. The basis functions are formed as

$$\phi_m(t) = \sum_{i=1}^{M} b_{mi}\psi_i(t).$$

Using the M basis functions denoted by

$$\phi_1(t), \ \phi_2(t), \ldots, \ \phi_M(t),$$

we can find the set of reconstruction coefficients $\{z_m\}$ corresponding to any given waveform $\psi(t)$:

$$z_m = \int \psi(t)\phi_m(t)dt.$$

Using these coefficients, we may represent $\psi(t)$ as a linear combination of the $\{\phi_m\}$ plus an error term:

$$\psi(t) = \sum_{m=1}^{M} z_m \phi_m(t) + \varepsilon(t).$$

The reconstruction coefficients are chosen to minimize the expected value of the square of the error term $\varepsilon(t)$. This can be shown to be equivalent to minimizing the entropy*

$$S_{KL} = -\sum_{m=1}^{M} \rho_m \log \rho_m,$$

where the density function is taken as

$$\rho_m = \sum_{i=1}^{M} z^2_{im},$$

and where z_{im} is obtained from the ith training waveform and the mth basis function:

$$z_{im} = \int \psi_i(t) \phi_m(t) dt.$$

The result is that the K-L basis coefficients are obtained as the eigenvalues $\{\lambda_m^{K-L}\}$ of the determinant

$$\|C_{ij} - \delta_{ij} \lambda_m^{K-L}\| = 0,$$

where the matrix elements C_{ij} are given by

$$C_{ij} = \int \psi_i(t) \psi_j(t) dt.$$

---

* S. Watanabe, "Karhunen-Loève Expansion and Factor Analysis, Theoretical Remarks and Applications," *Trans. of the Fourth Prague Conf. on Info. Theory, Statistical Decision Functions, Random Processes,* Prague, Aug. 31–Sept. 11, 1965, Academia, Publ. House of the Czechoslovak Academy of Sciences, Prague, 1967, pp. 635–660.

For further details on K-L expansion feature extraction, see Watanabe, Tou and Heydorn, Young and Calvert, and other general references on pattern recognition.

J.   Maximum Entropy Spectral Analysis

In the discussion so far in this chapter, we have regarded waveform measurements as containing independent variables information about a general dependent variable. For example, the electrocardiogram waveform contains information about the future health of a patient.

In many applications of interest, the dependent variable is specifically the shape of the same waveform but outside the region of available independent variables data. Thus, for example, if we sample a quantity at uniformly spaced points in the interval from t_1 to t_2, we may seek to predict the probable magnitude at an extrapolated time $t_3 > t_2$. We may also be interested in interpolation between observation points. In both cases, we seek a "smoothed" representation of a finite set of data points. One way of obtaining the smooth fit is by the use of entropy maximization.*

For example, let $F(t)$ be a stationary Gaussian process, and let $D(\omega)$ be its spectral density. Assume we have a sample of N values of $F(t)$, i.e., $F_1 = F(\Delta t)$, ..., $F_N = F(N\Delta t)$ where Δt is the uniform sampling interval. Then the spectral density can be expressed as

$$D(\omega) = \sum_{k=-\infty}^{\infty} \phi(k) \, e^{-i\omega k\Delta t},$$

where $\phi(k)$ is the autocorrelation for lag k, given by

$$\phi(k) = \frac{1}{N} \sum_{j} F_j \, F_{j+k}.$$

---

* J.P. Burg, "Maximum Entropy Spectral Analysis," paper presented at the 37th meeting Soc. Explor. Geophys., Oklahoma City, OK, Oct. 31, 1967.

From our sample, we know N values of the autocorrelation: $\phi(0),\ldots,$ $\phi(N-1)$. The entropy of the spectral density is given by

$$S = \frac{1}{4\omega_N} \int_{-\omega_N}^{\omega_N} \log D(\omega)\ d\omega\ ,$$

where ω_N is the Nyquist frequency.* To obtain an estimate $\hat{D}(\omega)$ of $D(\omega)$, we choose those values of the unknown autocorrelations for which S is a maximum. This gives

$$\hat{D}(\omega) = \frac{D_N}{\omega_N \left| 1 + \displaystyle\sum_{k=1}^{N-1} \alpha_k\ e^{-i\omega k\Delta t} \right|^2}\ ,$$

where D_N is a constant and the α_k are the prediction error coefficients determined from the data. For further details on maximum
entropy spectral analysis, see Burg, Smylie *et al.*, Akaike, and
Ulrych *et al.*

It is frequently stated that time is the independent variable and
amplitude is the dependent variable. This terminology, however,
fails to acknowledge that there is independent variables information in the amplitudes of the data points. The amplitude of a
point to be predicted is a dependent variable. The amplitudes of
the data points that have been observed are independent variables.

Thus the question arises: how do we group events in this independent variable space for probability estimation purposes? This
is an entropy *minimization* question. In this sense, there are important hidden entropy minimizing assumptions in the use of maximum
entropy spectral analysis.

---

* The inverse of the Nyquist interval (half the bandwidth), which is the maximum interval of uniform sampling to completely determine a waveform.

K.   Entropy Minimax Waveform Expansion

Fourier analysis is an efficient way of representing conditional information in cases in which periodicities, especially low or moderate frequency periodicities, in the independent variables are important to the dependent variable.

K-L analysis is efficient in cases in which the overall shape of the independent variable waveform is important.

In many cases of practical importance, key features of waveforms reside in small details not efficiently represented by either Fourier or K-L analysis. For example:

   o   In examining an EKG for abnormality in the heartbeat
       waveform for a patient, the physician must pay attention
       to many small details. Such details would often be omitted
       from either a Fourier or a K-L analysis unless the number
       of terms taken is so large as to virtually eliminate the
       data compression efficiency of the analysis.

   o   When interpreting the meaning of a human voice tone or
       facial expression, very slight changes can have considera-
       ble significance. In such cases, languages focusing upon
       overall shapes can be very inefficient for representing
       critical details.

These problems can be handled in a fashion analogous to K-L analy-sis, by defining linear combinations of the training waveforms which minimize error in conditional information.* The resulting entropy minimax basis functions can then be used as a means of data compression specifically tuned to enrichment of information about the dependent variable.

As with K-L analysis, we consider a set of N waveforms

$$\psi_1(t), \ \psi_2(t), \ldots, \psi_i(t).$$

---

* R.A. Christensen, "Entropy Minimax Expansions for Waveform Analysis," *Pattern Recognition*, *11*, 1979, pp. 41-49 (*Volume III*, pp. 239-247).

We wish to construct M basis functions $\{\phi_m\}$ by means of basis coefficients $\{b_{mi}\}$,

$$\phi_m(t) = \sum_{i=1}^{M} b_{mi} \, \psi_i(t),$$

so that we can represent an arbitrary waveform, $\psi(t)$, in terms of of the basis functions plus an error term:

$$\psi(t) = \sum_{m=1}^{M} z_m \, \phi_m(t) + \varepsilon(t).$$

We seek to minimize the expected conditional information loss. This is the entropy decrement

$$\Delta S = \int d\underset{\sim}{z} \; f(\underset{\sim}{z}) \sum_{k=1}^{K} p_k(\underset{\sim}{z}) \log \frac{p_k(\underset{\sim}{z})}{p_k'(\underset{\sim}{z})},$$

where $p_k(\underset{\sim}{z})$ is based on the full N terms, and $p_k'(\underset{\sim}{z})$ is based on the truncation to only M terms.

After some manipulations, it can be shown that the entropy minimax basis functions are given as the solution to two coupled eigenequations:

$$\sum_{i} C_{ji} \, G_{im} = \lambda_m \, z_{jm}$$

and

$$\sum_{i} A_{im} = 0,$$

where G_{im} and A_{im} are functions of the expansion coefficients $\{z_m\}$ and spreading parameters $\{a_m\}$ like the width vector components for potential functions defined in the next chapter.

L.   Feature Space Axes Selection

Entropy considerations provide useful tools to assist axes selection.   Suppose, for example, that we wish to conduct a computerized entropy minimax analysis with feature space limited, at most, to 100 dimensions, but that we have 1,000 candidate features.

A "mini-analysis" can be conducted on each candidate feature to assess its conditional information value singly as a guide to this selection.   One can compute the entropies for the primary, secondary, etc., thresholds for each candidate.   Lower entropy indicates more conditional information.

Next, we can assess the features pair-wise, three-at-a-time, and so forth.   Computer run time limitations usually restrict us to pairs or triplets when conducting exhaustive preliminary investigations in 100-dimensional feature spaces.   Preprocessing run times go as follows:

$$\binom{100}{1} = 100$$

$$\binom{100}{2} = 4.95 \times 10^3$$

$$\binom{100}{3} = 1.62 \times 10^5$$

$$\binom{100}{4} = 3.92 \times 10^6$$

$$\text{etc.}$$

When a subspace of a dozen or so dimensions is identified for a particular pattern, exhaustive examination of all possible binary combinations can be conducted for the data projected into this subspace.   Processing run times go as follows:

$$2^{12} = 4.10 \times 10^3$$

$$2^{14} = 1.64 \times 10^4$$

$$2^{16} = 6.55 \times 10^4$$

$$2^{18} = 2.62 \times 10^5$$

$$2^{20} = 1.05 \times 10^6$$

$$2^{22} = 4.19 \times 10^6$$

etc.

When assessing a candidate feature space axis, things to take into consideration include:

o   Entropy of primary, secondary, etc., thresholds. (How much conditional information does it contain as a feature singlet?)

o   Correlation with other candidate features. (Does it contribute independent information, or is it too redundant with another feature to warrant including it as an axis and spending computational resources on it?)

It is in feature selection that the most extensive human-computer interaction occurs in applications of SWAPDP to multivariate pattern discovery.

QUOTES

LEWIS (1962)

...let the N patterns to be recognized be denoted by Y_i, $i = 1$, ..., N and the M characteristics by C_j, $j = 1$, ..., M where the jth characteristic can take on n_j possible values.

* * * *

Because of the characteristic independence assumption, it is reasonable to attempt to determine, for each characteristic C_j, a number G_j which measures the "goodness" of that characteristic.

* * * *

Since only a single number is required for G_j, this suggests that G_j be the expected value of some function...

$$G_j = \sum_{i=1}^{N} \sum_{k=1}^{n_j} P[Y_i, C_j(k)] \log f\{P[Y_i, C_j(k)]\}$$

...f should be a measure of the correlation of the C_j with the Y_i. The particular f we shall use is

$$\frac{P[Y_i, C_j(k)]}{P(Y_i) \, P[C_j(k)]}.$$

Thus

$$G_j = \sum_{i=1}^{N} \sum_{k=1}^{n_j} P[Y_i, C_j(k)] \log \frac{P[Y_i, C_j(k)]}{P(Y_i) \, P[C_j(k)]}$$

$$= \sum_{i=1}^{N} \sum_{k=1}^{n_j} P[Y_i, C_j(k)] \log \frac{P[Y_i | C_j(k)]}{P(Y_i)}$$

$$= \sum_{i=1}^{N} \sum_{k=1}^{n_j} P[Y_i, C_j(k)] \log \frac{P[C_j(k) | Y_i]}{P[C_j(k)]}.$$

— Philip M. Lewis, II (1931–   )
"The Characteristic Selection Problem in Recognition Systems," *IRE Trans. on Info. Theory*, IT-8, Feb. 1962, pp. 172, 173, 176.

BLASBALG and VAN BLERKOM (1962)

The entropy-reducing [ER] transformation is an irreversible operation on the message which results in an "acceptable" reduction in fidelity. The vocoder (a device for compressing speech) technique[1] is an example of message degradation at the sending end which preserves speech intelligibility. The major emphasis in picture compression today is concerned with ER transformations. In both the vocoder and TV, the human observer generally determines the acceptable degree of message fidelity. In general, the ER transformations depend on the properties of the message source and its receiver. That is, an ER operation acceptable in one application can be unacceptable in another. Since the human sensors of speech and pictures are significantly different, one expects that the useful ER transformations for each case will be, and in fact are, significantly different.

$$* \quad * \quad * \quad *$$

The compression ratio is defined as

$$C_R = \frac{H_{max}}{H} = \frac{K \log_2 N}{-\sum_{i=1}^{D} P_i \log_2 P_i} \geq 1 \, .$$

> —Herman Blasbalg (1925–   ) and Richard Van Blerkom (1935–   )
> "Message Compression," *IRE Trans. on Space Electronics and Telemetry*, Sept. 1962, pp. 228, 229.

SEBESTYEN and EDIE (1964)

The average information gained about classes, in general, from the parameter set y, on the average, is...

$$I = \sum_{i=1}^{C} P(C_i) \int p(y|C_i) \, \log \frac{p(y|C_i)}{p(y)} \, dy$$

---

[1] "Proc. Seminar on Speech Compression and Processing," AF Cambridge Res. Ctr., L.G. Hanscom Field, Bedford, MA, Vol. 1, Rept. No. TR-59-198.

SEBESTYEN and EDIE (1964, cont.)

> ...P.M. Lewis*...employs this figure of merit (measure of "good-ness") to help to evaluate the relative utilities of different measurement subsets (for the case of Gaussian processes with *independent* variables).

* * * *

> A generalized measure of the distance or differences between classes (their probability densities in the property space) is, the "divergence" given by Kullback**... The divergence is a measure of "discriminability" between the classes C_i and C_j.

$$J(C_i, C_j) = \int \left[p(y|C_i) - p(y|C_j) \right] \log \frac{p(y|C_i)}{p(y|C_j)} \, dy$$

> —George Sebestyen (1931-   ) and Jay Edie (1937-   ) "Pattern Recognition Research, " AFCRL-64-821, Litton Systems, Inc., Waltham, MA, June 14, 1964, pp. 92, 93.

WATANABE (1965)

> The K-L-expansion is usually known as one which minimizes the average error committed by taking only a finite number of terms in the infinite series of an expansion when a given collection of functions is to be expressed as series in terms of some com-plete set of orthogonal functions...We shall point out that the the K-L-expansion can also be characterized as one which mini-mizes the entropy defined in terms of the average squared coef-ficients of an expansion...The degree of evenness of the dis-tribution of importance among base functions in this sense is expressed by the above mentioned entropy. In other words, to say that this entropy is small in the case of K-L-expansion amounts to saying, in somewhat loose use of language, that the importance in the above sense is concentrated heavily on very few base functions in this case.

> —Satosi Watanabe (1910-   ) "Karhunen-Loève Expansion and Factor Analysis, Theoretical Remarks and Applications," *Transactions of the Fourth Prague Conference on Information Theory, Statistical Decision Functions, Random Processes*, Prague, Aug. 31 - Sept. 11, 1965, Academia, Publ. House of the Czeckoslovak Academy of Sciences, Prague, 1967, p. 637.

---

*   P.M. Lewis, "The Characteristic Selection Problem in Recognition Systems," *IEEE Trans. on Info. Theory*, IT-8, 171-172, 1962.

** Kullback, Solomon, *Information Theory and Statistics*, J. Wiley & Sons, Inc., NY, 1959.

BURG (1967)

...the spectral estimate must be the most random or have the maximum entropy of any power spectrum which is consistent with the measured data. This new analysis technique gives a much higher resolution spectral estimate than is obtained by conventional techniques with a very little increase in computing time.

> —John Parker Burg (1931-   )
> "Maximum Entropy Special Analysis," paper present-
> ed at the 37th meeting Soc. Explor. Geophys.,
> Oklahoma City, OK, Oct. 31, 1967, p. ii.

SNEATH (1969)

...entropy measures disorder, and both a regularly spaced distribution and a clustered distribution are ordered and thus have low entropy.

> —P.H.A. Sneath (1923-   )
> "Evaluation of Clustering Methods," *Numerical
> Taxonomy*, ed. by A.J. Cole, Academic Press, NY,
> 1969, p. 264.

ULRYCH, SMYLIE, Jensen and CLARKE (1973)

A method of filtering and smoothing short records has been presented which depends on the prediction of the data by using the Burg algorithm for computing prediction filter coefficients and Burg's maximum entropy method for computing power spectra. Common methods of computing these parameters suffer from unrealistic assumptions about the extension of the data outside the known interval. These assumptions lead to spectral windows that severely limit the resolution of spectra of short records.

> —T.J. Ulrych (1935-   ), D.E. Smylie (1936-   ),
> O.G. Jensen (1905-   ) and G.K.C. Clark (1941-   )
> "Predictive Filtering and Smoothing of Short
> Records Using Maximum Entropy," *Journal of
> Geophysical Research, 78,* August 10, 1973, p. 4963.

TOU and GONZALEZ (1974)

The entropy concept can be used as a suitable criterion in the design of optimum feature selection. Features which reduce the uncertainty of a given situation are considered more informative than those which have the opposite effect. Thus, if one views entropy as a measure of uncertainty, a meaningful feature selection criterion is to choose the features which minimize the entropy of the pattern classes under consideration. Since this criterion is equivalent to minimizing the dispersion of the various pattern populations, it is reasonable to expect that the resulting procedure will have clustering properties.

> —Julius T. Tou (1926-   ) and
> Rafael C. Gonzalez (1942-   )
> *Pattern Recognition Principles,* Addison-Wesley
> Pub. Co., Reading, MA, 1974, p. 263.

REFERENCES

Akaike, H., "Fitting Autoregressive Models for Prediction," *Ann. Inst. Statist. Math., 21,* 1969, pp. 243-247.

_____, "Power Spectrum Estimation through Autoregressive Model Fitting," *Ann. Inst. Statist. Math., 21,* 1969, pp. 407-419.

_____, "Statistical Predictor Identifications," *Ann. Inst. Statist. Math., 22,* 1970, pp. 203-217.

Andrews, Harry C., "Feature Selection," *Intro. to Mathematical Techniques in Pattern Recognition,* Wiley-Interscience, NY, 1972, pp. 15-64.

Blasbalg, H. and R. Van Blerkom, "Message Compression," *IRE Trans. on Space Electronics and Telemetry,* Sept. 1962, pp. 228-238.

Bloomfield, Peter, *Fourier Analysis of Time-Series: An Introduction,* J. Wiley & Sons, Inc., NY, 1976.

Boyle, Brian E., "Symptom Partitioning by Information Maximization," Dept. of E.E., MIT, Cambridge, MA, 1972.

Burg, J.P., "Maximum Entropy Spectral Analysis," paper presented at the 37th meeting Soc. Explor. Geophys., Oklahoma City, OK, October 31, 1967.

_____, "All New Analysis Technique for Time-Series Data," paper presented at NATO Action Study Inst. on Signal Processing, Enschede, Netherlands, August, 1968.

Chien, Y.T. and K.S. Fu, "Selection and Ordering of Feature Observations in a Pattern Recognition System," *Inf. Control, 12,* 1968, pp. 395-414.

Christensen, R.A., "Applications of the Principles of Induction," Chapter 9 of *Foundations of Inductive Reasoning,* Berkeley, CA, 1964.

_____, "Entropy Minimax Method of Pattern Discovery and Probability Determination," Arthur D. Little, Inc., Camb., MA, March 7, 1972. [*Chapter 4 of Volume III.*]

_____, "Entropy Minimax, a Non-Bayesian Approach to Probability Estimation from Empirical Data," *Proc. of the 1973 Intl. Conf. on Cybernetics and Society,* IEEE Systems, Man and Cybernetics Society, Nov. 5-7, 1973, Boston, MA, 73 CHO 799-7-SMC, pp. 321-325. [*Chapter 5 of Volume III.*]

_____, "Entropy Minimax Analysis of Simulated LOCA Burst Data," report prepared for Core Performance Branch, Directorate of Licensing, U.S. Atomic Energy Commission, Contract No. AT(49-24)-0083, Entropy Limited, Belmont, MA, Dec. 31, 1974.

_____, "Entropy Minimax Determination of Threshold Values for Quantitative Variables," Carnegie-Mellon Univ., Pittsburgh, PA, Tech. Rept. No. 40.8.75, April 12, 1975. [*Chapter 7 of Volume III.*]

_____, "Entropy Minimax Rotations: The Extraction of Features from Sets of Quantitative Variables," Carnegie-Mellon Univ., Pittsburgh, PA, Tech. Rept. No. 40.9.75, April 28, 1975. [*Chapter 8 of Volume III.*]

Christensen, R.A., "Entropy Minimax Expansions for Waveform Analysis," *Pattern Recognition, 11,* No. 1, Jan. 1979, pp. 41-49. *[Chapter 15 of Volume III.]*

Christensen, R.A., R. Eilbert, O. Lindgren and L. Rans, "An Exploratory Application of Entropy Minimax to Weather Prediction: Estimating the Likelihood of Multi-Year Droughts in California," report to the Office of Water Research and Technology, U.S. Dept. of the Interior, Contract No. 14-34-001-8409, Entropy Limited, Lincoln, MA, Sept. 1980.

_____, "Successful Hydrologic Forecasting for California Using an Information Theoretic Model," *Journal of Applied Meterology,* (to be published).

Christensen, R.A. and A. Hirschman, "Automatic Phase Alignment for the Karhunen-Loève Expansion," *IEEE Trans. on Biomedical Engineering,* BME-26, Feb. 1979, pp. 94-99.

Fourier, Jean, *The Analytic Theory of Heat* (1822), Dover Pub., Inc., NY, 1966.

Franklin, D., "Analysis of Fission Gas Release Data by a Pattern Recognition Technique," IDG Note 2423, OECD Halden Reactor Project, Halden, Norway, Jan. 11, 1978.

Frantsue, A.G., "Some Problems in the Statistical Theory of Pattern Recognition," *JPRS, 35,* 1966, pp. 23-35.

Fu, K.S., "Feature Selection and Feature Ordering," *Sequential Methods in Pattern Recognition and Machine Learning,* Academic Press, NY, 1968, pp. 24-45.

Fukunaga, Keinosuke;" Feature Selection and Linear Mapping for One Distribution," *Intro. to Statistical Pattern Recognition,* Academic Press, NY, 1972, pp. 225-257.

Fukunaga, K. and W.L.G. Koontz, "Application of the Karhunen-Loève Expansion to Feature Selection and Ordering," *IEEE Trans. Comp. C-19,* 1970, pp. 311-318.

Karhunen, K., "Ueber lineare Methoden in der Wahrscheinlichkeitsrechnung," *Ann. Acad. Sci. Fenn., 37,* 1947.

Lanczos, Cornelius, *Discourse on Fourier Series,* Hafner Pub. Co., NY, 1966.

Leontiades, Kyriako, "Computationally Practical Entropy Minimax Rotations, Applications to the Iris Data and Comparison and Other Methods," Biomedical Engr. Program, Carnegie-Mellon Univ., Pittsburgh, PA, 1976.

Lewis, P.M., II, "The Characteristic Selection Problem in Recognition Systems," *IRE Trans. Information Theory, IT-8,* 1962, pp. 171-178.

Lim, E., "Coolant Channel Closure Modeling Using Pattern Recognition: A Preliminary Report," FRMPM21-2, SAI-175-79-PA, Science Applications, Inc., Palo Alto, CA, June 1979.

Loève, A., "Fonctions aleatoires de secondordre," *C.R. Acad. Sci., 220,* 1945; *222,* 1946 and *Rev. Sci., 83,* 1945; *84,* 1946:

_____, *Probability Theory* (1963), Springer-Verlag, NY, 1978.

Marill, T. and D.M. Green, "On the Effectiveness of Receptors in Recognition Systems," *IEEE Trans. on Information Theory, IT-9,* 1963, pp. 11-17.

Odell, Patrick, "A Model for Dimension Reduction in Pattern Recognition Using Continuous Data," *Pattern Recognition, 11,* No. 1, Jan. 1979, pp. 51-54.

Perez, Albert, "Information Theory Methods in Reducing Complex Decision Problems," *Trans. of the Fourth Prague Conf. on Information Theory, Statistical Decision Functions, Random Processes,* Prague, Aug. 31 - Sept. 11, 1965, Academia, Publ. House of the Czeckoslovak Academy of Sciences, Prague, 1967, pp. 58-87.

Reichert, Thomas A., "The Security Hyperannulus--A Decision-Assist Device for Medical Diagnosis," *Proc. of the Twenty-Seventh Annual Conference on Engineering in Medicine and Biology,* Philadelphia, PA, Oct. 6-10, 1974, p. 331.

Robinson, E.A., *Multichannel Time Series Analysis with Digital Computer Programs,* Holden-Day, San Francisco, CA, 1967.

Sebestyen, George and Jay Edie, "Pattern Recognition Research," AFCRL-64-821, Litton Systems, Inc., Waltham, MA, June 14, 1964.

Smylie, D.E., G.K.C. Clarke and T.J. Ulrych, "Analysis of Unequalization in the Earth's Rotation," *Methods in Computational Physics, 13,* Academic Press, NY, 1974, pp. 391-430.

Sneath, P.H.A., "Evaluation of Clustering Methods," *Numerical Taxonomy,* ed. by A.J. Cole, Academic Press, NY, 1969.

Tou, Julius T. and Rafael C. Gonzalez, *Pattern Recognition Principles,* Addison-Wesley Pub. Co., Reading, MA, 1974.

Tou, Julius T. and R.P. Heydorn, "Some Approaches to Optimum Feature Extraction," *Computer and Information Sciences-II,* ed. by J.T. Tou, Academic Press, NY, 1967, pp. 57-89.

Ulrych, T.J., "Maximum Entropy Power Spectrum of Threshold Sinusoids," *J. Geophys. Res., 77,* 1972, pp. 1396-1400.

Ulrych, T.J., D.E. Smylie, O.G. Jensen and G.K.C. Clarke, "Predictive Filtering and Smoothing of Short Records Using Maximum Entropy," *Journal of Geophysical Research, 78,* Aug. 10, 1973, pp. 4959-4964.

Ulrych, T.J. and T.N. Bishop, "Maximum Entropy Spectral Analysis and Autoregression Decomposition," *Reviews of Geophysics and Space, 13,* Feb. 1975, pp. 183-200.

Watanabe, Satosi, "Karhunen-Loève Expansion and Factor Analysis, Theoretical Remarks and Applications," *Transactions of the Fourth Prague Conference on Information Theory, Statistical Decision Functions, Random Processes,* Prague, Aug. 31 - Sept. 11, 1965, Academia, Publ. House of the Czechoslovak Academy of Sciences, Prague, 1967, pp. 635-660.

_____, "Minimum-Entropy Principles," Section 3.2 of *Advances in Information Systems Science,* ed. by Julius T. Tou, Plenum Press, NY, 1970, pp. 89-93.

Watanabe, Satosi, P.F. Lambert, C.A. Kulikowski, J.L. Buxton and R. Walker, "Evaluation and Selection of Variables in Pattern Recognition," *Computer and Information Sciences-II,* ed. by J.T. Tou, Academic Press, NY, 1967, pp. 91-122.

Young, Tzay Y. and Thomas A. Calvert, "Feature Extraction Theory," *Classification, Estimation and Pattern Recognition,* Am. Elsevier Pub. Co., NY, 1974, pp. 224-276.

CHAPTER 14

POTENTIAL FUNCTION ESTIMATION

CHAPTER 14

POTENTIAL FUNCTION ESTIMATION

A.   Continuity with Respect to Independent Variables

Scientists and engineers have in the past favored continuous rather than discrete models of physical processes.   This stems in part from a desire for formulae which serve all possible needs including possible needs for fine-structure detail, and partly from the contrary pre-digital-computer computational need for simple closed-form solutions even if they come from analyses that only roughly approximate reality.

It is evident, however, in the light of information theory that, with finite sample sizes, the assertion of arbitrarily fine-structured laws is neither logically justified nor informationally efficient.

Potential function analysis allows us to formulate laws that are continuous in their variables while retaining logical consistency and informational efficiency.   It achieves these desirable ends by empirical determination of the degree of fine-structuring in a continuum.

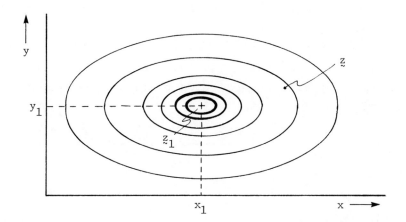

Fig. 66.   Equipotential Lines Representing Distributed Effect of z_1 in a Two-Dimensional Feature Space

Consider a single training sample data point (with outcome "+") at location $z_1 = (x_1, y_1)$ in a two-dimensional feature space. See Fig. 66. Consider also a test sample data point (of unknown outcome) at location $z = (x, y)$ in this space.

Now imagine that we define a potential function $\gamma(z, z_1)$ centered on the data point z_1. For example, we may choose a normal-shaped function:

$$\gamma(z, z_1) = \frac{1}{2\pi\sqrt{\sigma_1\sigma_2}} \, e^{-\frac{1}{2}\left[\left(\frac{x-x_1}{\sigma_1}\right)^2 + \left(\frac{y-y_1}{\sigma_2}\right)^2\right]}.$$

This function describes a "potential" felt by the test point. It is strongest near the data point and becomes weaker as the test point is moved farther from the data point.

Potential function analysis uses this basic concept to build models that describe the influence of each training sample data point on the outcome probabilities at arbitrarily chosen test points. This allows one to spread out, in a continuous fashion, the effects of the discrete data points. One may then compute the outcome probabilities at any point resulting from the combined intensities $\{\rho_k(z)\}$ of training sample data points distributed throughout the space. See Fig. 67.

Entropy maximization for this continuous case yields the result that the outcome probability function $p_k(z)$ is related to the outcome density by

$$p_k(z) = \frac{\rho_k(z) + w_k(z)}{\rho(z) + w(z)},$$

where

$$\rho(z) = \sum_k \rho_k(z)$$

and

$$w(\underset{\sim}{z}) = \sum_{k} w_k(\underset{\sim}{z}).$$

The weighting functions $w_k(\underset{\sim}{z})$ are discussed in Chapters 20 and 23.

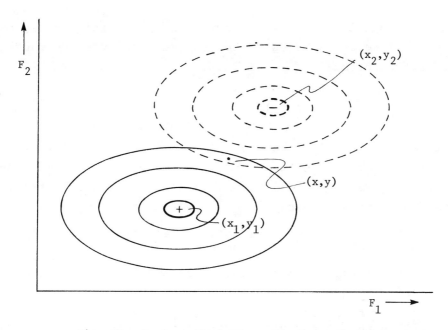

Fig. 67. Equipotential Lines for Two Data Points

There are many methods by means of which the effects of two or more data points can be combined to give outcome probabilities at a test point. Examples are density summation, wave superposition, and log-likelihood compounding.

In density summation, the potential functions are assumed to be real density functions. The outcome density $\rho_k(\underset{\sim}{z})$ is then defined as

$$\rho_k(\underset{\sim}{z}) = \sum_{i=1}^{n} \gamma_k(\underset{\sim}{z}, \underset{\sim}{z}_i).$$

In wave superposition, the potential functions are assumed to be complex wave functions. This enables modeling informational interference. The outcome density is defined as

$$\rho_k(\underset{\sim}{z}) = \left| \sum_{i=1}^{n} \gamma_k(\underset{\sim}{z},\underset{\sim}{z}_i) \right|^2 .$$

Finally, in log-likelihood compounding the potential functions are assumed to be real-valued, and the outcome density is defined as the support function:

$$\rho_k(\underset{\sim}{z}) = \log \prod_{i=1}^{n} \gamma_k(\underset{\sim}{z},\underset{\sim}{z}_i) .$$

There is a strong analogy between potential function analysis and parametric probability estimation. The former must address the same questions in independent variable space that the latter does in dependent variable space.

- o What criteria are to be used in selecting an underlying functional form for the potential function?

- o What are typical potential functions and their properties?

- o What criteria are to be used in assessing different estimates for the parameters of each potential function?

- o What are typical parameter estimation methods for potential functions and what are their properties?

- o What criteria are to be used in assessing different rules for combining the effects of observed data into estimates of outcome probabilities for test points?

- o What are typical rules for doing this and what are their properties?

The statement that a particular physical process can be modeled by a particular distribution in dependent variable space is a matter of physical fact, not a mathematical abstraction. We are interested in distributional forms that happen to be typically found to reasonably approximate the physical world. The contribution of mathematics is the study of the properties of such functional forms.

Similarly, a statement that a particular physical process can be modeled by a particular potential function in independent variable space is a matter of physical fact, not a mathematical abstraction. It is similarly the role of mathematics to study the properties of potential functions that happen to be typically found to reasonably approximate the physical world.

B. Potential Function Parameter Estimation

Consider, for example, a rectangular potential function. Within a specified interval (or hyper-rectangle) about a data point, it has a constant positive value. Outside this interval, the function is zero.

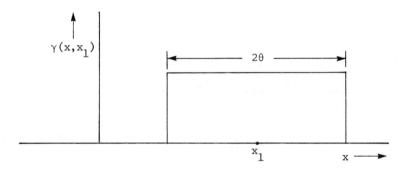

Fig. 68. Rectangular Potential Function

To simplify matters, consider a one-dimensional feature space. Fig. 68 shows such a rectangular potential function centered about a data point at x_1.

We can write this potential as

$$\gamma(x, x_1) = \begin{cases} 1 & \text{for } x_1 - \theta \leq x \leq x_1 + \theta \\ 0 & \text{otherwise.} \end{cases}$$

This function has one parameter, its half width θ. We seek a method for making an estimate of this parameter based on a data sample.

The entropy minimax estimator is defined as that value $\hat{\theta}$ for which the conditional entropy,

$$S = -\int dx\ f(x) \sum_k p_k(x) \log p_k(x),$$

is a minimum, where

$$p_k(x) = \frac{\rho_k(x) + w_k(x)}{\rho(x) + w(x)},$$

$$f(x) = \frac{(n/\rho)\ \rho(x) + v(x)}{n + v},$$

$$\rho(x) = \sum_k \rho_k(x),$$

$$\rho = \int dz\ \rho(x),$$

$$w(x) = \sum_k w_k(x) \text{ and}$$

$$v = \int dz\ v(x).$$

The weighting functions $w_k(x)$ and $v(x)$ are discussed in Chapter 20.

It is, in practice, a complex job to carry out the minimization of S with respect to θ in order to determine its estimator $\hat{\theta}$. However, some interesting observations can be made for special cases.

When all the data have the same outcome, then for an arbitrary number of events, the entropy minimax estimator is $\hat{\theta} = \infty$. In cluster terminology this amounts to drawing no boundaries and treating the entire space as a single cell.

With two data points of different outcome, separated by a distance Δx along the independent variable, the entropy minimax estimator can be shown to be $\hat{\theta} = \frac{1}{2}\Delta x$. Thus, entropy minimization draws a boundary midway between the two events.

C.   Continuity with Respect to the Dependent Variable

As we may use potential functions to introduce continuity with respect to the independent variable, so also we may introduce continuity with respect to the dependent variable.   To accomplish this
we define:

$f(k$ given $\underset{\sim}{z},\underset{\sim}{\theta})dk$ = probability of outcome between k and k+dk
given feature vector $\underset{\sim}{z}$ and parameters $\underset{\sim}{\theta}$, and

$f(\underset{\sim}{z}$ given $\underset{\sim}{\theta})d\underset{\sim}{z}$ = probability of feature vector in the
interval $\underset{\sim}{z}$ to $\underset{\sim}{z}+d\underset{\sim}{z}$ given the parameters $\underset{\sim}{\theta}$.

The conditional entropy is

$$S(\underset{\sim}{\theta}) = -\int [f(\underset{\sim}{z} \text{ given } \underset{\sim}{\theta})d\underset{\sim}{z}] \int [f(k \text{ given } \underset{\sim}{z},\underset{\sim}{\theta})dk]$$

$$\times \log[f(k \text{ given } \underset{\sim}{z},\underset{\sim}{\theta})dk] .$$

This quantity diverges because of the extra dk in the logarithm,
reflecting the infinite information content of infinitely fine-
structured information.   However, we may escape this problem by
defining a renormalized entropy from which the divergence is
eliminated.   This is given by

$$S_R(\underset{\sim}{\theta}) = \lim_{\Delta k \to 0} (S(\underset{\sim}{\theta}) + \log \Delta k)$$

$$= -\int d\underset{\sim}{z} \ f(\underset{\sim}{z} \text{ given } \underset{\sim}{\theta}) \int dk \ f(k \text{ given } \underset{\sim}{z},\underset{\sim}{\theta})$$

$$\times \log \ f(k \text{ given } \underset{\sim}{z},\underset{\sim}{\theta}) .$$

Because we have simply pulled out a constant, $S_R(\underset{\sim}{\theta})$ contains the
same information as $S(\underset{\sim}{\theta})$ concerning the relationship between feature space location and dependent variable value. Note that renormalized entropies can have negative as well as positive values.

This allows us to reformulate our probability density function as

$$f(k \text{ given } \underset{\sim}{z},\underset{\sim}{\theta}) = \frac{\rho(k,\underset{\sim}{z}) + w(k,\underset{\sim}{z})}{\rho(\underset{\sim}{z}) + w(\underset{\sim}{z})} .$$

Associated reformulations of the potential functions are given
for density summation by

$$\rho(k,\underset{\sim}{z}) = \sum_{i=1}^{n} \gamma(k,\underset{\sim}{z},\underset{\sim}{z_i}),$$

and for wave superposition by

$$\rho(k,\underset{\sim}{z}) = \left| \sum_{i=1}^{n} \gamma(k,\underset{\sim}{z},\underset{\sim}{z_i}) \right|^2 .$$

If we think of the dependent variable as a "parameter", we can
also define a new density function in terms of the evidence supporting the outcome

$$\rho(k,\underset{\sim}{z}) = - \frac{\partial^2}{\partial k^2} U(k,\underset{\sim}{z}),$$

where the support function is given by

$$U(k,\underset{\sim}{z}) = \log \prod_{i=1}^{n} \gamma(k,\underset{\sim}{z},\underset{\sim}{z_i}).$$

With this latter selection, potential function estimation uses the
machinery of statistical analysis of support and evidence.

It may be instructive to see how Newton's law of inertia can be
represented as a consequence of potential function data analysis.
We assume a one-dimensional space for simplicity. Fig. 69 shows
observations of a particle at two different times, t_1 and t_2, when
the positions were x_1 and x_2, respectively.

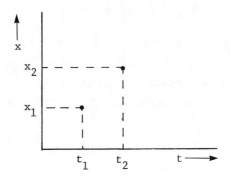

Fig. 69.   Measurements of Positions of
Particle at Two Different Times

Around each data point, we draw equipotential lines of a two-dimensional normal potential for finding a particle if we look at any chosen test point. See Fig. 70. Because space and time are not treated as equivalent in this example, we have generally different spreading constants in the space and time directions.

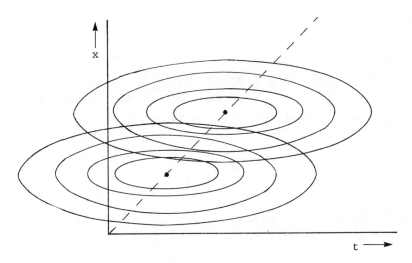

Fig. 70.   Potential Functions Spreading
Outward from Data Points

We form the outcome density by evidence compounding. The probability "ridge", i.e., local maximum, lies on the straight line connecting the two observation points. (This is evident from symmetry.) For any given t, it is most likely (based on the two data points and the choice of feature space and potential function) that we will find the particle to be on this straight line.

The slope of the line is the velocity of the particle. In the absence of any contrary data, the most probable trajectory in this space is a straight line with constant velocity.

Additional data points lying on this straight line will enhance the probabilities along this trajectory by sharpening the ridge. Data points scattered about the line, perhaps due to measurement error, will cause the ridge to spread. See Fig. 71.

Data exhibiting a consistent deviation from the straight line (say for an accelerating particle) will cause the ridge to curve. See Fig. 72. Note, however, that with a finite data set, the evidence will not support an unlimited curved trajectory in this space. As it is extrapolated farther (in either direction) from the outermost data point, it gradually straightens out. We cannot see any new corners in this space beyond the horizon of our finite data.

This does not mean that we cannot interpolate or extrapolate variable functions. By making a Fourier transformation, for example, from the time domain to the frequency domain we can conduct our potential function analysis on the frequency spectrum. Then when we transform back, we may appear to be making very complex and variable projections in the time domain. This is what is done in Kolmogorov and Wiener analysis for time-series interpolation and extrapolation. Karhunen-Loève and entropy minimax expansions provide alternate domains for interpolation and extrapolation, hence alternate types of stationarity and nonstationarity.

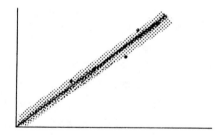

Fig. 71. Probability Ridge Broadening due to Data Scatter

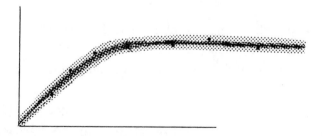

Fig. 72. Curved Probability Ridge Tracking
Data Points in a Smooth Arc

D.   Entropy Minimax Hazard Axes

Suppose we are concerned with some irreversible failure phenomena.
Examples include: light bulb burns out, aeroplane structural rib
cracks, auto brake lining wears out, human body dies.   We plot the
available data as failure status versus age.   See Fig. 73.

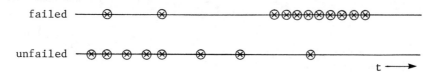

Fig. 73.   Failure Status as a Function of t

In general, unfailed individuals will be more dense at lower ages,
failed individuals more dense at higher ages.

We define the hazard function h(t) such that:

   h(t)Δt = probability that an individual unfailed at age t fails
            between t and t+Δt.

The maximum entropy estimator of h(x)Δx is then given by

$$h(t)\Delta t = \frac{n_f + w_f}{n + w} ,$$

where

   n_f = number failed during interval Δt ,

   n = number unfailed at beginning of interval Δt ,

   w_f = failure weight , and

   w = weight normalization .

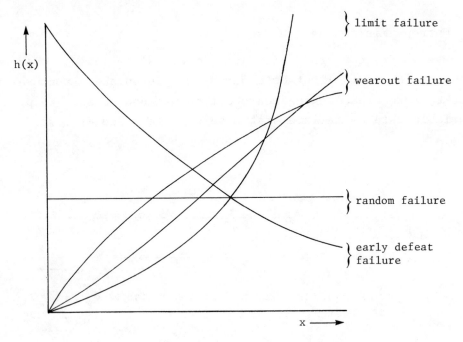

Fig. 74. Examples of Various Types
of Hazard Functions

Fig. 74 shows several commonly used hazard function shapes. The failure frequency function can be computed from the hazard function as follows:

$$F(t) = 1 - \exp\left\{-\int_{t_0}^{t} h(t')dt'\right\},$$

where t_0 is the starting point for each individual. See Fig. 75.

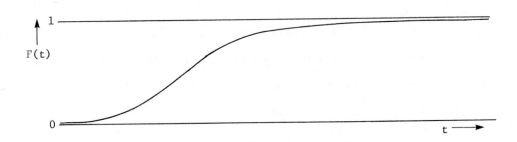

Fig. 75. Failure Frequency Function

The informativeness of F(t) as a failure indicator depends upon the sharpness of the rise from 0 to 1. If F(t) jumps discontinuously from 0 to 1 at a particular value $t = t_1$, then we have a categorical failure indicator. If, however, F(t) rises slowly over a wide and highly populated region of t, then we have a much less informative failure indicator.

We seek to improve the informativeness of our failure indicator by replacing t with a new variable x. Let $\{V_i(t), i=1,\ldots,m\}$ be a set of m monotonically nondecreasing time-dependent variables describing the failure process. Plausible V_i's include:

o   Light bulb: Accumulated watt-hours burned, time in various temperature ranges, number on-off cycles, maximum applied voltage.

o   Structural rib: Accumulated time in various stress ranges, equivalent creep, number take-offs and landings, peak stress reached.

o   Brake lining: Accumulated number of stops, time in various pressure ranges.

o   Human body: Accumulated number cigarettes smoked, amount of alcohol consumed, miles traveled in automobiles, ounces different foods consumed, time spent in various environments, time spent with blood pressure in various ranges.

Assume we have selected the variables so that each is nonnegatively correlated with failure. Then we can form a new hazard axis as

$$x(t) = X_o + \sum_{i=1}^{m} A_i V_i(t),$$

where X_o is a constant and the coefficients $\{A_i\}$ are nonnegative.

To maximize the predictive informativeness of the hazard axis x, we treat the coefficients as parameters to be determined by the entropy minimax criteria.

Let

   f(x)dx = probability of datum occurring between position
            x and x+dx.

Then the conditional failure entropy is

$$S = -\int_{-\infty}^{\infty} \{F(x) \log F(x) + [1-F(x)] \log [1-F(x)]\} f(x)dx.$$

The entropy minimax criteria may be stated as follows:

o   Use maximum entropy estimation to determine the fre-
    quency functions from the data.

o   Use minimum entropy linear combinations to determine
    the parameters $\{A_i\}$.

Examples of entropy minimax hazard axes for nuclear fuel failure
are given in Volume IV.

E.   Self-Consistent Failure Time Estimation from
     Ensemble Statistics

Another use of entropy minimax hazard axes, in addition to failure
prediction, is to estimate failure time in cases where the value
of the hazard axis is known but the time of failure is not known.
For example:

o   A population of patients.   The time of onset of the
    disease is unknown.

o   Materials in a mechanism subject to infrequent
    inspection.   The time of failure is unknown.

A self-consistent analysis is used to provide the failure time
estimates.

Fig. 76 plots failure data as a function of position on the hazard
axis at the time of examination.

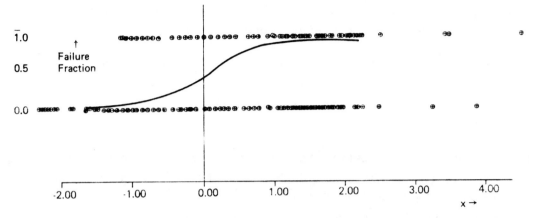

Fig. 76. Failure Fraction (1=Failed, 0=Not Failed) versus
Hazard Axis Position x at Examination Time

If we knew when each failure occurred, we could use these data to
compute the hazard function. Fig. 76 gives only the x-positions
at the time of examination. Failures, of necessity, must have occurred at earlier times and hence at lower x-positions.

If we knew the hazard function $h(x)$, we could compute the failure
probability distribution for each element observed to be failed
at $x = x_m$. For $x_0 < x < x_m$, the probability that the failure
occurred between x and x+dx would then be given by

$$f(x)\,dx = \frac{[1-F(x)]h(x)\,dx}{\displaystyle\int_{x_0}^{x_m}[1-F(x')]h(x')\,dx'}.$$

Thus, if we knew the "true" $h(x)$, and hence also the "true" $F(x)$,
we could distribute or smear each observed failure among points
earlier on the x-axis in accordance with the function $f(x)$. The
nonfailures would have to be similarly redistributed. The numer-
ators and denominators of the fractions $h(x)\Delta x$, defined in Section
14.D above, could then be corrected to account for the fractional
distribution of each element over points preceding the observation
point. The adjusted values of $h(x)\Delta x$ could then be used to estimate the functions $h(x)$ and $F(x)$. At the outset, we lack exact

knowledge of these two functions, but we do possess approximations
that can be used in a recursive process to estimate the true values.

As a natural starting point, we assume that all the failures occurred
at the observed end-of-cycle x-positions. This allows us to get
a zeroth order estimate for h(x) and F(x). We denote them as
$h_0(x)$ and $F_0(x)$. See Fig. 77.

Proceeding, we then distribute the data to earlier points along
the hazard axis on the basis of $h_0(x)$, $F_0(x)$. This permits us to
get a new estimate $h_1(x)$, $F_1(x)$. The process is repeated until
the change from $h_{i-1}(x)$ to $h_i(x)$ is insignificant. See Fig. 78.

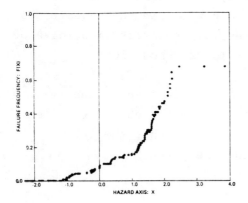
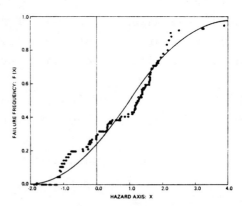

Fig. 77. Failure Frequency Function   Fig. 78. Estimated True Failure
at Examination Time   Frequency Function

QUOTES

TENNYSON (1869)

...I pluck you out of the crannies;—
Hold you here, root and all, in my hand,
Little flower—but if I could understand
What you are, root and all, and all in all,
I should know what God and man is.

> —Alfred Tennyson (1809-1892)
> "Flower in the Crannied Wall" (1869), *Great Poems of the English Language, An Anthology,* compiled by W.A. Briggs, Tudor Pub. Co., NY, 1936, p. 801.

WIENER (1949)

...the methods of prediction contemplated are to be linear, invariant with respect to the choice of an origin in time, and dependent only on the past and present of the function under investigation.

...We concern ourselves, only with the prediction of ensembles for which nearly all functions have the same auto-correlation coefficient.

...we shall assume that the integrated spectrum $\Lambda(\omega)$ is absolutely continuous...

...the simplest exception to the absolute continuity of $\Lambda(\omega)$ is the case in which $\Lambda(\omega)$ has one or more jumps...For such a function, the phase relations between the different components at present or in the past determine with perfect strictness the phase relations into the indefinite future. ...It is only such terms as depend on an external rigidly periodic influence, like the day or the year, which show even an approximation to such behavior. A periodicity in which the phase relations gradually alter through the course of the ages is not in fact a true periodicity and does not correspond to a sharp jump of $\Lambda(\omega)$, but to a very rapid rise...

If the jump spectrum is an idealization, never perfectly realized in practice, this is even more the case with the continuous spectrum which is not absolutely continuous. In both cases, to establish the existence of such a spectrum presupposes an infinitely long run of observations. In both cases, according to the work of Kolmogoroff,* the past of a function with a spectrum crowded into a set of frequencies of zero measure determines its future for an infinite time.

* * * *

---

* A.N. Kolmogoroff, Interpolation und Extrapolation, *Bulletin de l'academie des sciences de USSR,* Ser. Math. 5, 1941, pp. 3-14.

WIENER (1949, cont.)

> Let $f(t)$ and $g(t)$ be two complex time-series. Let $f(t)$ represent a message and $g(t)$ a disturbance...We shall suppose that sufficient auxiliary conditions are satisfied to permit the free shifting of the time origin of the series...
>
> ...In general the spectra of $f(t)$ and $g(t)$ will overlap, and when this is the case, not even an infinite lag in filtering will give perfect discrimination.
>
> ...our criterion of performance is the mean square distortion of the message over the time...
>
> > —Norbert Wiener (1894-1964)
> > *Extrapolation, Interpolation, and Smoothing of Stationary Time-Series, With Engineering Applications*, J. Wiley & Sons, Inc., NY, 1949, pp. 59-60, 81-82.

ROSENBLATT (1956)

> All estimates of the density function satisfying relatively mild conditions are shown to be biased.
>
> > *   *   *   *
>
> An obvious estimate of $f(y)$ is the difference quotient
>
> $$S(y; X_1, \ldots, X_n) = f_n(y) = \frac{F_n(y+h) - F_n(y-h)}{2h}$$
>
> of the sample distribution function $F_n(y)$, where $h = h_n$ is a function of the sample size n and approaches zero as $n \to \infty$ ... Fix and Hodges have used an estimate of this form in their discussion of a nonparametric discrimination problem [1].
>
> > —Murray Rosenblatt (1926-   )
> > "Remarks on Some Nonparametric Estimates of a Density Function," *The Annals of Mathematical Statistics*, *27*, 1956, pp. 832, 833.

---

[1] E. Fix and J.L. Hodges, Jr., *Discriminatory Analysis, Nonparametric Discrimination: Consistency Properties*, USAF School of Aviation Medicine, Project No. 21-49-004, Report No. 4, Contract AF41(128)-31, Feb. 1951.

WHITTLE (1958)

...the regularity hypothesis..."that the curve $f(x)$ being estimated is one of a whole population of curves, and that the "population correlation coefficient" of $f(x)$ and $f(x+\varepsilon)$ tends to unity as ε tends to zero". This hypothesis introduces the idea of an *a priori* distribution of the ordinate $f(x)$, an idea that often encounters a mixed reception...

(a) It is by no means a stringent demand, and allows the curve $f(x)$ to have many sorts of discontinuity and irregularity.
(b) It leads to weighting functions which make automatic allowance for sample size.
(c) If one is interested in only a finite interval of x, then the weighting functions will make automatic allowance for the end-effects of the finite interval.
(d) The weighting functions derived from the above regularity hypothesis are invariant to certain arbitrary transformations of the data...

The methods are formally almost identical with those used by Wiener (Wiener, 1949, p. 81) to recover the true signal from a signal disturbed by noise. The sampling variation of the curve corresponds to noise, the prior variation of curve ordinates to statistical variation of the signal. A complicating feature of the present case is the variation of "signal" and "noise" is not stationary in "time", and that one usually considers a finite rather than an infinite interval of "time". On the other hand, in this application the smoothed estimate may utilize the information contained in both "past" and "future".

—Peter Whittle (1931-   )
"On the Smoothing of Probability Density Functions,"
J. of the Royal Statistical Society, Ser. B, *20*, 1958, pp. 334-335.

CUTLER and EDERER (1958)

A principal advantage of the life table method is that it makes possible the use of all survival information accumulated up to the closing date of the study.

—Sidney J. Cutler (1917-   ) and
Fred Ederer (1926-   )
"Maximum Utilization of the Life Table Method in Analyzing Survival," *J. Chron. Dis.*, Dec. 1958, p. 699.

PARZEN (1962)

The problem of estimation of a probability density function $f(x)$ is interesting for many reasons. As one possible application we mention the problem of estimating the hazard, or conditional rate of failure, functions $f(x)/\{1-F(x)\}$ where $F(x)$ is the distribution function corresponding to $f(x)$.

* * * *

Various possible estimates of the probability density functions suggest themselves but none of them appear to be naturally superior. For example, as an estimate of $f(x)$ one might take

$$f_n(x) = \{F_n(x+h) - F_n(x-h)\}/2h$$

where h is a suitably chosen positive number. However, how should one choose h? It is clear that h should be chosen as a function of n which tends to 0 as n tends to ∞. But how fast should h tend to zero?

> —Emanuel Parzen (1929-   )
> "On Estimation of a Probability Density Function and Mode," *Ann. Math. Statist. 33*, 1962, pp. 1065, 1066.

MEISEL (1969)

How can the parameter of the potential function which determines its peakedness and rate of decline, such as the standard deviation of an exponential approximated to, be chosen? ...

...Values of the parameter within a fairly wide range are usually satisfactory; the parameter need not be determined with a high degree of accuracy.

...to test their efficacy, the resulting discriminant functions should be applied to samples not used in their generation.

> —William S. Meisel (1942-   )
> "Potential Functions in Mathematical Pattern Recognition," *IEEE Trans. on Computers, C-18*, Oct. 1969, pp. 916.

SNEATH (1969)

...it is not widely recognized that the degree of clustering depends on the scale of observation of the space.

> —P.H.A. Sneath (1923-   )
> "Evaluation of Clustering Methods," *Numerical Taxonomy*, ed. by A.J. Cole, Academic Press, NY, 1969, pp. 263-264.

REFERENCES

Aizerman, M.A., E.M. Braverman and L.I. Rozonoev, "Theoretical Foundations of Potential Function Method in Pattern Recognition," *Automat. i Telemekh, 25,* June 1964, pp. 917-963.

_____, "The Probability Problem of Pattern Recognition Learning and the Method of Potential Functions," *Automation and Remote Control, 25,* 1964, pp. 1175-1193.

Bashkerov, O.A., E.M. Braverman and I.B. Muchnik, "Potential Function Algorithms for Pattern Recognition Learning Machines," *Automat. i Telemekh, 25,* May 1964, pp. 692-695.

Braverman, E.M., "On the Method of Potential Functions," *Automat. i Telemekh, 26,* Dec. 1965, pp. 2130-2139.

Christensen, R.A., "Entropy Minimax Method of Pattern Discovery and Probability Estimation from Empirical Data," Arthur D. Little, Inc., Camb., MA, March 7, 1972. [*Chapter 4 of Volume III, pp. 100-103.*]

_____, "Entropy Minimax, A Non-Bayesian Approach to Probability Estimation from Empirical Data," *Proc. of the 1973 Intl. Conf. on Cybernetics and Society,* IEEE Systems, Man and Cybernetics Society, Nov. 5-7, 1973, Boston, MA, 73 CHO 799-7-SMC, pp. 321-325. [*Chapter 5 of Volume III, pp. 111-112.*]

_____, "Talk on Entropy Minimax Parameterization of Potential Functions in Pattern Discovery," presented to the Boston section of the IEEE, MIT, Cambridge, MA, Nov. 16, 1972. [*Chapter 12 of Volume III.*]

_____, "Entropy Minimax Parameterization of Potential Functions in Pattern Discovery," Tech. Rept. No. 40.11.75, Carnegie-Mellon Univ., Pittsburgh, PA, May 16, 1975. [*Excerpts in Chapter 13 of Volume III.*]

_____, "Entropy Minimax Expansion for Waveform Analysis," *Pattern Recognition, 11,* No. 1, Jan. 1979, pp. 41-49. [*Chapter 15 of Volume III.*]

_____, "Nuclear Fuel Rod Hazard Axes," *Fuel Rod Mechanical Performance Modeling, Task 3: Fuel Rod Modeling and Decision Analysis,* FRMPM33-2, Aug. - Oct. 1979, pp.7.1 - 7.11 and FRMPM34-1, pp. 40-50, 104-105, Entropy Limited, Lincoln, MA, 1979, 1980.

Christensen, R.A. and E. Duchane, "Element Specific Failure Time Estimation From Ensemble Statistics," *Fuel Rod Mechanical Performance Modeling, Task 3: Fuel Rod and Decision Analysis,* FRMPM32-2, Entropy Limited, Lincoln, MA, May - July 1979, pp. 6.34-6.50.

Christensen, R.A., R. Eilbert and S. Oldberg, "Entropy Minimax Hazard Axes For Failure Analyses," *Fuel Rod Mechanical Performance Modeling, Task 3: Fuel Rod Modeling and Decision Analyses.* FRMPM31-2, Entropy Limited, Lincoln, MA, May - July 1979, pp. 6.20-6.33.

Cutler, Sidney J. and Fred Ederer, "Maximum Utilization of the Life Table Method in Analyzing Survival," *J. Chron. Dis.,* Dec. 1958, pp. 699-712.

Fix, E. and J.L. Hodges, Jr., *Discriminatory Analysis, Nonparametric Discrimination: Consistency Properties,* USAF School of Aviation Medicine, Randolf Field, TX, Project 21-49-004, Rept. 4, Contract AF41(128)-31, Feb. 1951.

Liptser, R.S. and A.N. Shiryayev, *Statistics of Random Processes* (2 Vols.), Springer-Verlag, NY, 1978.

Meisel, W.S., "Potential Functions in Mathematical Pattern Recognition," *IEEE Trans. on Computers, C-18,* Oct. 1969, pp. 911-918.

Parzen, E., "On Estimation of a Probability Density Function and Mode," *Ann. Math. Stat., 33,* Sept. 1962, pp. 1065-1076.

Rosenblatt, M., "Remarks on Some Nonparametric Estimates of a Density Function," *Annals. Math. Statist., 17,* 1956, pp. 823-837.

Sneath, P.H.A., "Evaluation of Clustering Methods," *Numerical Taxonomy,* ed. by A.J. Cole, Academic Press, NY, 1969.

Tennyson, Alfred, "Flower in the Crannied Wall" (1869), *Great Poems in the English Language, An Anthology,* compiled by W.A. Briggs, Tudor Pub. Co., NY, 1936, p. 801.

Whittle, P., "Curve and Periodogram Smoothing," *J. Roy. Statist. Soc.,* Ser. B, *19,* 1957, pp. 38-47.

_____, "On the Smoothing of Probability Density Functions," *J. Roy. Statist. Soc.,* Ser. B, *20,* 1958, pp. 334-343.

Wiener, Norbert, *Extrapolation, Interpolation, and Smoothing of Time-Series,* J. Wiley & Sons, Inc., NY, 1949.

CHAPTER 15

UNCERTAINTY, IRRELEVANCY, AMBIGUITY
AND NON-BAYESIAN ASPECTS

CHAPTER 15

UNCERTAINTY, IRRELEVANCY, AMBIGUITY AND NON-BAYESIAN ASPECTS

A. Uncertainty

In this chapter we turn to some pattern analysis difficulties with which we must often cope in practice. The first of these is the problem of determining pattern matches for cases in which the value of an independent variable is uncertain.

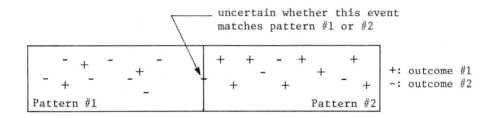

Fig. 79. Example in which it is Uncertain on which Side of a Boundary an Event Falls

Fig. 79 illustrates a situation in which we wish to estimate the outcome probabilities for two patterns based on event counts within a training sample, but are uncertain which pattern is matched by one of the events. We have insufficiently well-defined independent variable information to assign this event definitely to one pattern or the other.

In such cases, we proceed by assigning a fractional event count to each pattern, corresponding to the estimated probability of the event matching the pattern. This probability is estimated in the usual fashion from a combination of information about the particular event and the information for all such events.

The most extreme example of uncertainty is when the independent variable is completely unknown. In this case, we distribute the fractional event counts solely in accord with the overall sample distribution for events with the same outcome class.   This preserves
the training sample outcome information (for overall count purposes)
while retaining the known independent variable distributions.

The following two examples illustrate cases of pattern matching under uncertainty in an independent variable:

1.   Outcome class: Room is comfortable or not, as judged by
                           a specific set of people.

     Pattern #1:   Temperature is $30 \pm 1^{\circ}$C.

     Pattern #2:   Temperature is greater than 31°C.

     Feature: Temperature.

     Classes: ..., $28 \pm 1^{\circ}$C, $30 \pm 1^{\circ}$C, $32 \pm 1^{\circ}$C, ...,

     Event: Temperature measured by thermometer is 30°C
                           with normal error distribution and stan-
                           dard deviation of $\pm 1^{\circ}$C.

     Classification: Temperature is uncertain for this event.

     Pattern matches: Pattern #1 matched with probability of 0.68.
                           Pattern #2 matched with probability of 0.17.

2.   Outcome class: Did or did not survive two years after
                           treatment.

     Pattern #1:   Patients of age 30 ± 5 years.

     Pattern #2   Patients of age 50 ± 5 years.

     Feature: Age.

     Classes: ..., 30 ± 5, 40 ± 5, 50 ± 5, ..., years.

     Event: Patient who is known to have died one year
                           after treatment, but whose age is not known
                           and for whom no records are available from
                           which age can be determined.

     Pattern matches: Pattern #1 matched with probability determined
                           by frequency of age 30 ± 5 years among patients
                           who did not survive two years after treatment.
                           Pattern #2 matched with probability determined
                           by frequency of age 50 ± 5 years among patients
                           who did not survive two years after treatment.

There are, of course, a wide variety of complications that can enter into the analysis of situations with uncertain or unknown values for independent variables.   Distributions, intercorrelations, and generally available background information can all be relevant.   For example, one might acquire evidence that age distributions are different for patients with incomplete records than for those with complete records.   The researcher must take care to consider all the available information, and make the probability estimates accordingly.

B.   Irrelevancy and Ambiguity

Whereas uncertainty arises from lack of feature information about a particular event, irrelevancy and ambiguity arise from inadequacies in the feature definitions themselves.

It is an inherent limitation of human knowledge that all terms referencing physical objects and events are incompletely defined.   In observational and experimental science, we constantly gather data about new events after a fixed event classification scheme has been defined.   However, in fixing the classification scheme we are subject to the problem that a new event may be encountered with characteristics which we did not anticipate and which does not fit into our scheme.

We might cope with this problem by simply ignoring all such events. By use of this policy, however, our analysis may be prone to failure to learn from new information.   On the other hand, we could revise the classification scheme every time such a case arises; but a continual succession of such revisions would be unworkable, both because of the difficulty of re-expressing prior events in the new system when some of the data may no longer be available, and because of the computational requirements of a repetitious sequence of complete re-analyses.   True, in some cases redefinition of the classification may eventually be warranted when new events reveal important findings.   However, we need a procedure for processing data in the meantime. This is the task of irrelevancy and ambiguity analysis.

When the feature values are merely uncertain or unknown, the total event counts over feature space are unaffected. An event with an uncertain or unknown feature value still counts as one event. It is merely distributed fractionally among the patterns.

The situation in cases of irrelevancy and ambiguity is, however, different. Here the total event count is affected.

When a feature class is irrelevant to a particular event, the event contributes zero count to any pattern requiring this class. The total event count in such cases is less than the number of events.

When a feature class is ambiguous, the event is given a full count for each pattern over which it is ambiguous. Thus, each event with such an ambiguity contributes more than one count to the total feature space.

Irrelevancy arises when a feature condition specifies a threshold on a variable which does not logically pertain to the particular event in question.

Ambiguity arises when a feature condition specifies a distinction which does not logically pertain to the particular event in question, even though the variable referenced does pertain.

The following examples illustrate event classification in cases involving irrelevancy and ambiguity:

1. Pattern #1:   Cats with one or more fleas on their left ears.
    Pattern #2:   Cats with no fleas on their left ears.
    Feature: Fleas on left ear.
    Partition: Some, none.
    Event: Cat with no left ear.

Analysis: Left ear flea count is irrelevant for this event.

Pattern matches: Neither #1 nor #2.

2. Pattern #1:   Auto trips at speed 70±5 kph.
    Pattern #2:   Auto trips at speed 60±5 kph.
    Feature: Speed.
    Partition: ..., 50±5, 60±5, 70±5, ...kph.
    Event: Auto trip at speeds ranging from 40 to 80 kph.
    Analysis: Speed is ambiguous for this event.
    Pattern matches: Both #1 and #2.

3. Pattern #1:   Sugar molecules with a carbon atom at the fifth position on their carbocyclic ring.
    Pattern #2:   Sugar molecules with something other than a carbon atom at the fifth position.
    Feature: Carbocyclic 2" ring position substituent.
    Partition: C, H, O, etc.
    Event: A furanose sugar molecule with no 2" position on its carbocyclic ring.
    Analysis: The 2" position substituent is irrelevant for this event.
    Pattern matches: Neither #1 nor #2.

4. Pattern #1:   Tables of width 1.4±0.1 meter.
    Pattern #2:   Tables of width 1.8±0.1 meter.
    Feature: Width.
    Partition: ..., 1.0±0.1, 1.2±0.1, ...meter.
    Event: A trapezoidal table of width 1 meter at one side and 2 meters at the other.
    Analysis: Width is ambiguous for this event.
    Pattern matches: Both #1 and #2.

5. Pattern #1:   Murders using a gun as the weapon.
    Pattern #2:   Murders using a weapon other than a gun.
    Feature: Murder weapon.
    Partition: Gun, knife, poison, etc.
    Event: A murder by starvation.
    Analysis: Murder weapon is irrelevant for this event.
    Pattern matches: Neither #1 nor #2.

6. Pattern #1:   Liquids which boil at $100 \pm 1°C$.
    Pattern #2:   Liquids which boil at $98 \pm 1°C$.
    Feature: Boiling point.
    Partition: ..., $96 \pm 1°C$, $98 \pm 1°C$, $100 \pm 1°C$, $102 \pm 1°C$,...
    Event: Water in river running from a certain eleva-
                       tion (B.P. 98°C) to sea level (B.P. 100°C).
    Analysis: Boiling point is ambiguous for this event.
    Pattern matches: Both #1 and #2.

7. Pattern #1:   Observations of blue-colored objects.
    Pattern #2:   Observations of non-blue-colored objects.
    Feature: Observed color.
    Partition: Violet, blue, green, yellow, orange, red,...
    Event: Observation of a clear pane of glass.
    Analysis: Observed color is irrelevant for this event.
    Pattern matches: Neither #1 nor #2.

8. Pattern #1:   Observations of blue-colored objects.
    Pattern #2:   Observations of non-blue-colored objects.
    Feature: Observed color.
    Partition: Violet, blue, green, yellow, orange, red,...
    Event: Object identified as blue by a person known
                       to have blue-green color blindness.
    Analysis: Observed color is ambiguous for this event.
    Pattern matches: Both #1 and #2.

9. Pattern #1:   Good takings of life.
    Pattern #2:   Evil takings of life.
    Feature: Morality.
    Partition: Good, evil.
    Event: Person taking all conceivable precautions
                          accidentally shoots someone while cleaning
                          a gun.
    Analysis: Morality is irrelevant for this event.
    Pattern matches: Neither #1 nor #2.

10. Pattern #1:   Good takings of life.
    Pattern #2:   Evil takings of life.
    Feature: Morality.
    Partition: Good, evil.
    Event: Execution of a murderer by the State.
    Analysis: Morality is ambiguous for this event.
    Pattern matches: Both #1 and #2.

Irrelevancy is generally easy to identify. However, distinguishing between ambiguity and uncertainty can be difficult. The critical difference is that uncertainty can be resolved by obtaining
more definitive information about the event, whereas ambiguity can
only be resolved by refining the definition of the feature space
classes.

C. Non-Bayesian Aspects of Maximum Entropy Probabilities

Bayes' inverse probability theorem (see Section 7.E) is a deductive
consequence of certain basic axioms of probability theory plus some
additional, often unstated, assumptions. Analyses labeled *non-
Bayesian* are ordinarily non-Bayesian only in the sense that they do
not formally use Bayes' theorem in probability estimation, not because they fail to comply with the theorem. The need to employ
non-Bayesian methods typically arises from an inability to obtain
values for one or more of the probabilities required by Bayes' theorem.

Maximum entropy probabilities, on the other hand, do not always
satisfy the inverse probability formula. This situation arises
under conditions in which one or more of its axioms or assumptions
is violated. Bayes' theorem implicitly assumes the same base of
events is used for all of the probabilities among which it expresses the inverse relation. In contrast, entropy minimax analysis
estimates each probability on the basis of *all* events, both real
and virtual, that are available for that probability. When different probabilities are based on different sets of events, Bayes'
theorem is inapplicable.

Consider, for example, the hypothetical training data set given in
Table 39.

Table 39. Hypothetical Training Data

Condition	Outcome	# Training Events
C_1	O_1	12
C_1	O_2	11
C_1	unknown	10
C_2	O_1	9
C_2	O_2	8
C_2	unknown	7
unknown	O_1	6
unknown	O_2	5
irrelevant	O_1	4
irrelevant	O_2	3
ambiguous	O_1	2
ambiguous	O_2	1

Total = 78

Suppose that we attempt to use the inverse probability formula:

$$P(O_k \text{ given } C_i)\Big|_{Bayes} = \frac{P(C_i \text{ given } O_k)P(O_k)}{\sum_j P(C_i \text{ given } O_j)P(O_j)}.$$

Assume, for simplicity, all the weight normalizations (see Chapter 20) to be $w = 15$. Distributing the unknowns in proportion to the knowns, the maximum entropy probabilities for the right-hand side are computed as follows:*

$$P(O_1) = \frac{33 + 13\left(\frac{33}{61}\right) + 1}{61 + 15} = 0.539905$$

$$P(O_2) = \frac{28 + 13\left(\frac{28}{61}\right) + 1}{61 + 15} = 0.460095$$

$$P(C_1 \text{ given } O_1) = \frac{14 + 10\left(\frac{14}{23}\right) + 13\left(\frac{36}{60}\right) + 1}{23 + 17\left(\frac{33}{61}\right) + 15} = 0.612054$$

$$P(C_2 \text{ given } O_1) = \frac{11 + 7\left(\frac{11}{23}\right) + 13\left(\frac{27}{60}\right) + 1}{23 + 17\left(\frac{33}{61}\right) + 15} = 0.449138$$

$$P(C_1 \text{ given } O_2) = \frac{12 + 10\left(\frac{12}{20}\right) + 13\left(\frac{36}{60}\right) + 1}{20 + 17\left(\frac{28}{61}\right) + 15} = 0.626120$$

$$P(C_2 \text{ given } O_2) = \frac{9 + 7\left(\frac{9}{20}\right) + 13\left(\frac{27}{60}\right) + 1}{20 + 17\left(\frac{28}{61}\right) + 15} = 0.443891$$

---

* As an example of this computation, the terms in the expression for $P(C_1 \text{ given } O_1)$ are the following:

$$14 = 12[C_1,O_1] + 2[A,O_1]$$

$$10\left(\frac{14}{23}\right) = 10[C_1,U]\frac{12[C_1,O_1] + 2[A,O_1]}{12[C_1,O_1] + 9[C_2,O_1] + 2[A,O_1]}$$

$$13\left(\frac{36}{60}\right) = 13[w-t-f]\frac{12[C_1,O_1] + 11[C_1,O_2] + 10[C_1,U] + 2[A,O_1] + 1[A,O_2]}{78[T] - 6[U,O_1] - 5[U,O_2] - 4[I,O_1] - 3[I,O_2]}$$

$$1 = t$$

$$23 = 12[C_1,O_1] + 9[C_2,O_1] + 2[A,O_1]$$

$$17\left(\frac{33}{61}\right) = \left(10[C_1,U] + 7[C_2,U]\right)\frac{12[C_1,O_1] + 9[C_2,O_1] + 6[U,O_1] + 4[I,O_1] + 2[A,O_1]}{78[T] - 10[C_1,U] - 7[C_2,U]}$$

$$15 = w$$

The denominators in Bayes' theorem

$$D_i = \sum_j P(C_i \text{ given } O_j) P(O_j),$$

are given by:

$$D_1 = (0.612054)(0.539905)+(0.626120)(0.460095) = 0.618526$$

$$D_2 = (0.449138)(0.539905)+(0.443891)(0.460095) = 0.446724 .$$

The final results given by Bayes' theorem are:

$$P(O_1 \text{ given } C_1)\Big|_{\text{Bayes}} = (0.612054)(0.539905)/D_1 = 0.534256$$

$$P(O_2 \text{ given } C_1)\Big|_{\text{Bayes}} = (0.626120)(0.460095)/D_1 = 0.465744$$

$$P(O_1 \text{ given } C_2)\Big|_{\text{Bayes}} = (0.449138)(0.539905)/D_2 = 0.542822$$

$$P(O_2 \text{ given } C_2)\Big|_{\text{Bayes}} = (0.443891)(0.460095)/D_2 = 0.457177 .$$

These same probabilities, as computed according to the maximum entropy criterion, are different. The maximum entropy probabilities are computed as follows:

$$P(O_1 \text{ given } C_1) = \frac{14 + 6\left(\frac{14}{26}\right) + 13\left(\frac{33}{61}\right) + 1}{26 + 11\left(\frac{36}{60}\right) + 15} = 0.530747$$

$$P(O_2 \text{ given } C_1) = \frac{12 + 5\left(\frac{12}{26}\right) + 13\left(\frac{28}{61}\right) + 1}{26 + 11\left(\frac{36}{60}\right) + 15} = 0.446952$$

$$P(O_1 \text{ given } C_2) = \frac{11 + 6\left(\frac{11}{20}\right) + 13\left(\frac{33}{61}\right) + 1}{20 + 11\left(\frac{27}{60}\right) + 15} = 0.559018$$

$$P(O_2 \text{ given } C_2) = \frac{9 + 5\left(\frac{9}{20}\right) + 13\left(\frac{28}{61}\right) + 1}{20 + 11\left(\frac{27}{60}\right) + 15} + 0.456000 .$$

Direct comparison reveals that the maximum entropy probabilities
are non-Bayesian:

Maximum Entropy		Bayes' Theorem
0.530747	$\neq$	0.534256
0.446952	$\neq$	0.465744
0.559018	$\neq$	0.542822
0.456000	$\neq$	0.457177

In fact, because of the presence of irrelevancies and ambiguities
in the data, these maximum entropy probabilities violate unit
measure:

$$0.530747 + 0.446952 = 0.977699 \neq 1.$$

$$0.559018 + 0.456000 = 1.015018 \neq 1.$$

The entropy minimax values are non-Bayesian because of the effects
of probability shrinkage in accord with the weight normalization,
and the effects of accounting for irrelevancies and ambiguities.
With shrinkage alone, they would still violate the inverse probability formula, but they would sum to unity. Irrelevancy and
ambiguity are what introduce unit measure violation. (Though
usually small, this is a real effect, since with irrelevance and
ambiguity, frequency counts also violate unit measure.)

QUOTES

ARISTOTLE (c. 335-323 B.C.)

...there cannot be an intermediate between contradictories, but of one subject we must either affirm or deny any one predicate...

* * * *

While the doctrine of Heraclitus, that all things are and are not, seems to make everything true, that of Anaxagoras, that there is an intermediate between the terms of a contradiction, seems to make everything false; for when things are mixed, the mixture is neither good nor not-good, so that one cannot say anything that is true.

> —Aristotle (384-322 B.C.)
> "The Law of Excluded Middle Defended," *Metaphysics* (c. 335-323 B.C.), tr. by W. Ross, *The Basic Works of Aristotle*, ed. by R. McKeon, Random House, NY, 1941, pp. 749-750.

For it is by stating the species or the genus that we appropriately define any individual man; and we shall make our definition more exact by stating the former than by stating the latter. All other things that we state, such as that he is white, that he runs, and so on, are irrelevant to the definition.

> —Aristotle (384-322 B.C.)
> *Categoriae* (c. 335-323 B.C.), tr. by E. Edghill, *The Basic Works of Aristotle*, ed. by R. McKeon, Random House, NY, 1941, p. 10.

LEIBNIZ (1696)

...men speak of this same subject, forming very different ideas of it, both by the different combination of the simple ideas they make and because the greater part of the qualities of bodies are the powers which they have of producing changes in other bodies and receiving them...one is contented with weight and color as criteria for a knowledge of gold; another includes ductility, fixedness, and the third desires to make us take into consideration its solubility in aqua regia.

* * * *

LEIBNIZ (1696, cont.)

Men will say that instead of imputing these imperfections to the words, we should rather put them to the account of our understanding; but I reply that words interpose themselves to such an extent between our mind and the truth of things, that we may compare them with the medium, across which pass the rays from visible objects, and which often spreads a mist before our eyes; and I have tried to think that, if the imperfections of language were more thoroughly examined, the majority of the disputes would cease of themselves, and the way to knowledge and perhaps to peace would be more open to men.

> —Gottfried Wilhelm Leibniz (1646-1716)
> "On Locke's Essay on Human Understanding" (1696),
> *New Essays Concerning Human Understanding*, tr. by
> A. Langley, Open Court Pub. Co., La Salle, IL,
> 1949, pp. 374, 375.

KEYNES (1921)

...h_1 is irrelevant to x on evidence h, if there is no proposition inferrible from $h_1 h$ but not from h, such that its addition to evidence h affects the probability of x.

> —John Maynard Keynes (1883-1946)
> *A Treatise on Probability* (1921), Macmillan & Co.,
> London, 1957, p. 55.

REFERENCES

Aristotle, "The Law of Excluded Middle Defended," *Metaphysics*, tr. by W. Ross, *The Basic Works of Aristotle*, ed. by R. McKeon, Random House, NY, 1941.

_____, *Categoriae*, tr. by E. Edghill, *The Basic Works of Aristotle*, ed. by R. McKeon, Random House, NY, 1941.

Christensen, R.A., *Foundations of Inductive Reasoning*, Berkeley, CA, 1964.

_____, "A General Approach to Pattern Discovery," Tech. Rept. No. 20, Computer Center, Univ. of Calif., Berkeley, CA, June 29, 1967 (revised Nov. 15, 1967). [*Chapter 1 of Volume III, p. 22.*]

_____, "Entropy Minimax, A Non-Bayesian Approach to Probability Estimation from Empirical Data," *Proc. of the 1973 International Conference on Cybernetics and Society*, IEEE Systems, Man and Cybernetics Society, Nov. 5-7, 1973, Boston, MA, 73 CHO 799-7-SMC, pp. 321-325. [*Chapter 5 of Volume III.*]

_____, "Entropy Minimax Processing of Incomplete Data Sets," *Proceeding of the 1973 International Conference on Cybernetics and Society*, IEEE Systems, Man and Cybernetics Society, Nov. 5-7, 1973, Boston, MA, 73 CHO 799-7-SMC, pp. 326-327. [*Chapter 6 of Volume III.*]

_____, "Entropy Minimax Analysis of Simulated LOCA Burst Data," report prepared for Core Performance Branch, Directorate of Licensing, U.S. Atomic Energy Commission, Contract No. AT(49-24)-0083, Entropy Limited, Belmont, MA, Dec. 31, 1974.

Christensen, R.A. and K. Leontiades, "Data Qualification for Entropy Minimax Analysis," EPRI 508-2 Project Memorandum, Entropy Limited, Lincoln, MA, Jan. 21, 1977.

Christensen, R.A. and T. Reichert, "A Preliminary Entropy Minimax Search for Patterns in Structural, Physio-Chemical and Biological Features of Selected Drugs That May Be Related to Activity in Retarding Lymphoid Leukemia, Lymphocytic Leukemia and Melanocarcinoma in Mice," a report to the National Cancer Institute under subcontract No. A10373 of Contract No. NO1-CM-23711, Entropy Limited, Pittsburgh, PA, June 30, 1975.

_____, "Unit Measure Violations in Pattern Recognition, Ambiguity and Irrelevancy," *Pattern Recognition, 8*, 1976, pp. 239-245.

Keynes, John Maynard, *A Treatise on Probability* (1921), Macmillan & Co., London, 1957.

Leibniz, G.W., "On Locke's Essay on Human Understanding" (1696), *New Essays Concerning Human Understanding*, tr. by A. Langley, Open Court Pub. Co., La Salle, IL, 1949.

Lyons, John, *Intro. to Theoretical Linguistics*, Camb. Univ. Press, Cambridge, 1968.

Miller, Arthur S., "Statutory Language and the Purposive Use of Ambiguity," *Yale Law Journal, 42*, 1956, pp. 23-39.

Morris, Charles, *Signs, Language and Behavior*, George Braziller, Inc., NY, 1946.

Reichert, T.A., "The Security Hyperannulus—A Decision Assist Devise for Medical Diagnosis," *Proc. of the Conf. on Engr. in Medicine and Biology*, AEMB, October 6-10, 1974, p. 331.

Stalnaker, Robert C., "Pragmatic Ambiguity," in "A Theory of Conditionals," *Studies in Logical Theory*, ed. by N. Rescher, APQ Monograph No. 2, Blackwell, Oxford, 1968.

PART IV

ENTROPY MINIMAX PREDICTIVE MODELS

CHAPTER 16

PREDICTIVE RELIABILITY AND DEFINITIVENESS

CHAPTER 16

PREDICTIVE RELIABILITY AND DEFINITIVENESS

A. Assessment of Predictions

It is easy to demonstrate by example that no single number can
suffice to assess a prediction. Suppose we are dealing with a
case of categorical prediction. A friend wishes us to predict
whether or not he will get hit by a car when he next walks across
a street. After he next walks across a street, there will be four
possibilities:

Type I Verity: (true negative)	We predicted that he would not get hit and he did not.
Type II Verity: (true positive)	We predicted that he would get hit and he did.
Type I Error: (false negative)	We predicted that he would not get hit and he did.
Type II Error: (false positive)	We predicted that he would get hit and he did not.

We can represent these four possibilities as the corners of an
assessment diagram. See Fig. 80.

Our concern here is with that information which must be provided
to enable assessment of predictions. Although they are also essential to assessment, we will not consider value judgment aspects,
except as they dictate informational requirements. (See Chapter
25 for a discussion of the relation of the problem of value judgment to predictive reliability and definitiveness.)

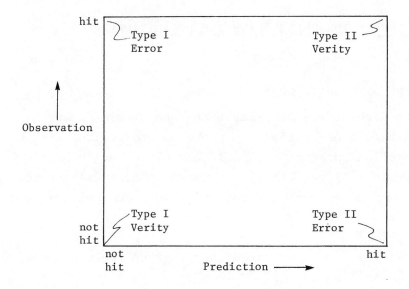

Fig. 80. Categorical Prediction Assessment Diagram

Assume that we are tabulating data upon which to assess our predictions. Assume, further, that we have made similar predictions
a total of 100 times and the results were as follows:

Verity Type I	80	Error Type I	10
Verity Type II	5	Error Type II	5
Total Verities	85	Total Errors	15

It is clear that we cannot construct a single number (such as the
error rate of 15%) to assess the prediction. This is because of
the severe difference between the disutility to our friend of a
Type I Error and of a Type II Error. Thus we must retain more detailed information in order for such differential disutilities to
be taken into consideration. In fact, we must present the entire
prediction assessment diagram in order to enable a complete assessment by an arbitrary user. See Fig. 81. (The utilities entered
are the conditional utilities of the outcome given the prediction.)

rate = 0.05   rate = 0.05
utility = -1000   utility = -1000

rate = 0.80   rate = 0.10
utility = 1   utility = -2

Fig. 81.   Data Entered onto Categorical
Prediction Assessment Diagram

Treating probabilistic predictions as a generalization of categorical predictions, we can formulate a similar assessment information diagram for this case.   We accomplish this by ordering the
events according to the magnitude of the probability.   We then
plot the individual events or groups of events as observed versus
predicted frequency.

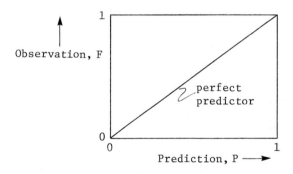

Fig.   82.   Probabilistic Prediction Assessment Diagram
Showing the Perfect Prediction Line

Fig. 82 shows observed frequency, F, versus probability, P.   The
$45°$ line from the lower left corner (0,0) to the upper right corner (1,1) represents a perfect predictor.

The user will map a utility function onto this space.   It will
generally vary with F.   If the value of the outcome is independent

of the prediction error, then utility will depend upon F alone.
It may also vary with P, though, if state of expectation is important.   In general, to enable comprehensive assessment of the
predictions, we must display the complete functional dependence
of F upon P.

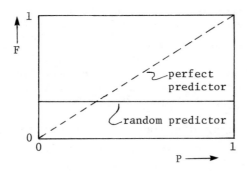

Fig. 83.   Random Predictor (solid line)

Fig. 83 adds a horizontal line representing a *random predictor*. Regardless of what probability is used as the prediction, the observed frequency is the same.

The point at which the random predictor intersects the perfect predictor is the overall sample average.

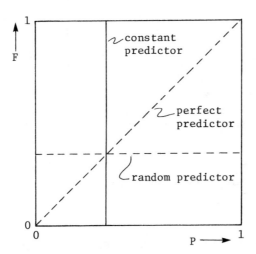

Fig. 84.   Constant Predictor (solid line)

Fig. 84 adds a vertical line representing a *constant predictor*. This predictor always gives the same estimate, regardless of the circumstances.
The best a constant predictor can do is pass through the intersection of the perfect and random predictors, i.e., the sample average.
It is the most conservative predictor possible in that it is a
single prediction based upon the entire set of training data.

An unshrunk predictor extracted from real information can be expected to produce an inverted and flattened S-shaped curve in assessment space.* See Fig. 85.

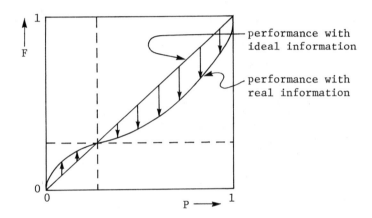

Fig. 85. Effect of Unknown Information Is to Pull
          Performance Line Away from Perfect Predic-
          tor toward Random Predictor

This is because, for events in any interval P to P + dP, the unknowns in the situation will tend to pull the observed frequency
F away from the perfect predictor line toward the random predictor line. This is the way in which predictive performance will
degrade on test samples for models designed to match training
data without compensating for the likelihood of chance correlations.

---

*R.A. Christensen, "Assessment of Statistical Models," Sec. 8.3 of *SPEAR Fuel
Reliability Code System: General Description*, NP-1378, Electric Power Research
Institute, Palo Alto, CA, March 1980, p. 8-22.

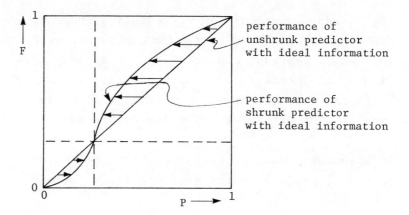

Fig. 86.   A Measured Amount of Conservatism Introduced into the
Prediction on Ideal Training Data by Shrinking in the
Direction of the Constant Predictor

The rationale for using a shrunk estimator such as $p = (x+t)/(n+t+f)$
can be seen in terms of the assessment diagram.   The performance
of an unshrunk predictor with ideal information (i.e., assuming
that the future will be exactly like the past) lies on the perfect
predictor 45° line.   Shrinkage moves the estimator toward the constant predictor vertical line.   See Fig. 86.

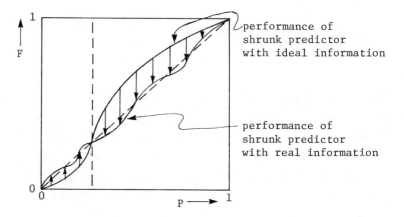

Fig. 87.   Performance on Test Sample of Shrunk Predictor with
Ideal Training Sample.   (If we have introduced just
the right amount of conservatism into our prediction
via shrinkage, we will, on the average, end up near
the perfect predictor line when pulled toward the
random predictor by the unknowns.)

The effect of the unknown aspects of the real information is to pull the points toward the random predictor line. If the shrinkage magnitude is correct, it will, on average, be just enough to compensate for this randomization due to the unknowns, resulting in a final performance on real test (future) data that is close to the perfect predictor line. See Fig. 87.

The S-shaped curve on the training sample is a purposeful "error" introduced into our representation of the historical data. It is like the skeet shooter who aims where he expects the target to be when the bullet gets there rather than where he sees it when pulling the trigger. Entropy minimax does the same sort of thing statistically. We could "aim" at the frequencies we see in the training data, i.e., we could *explain* the past very accurately. But then our prediction of the future would miss the target which will move toward the random region because of the imperfection and incompleteness of our knowledge. To improve prediction accuracy, we give the target a statistical "lead" by aiming somewhat away from our past experience and toward the constant predictor line.

Of course, shrinkage tends to cluster the points closer to the sample average point. Shrinkage makes a tradeoff by sacrificing definitiveness to improve accuracy.

Having reviewed some of the basic ideas of assessment space, let us now return to our test sample. Assume we are assessing probability estimates for outcomes that either occur or do not occur. We get results such as Fig. 88 if we plot the points individually. Every point is either on the F=0 line or on the F=1 line.

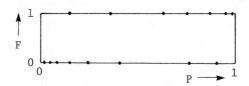

Fig. 88.   Individual Raw Data Points with Frequencies of 0 or 1

Suppose that, for assessment purposes, we group them into N equi-
populated segments along the P-axis. The groups now are spread
across the range of F. See Fig. 89.

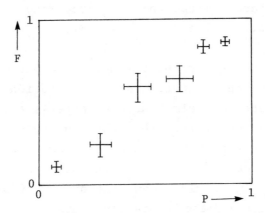

Fig. 89. Lumping Raw Data to Obtain Means and Uncertainty Spreads
(We want to put enough data in each lump to be reasonably
confident of the location of its mean.)

These groupings are made for each outcome class (dependent variable
value). If there are only two outcome classes, then the ordering
of the events along one is simply the reverse of the ordering of
the other. However, for three or more outcome classes, one ordering will, in general, be shuffled relative to another. The data
groups corresponding to the various segments will then be different
for different classes.

The central point in each segment is set at the average for test
points in the segment. The lengths of the horizontal and vertical
bars are adjusted to represent one standard deviation on either
side of the central point. If most of the data tend to be clustered near the (0,0) and (1,1) points, then the bars are shorter
for segments near the extremes and broader for segments in the
middle of the assessment diagram. The bars are not exactly centered on their points of intersection. Rather, they are skewed in
a fashion similar to the skewing of the beta distribution.

The coordinates of the central assessment point are given by

$$\overline{P}_i = \frac{1}{N_i} \sum_{j=1}^{N_i} P_j \;, \text{ and}$$

$$\overline{F}_i = \frac{1}{N_i} \sum_{j=1}^{N_i} F_j \;,$$

where

N_i = number of test sample events in i^{th} P-axis segment,

P_j = outcome probability for j^{th} of these N_i events, and

F_j = outcome frequency for j^{th} of these N_i events (F_j = 0 or 1 for well-defined individual event observations).

See Fig. 90. The horizontal bars represent the fact that we are lumping together predictions over an interval of probabilities. Actually, there is an additional complication here. Each prediction is not just a single probability but rather a probability together with its associated uncertainty, for example

$$P = 0.11 \begin{smallmatrix} +0.05 \\ -0.03 \end{smallmatrix} \;.$$

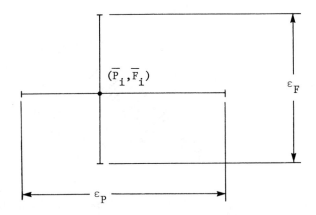

Fig. 90. Detail of Assessment Point and Horizontal and Vertical Bars

Thus, the horizontal bars also represent the fact that there are uncertainties in the individual predictions. The formula for combining these two to give the width ε_P of the horizontal bars is

$$\varepsilon_P^2 = \frac{1}{N_i^2} \sum_{j=1}^{N_i} U_j^2 + \frac{1}{N_i(N_i-1)} \sum_{j=1}^{N_i} (P_j-\overline{P}_i)^2 ,$$

where

$$U_j^2 = \frac{1}{M_j} P_j(1-P_j) , \text{ and}$$

M_j = number training data events supporting j^{th} test sample point prediction.

The vertical bars ε_F represent the statistical scatter in outcome frequencies for the number of events represented by each point. They are given by

$$\varepsilon_F^2 = \frac{1}{N_i(N_i-1)} \sum_{j=1}^{N_i} (F_j-\overline{F}_i)^2 .$$

This vertical scatter occurs no matter how nearly perfect or how nearly random are the predictions. Thus, for proper comparison with observed results, our assessment diagram should show the scatter bands for the perfect and the random predictors. See Fig. 91.

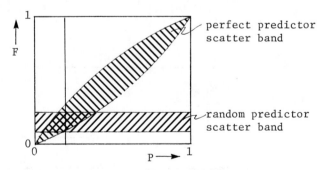

Fig. 91.   Statistical Scatter of Perfect and Random Predictors (Scatter bands are shown as the shaded regions. They are narrower the greater the number of data points in each group.)

Although a complete assessment requires, in general, the full assessment diagram, summary measures of certain aspects of the assessment information may be useful for some purposes.

These summary measures can be grouped into two categories. First, there are measures associated with the *reliability* of the predictions. Reliability is assessed by comparing predictions to observations. There are two aspects to such a comparison. One is the *accuracy* of the predictions. The other is the statistical *confidence* that the accuracy attained was not merely a chance occurrence. Accuracy is determined by how close we are to the 45° perfect predictor line. Confidence is determined by how many events we have in our test sample, hence, how narrow the uncertainty bars are on our predictions compared to the scatter bands for perfect or random predictors.

Second, there are summary measures concerned with the *definitiveness* of the predictions. These define the extent to which the predictions tend to cluster about the extremes (0,0) and (1,1). Accuracy, confidence and definitiveness are discussed in the following sections.

B.   Accuracy Assessment

Accuracy assessment is best introduced by an example. Suppose our prediction of the probability of rain tomorrow is

$$P = 0.11 \, {}^{+0.05}_{-0.03} \ @ \ 70\%.$$

We should begin our assessment of accuracy with a clear understanding of just what is being predicted. We are not predicting that the frequency of rain tomorrow will be 0.11. This would certainly be wrong. It will be either 0 or 1.

Also, we are not predicting that, if we repeat this situation indefinitely, the next-day-rain frequency will be 0.11. This would

almost certainly be wrong. The likelihood of hitting exactly this
frequency is miniscule.

What our prediction represents, rather, is the statement that the
probability is about 70% that, if we repeat this situation indefinitely, the next-day-rain frequency will be found to lie in the
interval between 0.08 and 0.16.*

But what does this mean? We have not completely defined a prediction until we have clearly described what could happen to make it
be wrong. Suppose we observe ten such cases and the next-day-rain
frequency for them turns out to be 0.2. Can our predictor not
simply reply, "Well, that must be one of the 30% exceptional cases"?

With probabilistic predictions, one is never definitely right or
definitely wrong. There are, nonetheless, gradations of rightness
and wrongness which can be assessed. It is at the discretion of
the decision-maker using the predictive information to set the
threshold for acceptance.

Any reasonable measure of predictive accuracy should have the following properties:

o   The measure is always a nonnegative number.

o   The measure is zero if and only if $P_i = F_i$ for all $i = 1, \ldots, n$.

o   If all the $\{p_i\}$ are held constant except an arbitrary two
    of them, say p_j and p_k where $p_j \geq f_j$ and $p_k \leq f_k$, then the
    measure will grow larger if p_j is increased and p_k is cor-
    respondingly decreased.

There are a number of accuracy measures meeting these criteria.
Commonly used examples are described in the next section.

---

* Of course, it is equally unlikely that we have hit upon exactly the right
number if we say this probability is 70%. To completely define a prediction,
we would need to specify an infinite number of moments. However, for most practical purposes, it is sufficient to specify these two (which are the prediction
equivalents to the mean and standard width of an observed sample).

C. Measures of Independence of Observation and Prediction

Let us display the assessment space as a contingency table. See Fig. 92.

Fig. 92. Contingency Table for Observation (F) versus Prediction (P)

The columns are defined to correspond to the groups along the P-axis, the rows to groups along the F-axis. The numbers n_{ij} give the data counts in each cell. We define also

R = # of rows (# groups along F),

C = # of columns (# groups along P),

$$r_i = \sum_{j=1}^{C} n_{ij} = \text{number of events in } i^{th} \text{ row },$$

$$c_j = \sum_{i=1}^{R} n_{ij} = \text{number of events in } j^{th} \text{ column, and}$$

$$n = \sum_{ij} n_{ij} = \text{total number of events.}$$

We now ask the question: Do our predictions contain any information about the outcome? If not, then the observed frequency and
the predicted probability should be independent. We can test this
hypothesis using the assessment contingency table.

Numerous test-statistics for contingency tables have been devised.
They include Pearson's T, Yates' Corrected T, Fisher's Hypergeo-
metric P, Cramer's V, Pearson's Contingency Coefficient, Kendall's
Tau, Somer's D, and Goodman-Kruskal's Lambda test-statistics.

Pearson's T-test

Conventional T-testing uses sample average estimation of the row
and column probabilities r_i/n and c_j/n and takes the assumption of
row-column independence $p_{ij} = (r_i/n)(c_j/n)$ as the null hypothesis.
The T-test-statistic then becomes

$$T = \sum_{i=1}^{R} \sum_{j=1}^{C} \frac{\left(n_{ij} - \frac{r_i c_j}{n}\right)^2}{\frac{r_i c_j}{n}} .$$

T is approximately χ^2 distributed with $\nu = (R-1)(C-1)$ degrees of
freedom. Use of T involves integrating its distribution over
values equal to and greater than the value on the observed data.

Yates' Corrected T-test

$$Y = \sum_{i=1}^{R} \sum_{j=1}^{C} \frac{\left(\left|n_{ij} - \frac{r_i c_j}{n}\right| - \frac{1}{2}\right)^2}{\frac{r_i c_j}{n}}$$

Y is closer to χ^2 behavior than T for small samples. This is
especially true for $R \times C = 2 \times 2$ contingency tables.

Fisher's Hypergeometric P-test *

If $R = C = 2$, then

$$P = \frac{r_1! \ r_2! \ c_1! \ c_2!}{n! n_{11}! \ n_{12}! \ n_{21}! \ n_{22}!}$$

is the probability of the cell distribution $\{n_{ij}\}$ assuming the cell probabilities to be

$$p_{ij} = \left(\frac{r_i}{n}\right)\left(\frac{c_j}{n}\right).$$

Use of P involves summing its values over the set of distributions including the observed distribution and all more extreme distributions.

Cramer's V-test

$$V_c = \sqrt{\frac{T}{n \ \min\{(R-1),(C-1)\}}}$$

V_c always has values in the range from 0 to 1.

Pearson's Contingency Coefficient-test

$$C_p = \frac{T}{T+n}$$

For 2×2 tables C_p ranges from 0 to 0.707. The upper limit varies with R and C.

Kendall's Tau-test

Define

P = # of concordant pairs (order same on both variables)

Q = # of discordant pairs (order opposite on both variables)

T = # of tied pairs (equality for at least one variable)

Note that $P + Q + T = \frac{1}{2} n \ (n-1) = $ total/number of pairs.

---

* Also referred to as Fisher's "exact" test.

Kendall defined the Tau-test in two forms, one for contingency tables with the same numbers of rows and columns

$$\text{Tau B} = \frac{P-Q}{\frac{1}{2}n\,(n-1)} \quad,$$

and one with unequal numbers of rows and columns for tables

$$\text{Tau C} = \frac{2m(P-Q)}{n^2(m-1)} \text{ when } m = \min\{R,C\}.$$

A related ordinal symmetric measure is

$$\text{Gamma} = \frac{P-Q}{P+Q} \quad.$$

Somer's D-test

Somer's D-test employs two test-statistics:

$$D_{\text{symmetric}} = \frac{4(P-Q)}{[\,(n^2-\sum_j c_j^2) + (n^2-\sum_j r_i^2)\,]} \quad, \text{ and}$$

$$D_{\text{asymmetric}} = \frac{2(P-Q)}{n^2-\sum_j c_j^2} \quad,$$

where we have defined $D_{\text{symmetric}}$ assuming the row-variable F as the dependent variable.

Goodman-Kruskal's Lambda-test

The Lambda test also employs both symmetric and asymmetric test statistics.

$$L_{\text{Symmetric}} = \frac{\sum_j \max_i \{n_{ij}\} + \sum_i \max_j \{n_{ij}\} - \max_i \{r_i\} - \max_j \{c_j\}}{2n - \max_i \{r_i\} - \max_j \{c_j\}} \quad, \text{ and}$$

$$L_{\text{Asymmetric}} = \frac{\sum_j \max_i \{n_{ij}\} - \max_i \{r_i\}}{n - \max_i \{r_i\}} \quad,$$

where we have again defined the row as the dependent variable for the asymmetric test.

D. Least Squared Error Measures of Accuracy

We next turn to measures of predictive accuracy. The most frequently used of these are based on the squared error criterion
introduced by Legendre in 1805 and Gauss in 1809. Two examples
are:

   Pearson's T-test:

$$T = n \sum_{i=1}^{L} \frac{(\bar{P}_i - \bar{F}_i)^2}{\bar{P}_i} , \text{ and}$$

   Dispersion Ratio R^2-test:

$$R^2 = \sum_{i=1}^{L} \left(\frac{\bar{P}_i - \bar{F}_i}{U_i} \right)^2 ,$$

where

   n_i = number of events in the ith group, i=1,...,L,

   $\bar{F}_i$ = average frequency of observation of the outcome in
        question among events in the ith group of events,

   $\bar{P}_i$ = average predicted probability for this outcome for
        events in this group,

   U_i = uncertainty in $\bar{P}_i$, and

   n = $\sum n_i$.

The T-test assesses the accuracy of our estimates $\{\bar{P}_i\}$ of the mean
frequencies $\{\bar{F}_i\}$, relative to the magnitude of the estimates themselves. The R^2-test assesses the accuracy of our estimates relative to the magnitude of the uncertainties $\{U_i\}$.

For large n_i, the measure T is asymptotically chi-squared distributed. This allows one to use the tabulated tail probabilities for
the chi-squared distribution to conveniently assign confidence
levels as a function of T and ν, where $\nu = L - 1 - n_\theta$ and n_θ is the
number of parameters estimated.

The expected value of R^2 for a perfect predictor is the number of degrees of freedom ν. (With no parameters estimated, this is $\nu = L - 1$.)

If we assume the process is binomially distributed, then

$$U_i = \sqrt{\frac{\overline{P}_i(1-\overline{P}_i)}{n_i}} \ .$$

This gives us

$$R^2 = \sum_{i=1}^{L} \frac{n_i(\overline{P}_i - \overline{F}_i)^2}{\overline{P}_i(1-\overline{P}_i)} \ ,$$

which is proportional to the dispersion squared $Q^2 = \frac{R^2}{L-1}$, similar in form to that defined by Lexis in 1879. Normal dispersion is $Q = 1$, supernormal is $Q > 1$, and subnormal is $Q < 1$. (Lexis dealt with a case in which the underlying probability was the same for all groups, and he estimated $\overline{P}_i$ by the mean frequency $F = \frac{1}{L}\sum \overline{F}_i$.)

If we assume the process is Poisson distributed (say as a small $\overline{P}_i$ approximation to the binomial), then

$$U_i = \sqrt{\frac{\overline{P}_i}{n_i}} \ .$$

If we let the U_i represent the scatter in the individual events rather than the uncertainty in the mean $\overline{P}_i$, then R^2 becomes simply the Phi-statistic: $\phi = T/n$.

E.   Informational Disagreement Measures of Accuracy

Although the least-squared error measures of accuracy are suitable for many circumstances, they have a defect which makes them undesirable for small sample sizes.   This is their extreme sensitivity to outliers. A few large errors can make T, for example, "unreasonably" great because of the squared dependence.

One way to soften the dependence is to use a criterion based on the sum of the absolute errors.   Absolute error measures were used by Boscovich, Laplace and others prior to the introduction of error squared measures by Legendre.   They, however, are too weak and do not have a natural relationship to the normal distribution as does the squared error measure.   They also suffer from the complications introduced into the analysis by the lack of smoothness of the absolute value, although this is less of a consideration with digital computers.

With the use of information theory, it is possible to define accuracy measures which lie between the absolute error and the error squared measures in sensitivity to outliers, and which are smooth throughout the range of the variables.

One such measure is the informational disagreement:[*]

$$\Delta S_e = \sum_{i=1}^{L} \bar{F}(C_i) \sum_{k=1}^{K} \bar{F}(O_k \text{ given } C_i) \log[\bar{F}(O_k \text{ given } C_i)/\bar{F}(O_k \text{ given } C_i)].$$

It is used extensively in communications theory in studying the probability of error for codes generated by various procedures.

This measure has a form similar to that of an expectation value of a log-likelihood ratio.   Consider the values for any chosen k, i: $P_{ki} = \bar{F}(O_k \text{ given } C_i)$ and $F_{ki} = \bar{F}(O_k \text{ given } C_i)$.   Since log 1 = 0, it is obvious that $\Delta S_e = 0$ if the predictions are exact for each

---

[*]   See S. Kullback, *Information Theory and Statistics*, J. Wiley & Sons, Inc., NY, 1959 for historical references.

group, i.e., $P_{ki} = F_{ki}$ for all k, i. It is not difficult to show
that $\Delta S_e \geq 0$ in all cases, and the $\Delta S_e = 0$ *only* when $P_{ki} = F_{ki}$ for
all k, i. Thus, ΔS_e has the correct general properties for an accuracy measure.

The fact that ΔS_e has the formal structure of entropy exchange is
seen by defining

$$\overline{P}(C_i \text{ and } D) = \overline{F}(C_i), \text{ and}$$

$$\overline{P}(O_k \text{ given } (C_i \text{ and } D)) = \overline{F}(O_k \text{ given } C_i),$$

where D represents the condition of having been already observed.
ΔS_e is thus the mutual entropy exchange between the condition D
and the set of observations $\{O_k \text{ given } C_i\}$.

Considering the hierarchy

$$1, \log x, x, x \log x, x^2, \ldots$$

we see that ΔS_e lies between linear error measures such as Laplace's
absolute difference or Kolmogorov's greatest absolute difference,
and quadratic error measures such as Lexis' dispersion or Pearson's
T. This gives it a lower sensitivity to outliers than quadratic
measures, but greater than linear measures.

Because $\overline{F}_i$ and $\overline{P}_i$ are restricted to the range 0 to 1, we do not
need to take absolute values, as we do with a linear measure to
avoid cancellation of errors with opposite signs.

For small sample sizes, tabulated values of tail probabilities for
the ΔS_e distributions can be used. For large sample sizes, it can
be shown that $-L \log \Delta S_e$ is asymptotically chi-squared distributed.

Like Pearson's T-test, the measure ΔS_e ignores the uncertainties
$\{U_i\}$. It is not difficult, however, to formulate an information-
theoretic accuracy measure which does take the uncertainties into
consideration.

Suppose we think of each of the L groups as a channel through which we acquire information about future outcome probabilities. Then U_i is the equivocation in the ith channel, and the difference $|P_i-F_i|$ is the received message indicating that P_i may be in error. (If we were dealing with the transmitted signal, then U_i would be the noise. See Chapter 7.)

After reception through the ith channel, our total uncertainty is $\sqrt{U_i^2+(P_i-F_i)^2}$. (Since the equivocation from the training data and the message from the test data are treated as uncorrelated, we use root-mean-square compounding.)* We cannot detect differences in the reception less than U_i. Thus, the number of detectably different receptions possible is

$$\frac{\sqrt{U_i^2+(P_i-F_i)^2}}{U_i} = \left[1+\left(\frac{P_i-F_i}{U_i}\right)^2\right]^{\frac{1}{2}}.$$

Using maximum entropy estimation, these receptions are equiprobable. Thus, the information content of each reception is given by

$$\frac{1}{2}\log\left[1+\left(\frac{P_i-F_i}{U_i}\right)^2\right].$$

The ith group has a bandwidth of $L=1$ reception per analysis. If a family of time functions is limited to a frequency of L cycles per unit time, then a time interval of length T can be represented, at most, by 2TL independent variables.

Summing over all L groups, we have

$$\Delta S = \sum_{i=1}^{L}\log\left[1+\left(\frac{P_i-F_i}{U_i}\right)^2\right]$$

as the total rate (per analysis) at which we can receive information that the $\{P_i\}$ are in error. This is nonnegative, being zero if

* If the test and training samples have different sizes, we adjust this to

$$\sqrt{\frac{n_i^{trng}U_i^2+n_i^{test}(P_i-F_i)^2}{n_i^{trng}+n_i^{test}}}.$$

and only if $F_i = P_i$ for all $i = 1, 2, \ldots, L$. If all the uncertainties U_i exactly equal the errors $|P_i - F_i|$, then ΔS becomes L bits. In the limit of small error-to-uncertainty ratios, it becomes equivalent to the generalized dispersion ratio test.

Note that, if the groups are randomly selected instead of being constructed by the entropy minimization criterion, the P_i, F_i and U_i are the same (approximately) for all groups. In this case we have simply

$$\Delta S = L \, \log \left[1 + \left(\frac{P-F}{U} \right)^2 \right],$$

where L is the total bandwidth (receptions per analysis) of our information extraction system.

F. Hypothesis Testing and Critical Values

In order for a measure of error such as T, R^2, or ΔS_e to be useful, we must have a means of determining what is a large and what is a small error.

We posit the following hypothesis H_o: the frequencies $\{F_i\}$ are drawn from a population with the probabilities $\{P_i\}$. To test the accuracy of the hypothesis we select a measure of error. Assume, for example, that we use the ΔS_e measure. Let $f_o(\Delta S)$ represent the distribution of ΔS for a random set of test data assuming the probabilities $\{P_i\}$. See Fig. 93.

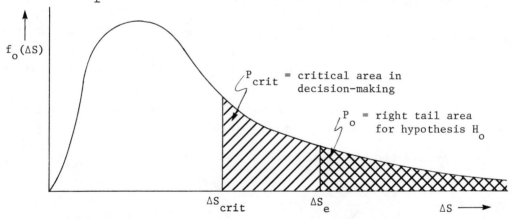

Fig. 93. Hypothesis Testing with Entropy
Decrement Distribution

For any observed value ΔS_e for ΔS, we can compute the probability of the data having this amount or more informational disagreement with the predictions, assuming the predicted probabilities $\{P_i\}$:

$$P_o = \int_{\Delta S_e}^{\infty} f(\Delta S)\, d\Delta S.$$

If this probability P_o is below some critical value P_{crit} established for decision-making purposes, then the decision-maker rejects the hypothesis that the probabilities $\{\overline{P}_i\}$ are accurate.

If P_o is above this limit, the hypothesis is not rejected. This does not mean, however, that it is necessarily accepted as proven. For example, an alternative may produce an even higher P_o.

G. Xenophobia and Statistical Confidence

The confidence level rationale for hypothesis rejection is that accepting the hypothesis in the face of low P_o requires believing that the observed frequencies $\{F_i\}$ happened by chance to be in a highly improbable class of surprisingly great disagreement with the probabilities $\{P_i\}$.

One should be careful to understand precisely what this means. Low P_o does *not* prove that the hypothesis has a low probability of being true. If we could make this strong a statement, then probability theory and statistics could have avoided what is now running over two centuries of difficulty.

All that statistical hypothesis testing can prove is that, if we assume the hypothesis is true, then the observed events are more or less surprising.

In order to reach a conclusion about the probable truth of the hypothesis itself, we would need to have additional information, such as the a priori probabilities required by Bayes' theorem. Unfortunately, these are often not available, even in an approximate sense.

Furthermore, a "probability of truth of the hypothesis" interpretation fails to consider the degree to which hypotheses are generally only approximations of reality. Most hypotheses, even those with very high P_o, are, strictly speaking, false.

Even if we could, in some sense, reach a conclusion about the probable truth of hypotheses, it would not necessarily be justifiable for the researcher to reject them for low probability. Suppose the data are surprising at the 5% level based on the hypothesis. Then we might logically wish to accept 5% of all hypotheses conventionally rejected at this level! But which 5%? To adopt a fixed policy of rejecting 100% of hypotheses for which the data are surprising at this level may be to miss many important discoveries.

The researcher should report the surprisal level, but neither "reject" nor "accept" the hypothesis. This is the province of the decision-maker. To do otherwise is to misrepresent the extent of the evidence and to impose unwarranted coherence among users of the information.

H.   Confidence Against Competing Hypothesis - Overall Sample Averages

In order to assess the significance of our predictions, we need more than a measure of the accuracy of our predictions $\{P_i\}$. We need also a measure of the likelihood that this accuracy was not due to chance. If there is too great a likelihood of achieving the observed accuracy by random chance, then we should not be impressed by the predictive success of our theory no matter how close to a perfect match is indicated by an accuracy test. Since randomness is relative to our information, this is another way of saying we are not surprised by what we already knew.

Assessment of this likelihood requires testing a competing hypothesis. The most commonly used hypothesis is the set of overall average training sample frequencies $\{\overline{A}_i\}$. For example, suppose someone espouses a theory on the basis of which we can predict the batting averages for baseball players based on their experience, physical attributes, etc. Suppose this theory is tested on a sample of 100 players and is claimed to have "high" accuracy. If, however, an overall player average would have equivalent accuracy on this sample, we would not regard these data as evidence that the theory has any particular significance.

To conduct such a test, for whichever accuracy measure we have chosen, we simply repeat the accuracy calculations using the hypothesis H_a: The observed frequencies $\{F_i\}$ are drawn from a population with probabilities $\{A_i\}$. Suppose we used ΔS_e to measure accuracy. Then, to measure confidence against the competing hypothesis H_a, we compute

$$\Delta S_a = \sum_{i=1}^{L} \overline{F}(C_i) \sum_{k=1}^{K} \overline{F}(O_k \text{ given } C_i) \log[\overline{F}(O_k \text{ given } C_i)/\overline{A}(O_k \text{ given } C_i)].$$

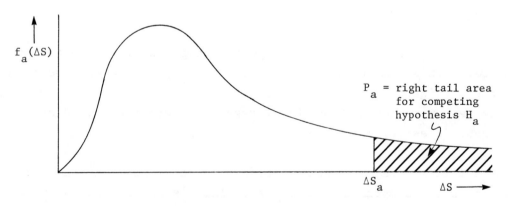

Fig. 94.   Area under Tail beyond Critical Value

The entropy decrement ΔS_a is simply ΔS_e with our predicted probabilities replaced by the sample averages. The computation is made
in accord with the following prescription:

    o The event classes (in independent variable space) used
       in calculating the individual averages are the same (mini-
       mum entropy) classes used in computing the individual
       probabilities and frequencies in ΔS_0. Thus ΔS_a *does* de-
       pend upon the entropy minimax analysis in this sense.

    o The predictions are now ordered according to increasing
       sample average for purposes of forming the L groups for
       which the $\overline{A}$ averages are computed.

Let $f_a(\Delta S)$ be the distribution of ΔS for a random set of test data
assuming the probabilities $\{A_i\}$.

Fig. 94 shows the right tail area P_a beyond the observed ΔS for
this distribution

$$P_a = \int_{\Delta S_a}^{\infty} f_a(\Delta S) \, d\Delta S \ .$$

Suppose we used a particular standard P_{crit} to assess the hypothesis H_o and decided we could not reject it because $P_o > P_{crit}$.

If we also cannot reject the competing sample average hypothesis
H_a using this same criterion because $P_a > P_{crit}$, then we have insufficient evidence to decide between the two. However, if $P_a < P_{crit}$,
then we can reject the sample average in favor of the predicted
probabilities $\{\overline{P}_i\}$.

The ability to reject the competing hypothesis will generally improve as the sample size increases. This is because the distribution tends to become more sharply peaked at lower ΔS as sample
size increases.

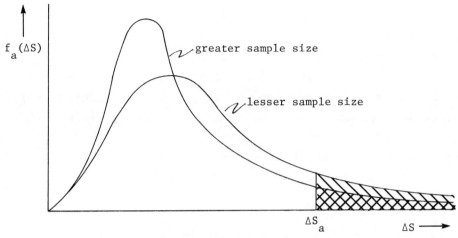

Fig. 95.   Effect of Increased Sample Size Is to
Make $f_a(\Delta S)$ More Peaked, thereby
Reducing the Area beyond ΔS_a

Fig. 95 shows this effect. For any given ΔS_a, a greater sample size
will mean a lower P_a. Thus, so long as ΔS_a remains sufficiently
greater than ΔS_e, there will be a sample size for which the distinction
can be made. (Of course, one must actually obtain this larger sample to ensure that ΔS_a does indeed stay sufficiently greater than
ΔS_e.)

Even further increasing of the test sample size may eventually result in rejection of the hypothesis H_o itself. However, this would
amount to using an hypothesis based on a small training sample to
try to predict a large test sample. With this much data, it would
be better to remove some from the test sample and use them to en-
large the training sample, thereby enabling the pattern discovery
analysis to obtain a more refined set of conditional probabilities.

I. Definitiveness Assessment

It might appear that reliability constitutes a complete assessment
of a prediction. But this is not so. What does it mean to assess
whether or not a prediction needs improvement? It means determining whether or not the prediction would, in some sense, be better
if we were to modify it. One way to modify a prediction is to
change the numerical values of the stated probabilities or confidence intervals. This may improve its accuracy. Another way is
to change its domain, i.e., the wording of the conditions under
which it is stated to apply. This may improve another aspect of
the prediction: its definitiveness. (Of course, this generally
implies an accompanying change in the probabilities to maintain
the accuracy.)

Predictive definitiveness is assessed in terms of how close the
probability estimates are to extremal values of zero or unity.

The information decrement measure of predictive accuracy ΔS_e is a
measure of the informational difference between the performance
line for the actual predictor and that for a perfect predictor.
See Fig. 96.

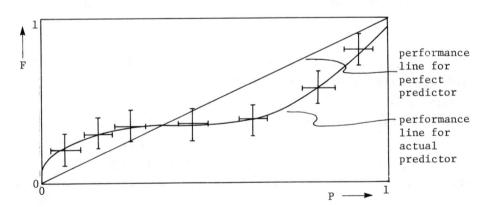

Fig. 96. ΔS_e Is a Measure of the Informational Difference between
Perfect and Actual Predictors on Test Data

The objective is to minimize ΔS_e. This can easily be accomplished simply by shrinking the prediction all the way down to the sample average, thus producing a constant predictor. This, however, would give us a prediction with no definitiveness. See Fig. 97.

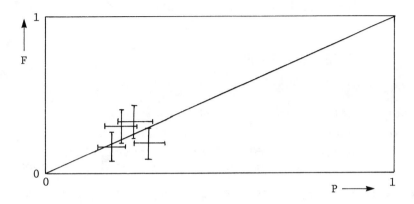

Fig. 97. ΔS_e Reduced (at the sacrifice of definitiveness) by Shrinking All Predictions to a Neighborhood of the Sample Average

What we would ideally like to achieve is a predictor which separates the groups such that they cluster near either the (0,0) point or the (1,1) point, with none in between. See Fig. 98.

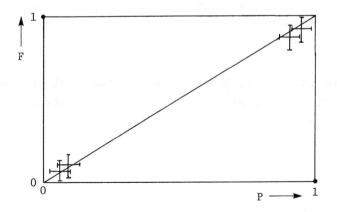

Fig. 98.   Ideal Predictor which Separates Test Data into a Cluster near (0,0) and a Cluster near (1,1)

426 ENTROPY MINIMAX SOURCEBOOK

J. Skill Ratio Measures of Definitiveness

In meteorology, predictive definitiveness is often measured in terms of a skill ratio.

For categorical predictions of a particular outcome, it is defined as:

$$K = \frac{correct - chance}{total - chance} ,$$

where

$$total = \text{total number of predictions made},$$
$$correct = \text{number of correct predictions, and}$$
$$chance = \text{expected number of correct predictions for a random predictor based on the overall sample average.}$$

A similar ratio can be defined for probabilistic prediction:

$$K = \frac{(N-|NF-NP|) - (N-|NF-NA|)}{(N) \qquad - (N-|NF-NA|)}$$

$$= \frac{|F-A| - |F-P|}{|F-A|},$$

where

$$F = \text{observed frequency of the outcome,}$$
$$P = \text{predicted frequency (probability) of the outcome,}$$
$$A = \text{expected frequency of the outcome from overall sample average, and}$$
$$N = \text{number of events.}$$

The skill ratio is zero if one uses an "unskillful" prediction based on the training sample average. It increases as P moves further from A and closer to F.

K.   Entropy Decrement Measures of Definitiveness

Just as we express a preference for prediction accuracy by minimum informational difference from a good predictor line, we can express a preference for definitiveness by maximum informational difference from a poor predictor line.

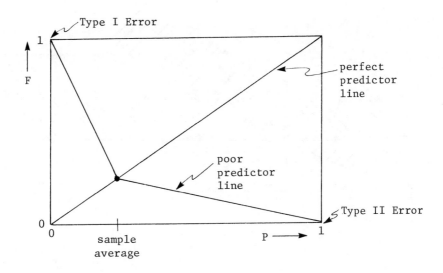

Fig. 99.   Poor Predictor Line Extending from the Type I Error
Point to the Type II Error Point   (If the sample
average is known, it should pass through this point.
Shown here is a two-segment piecewise linear poor
predictor line with its elbow at the sample average.)

A poor predictor line should, I propose, extend from the (0,1) Type I error point to the (1,0) Type II error point.   To avoid bias, it should also pass through the sample average.   Fig. 99 shows one such poor predictor line.

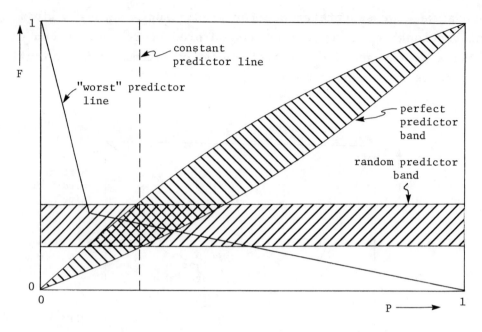

Fig. 100. "Worst" Predictor Line Defined as a Two-Segment
Piecewise Linear Poor Predictor Line with its
Elbow Midway between the Sample Average and the
Nearest Extremum

Fig. 100 shows another poor predictor line which I have labeled
the "worst" predictor. It is defined to extend further into the
random predictor band than the one in Fig. 99. The corner is set
half-way between the constant predictor line and the nearest extremum. This reflects an emphasis on penalization of predictions
in the random band relative to those along the constant predictor
line.

The two-leg worst predictor line in Fig. 100 can be thought of as
a piecewise linear approximation to a hypothetical smooth worst
predictor line as illustrated in Fig. 101. For most computational
purposes, the two-leg line is an adequate representation.

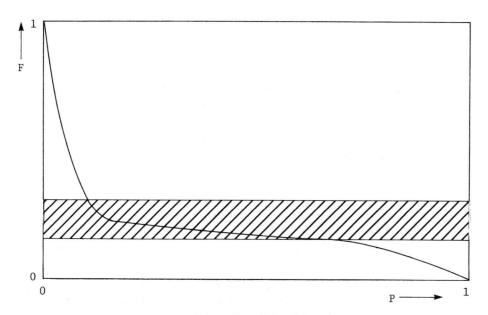

Fig. 101.   Smooth Worst Predictor Line

The informational difference between the actual predictor performance line and the worst predictor line is defined as*

$$\Delta S_d = \sum_{i=1}^{L} \left\{ (\overline{P}_i - \overline{D}_i) \log \left[\overline{P}_i / \overline{D}_i \right] + (\overline{D}_i - \overline{P}_i) \log \left[(1 - \overline{P}_i) / (1 - \overline{D}_i) \right] \right\} ,$$

where

$\overline{D}_i$ = horizontal distance from the ith group mean to the worst predictor line.

The worst predictor line for a particular user is normal to the
user's equiutility lines in assessment space.

---

* R.A. Christensen, "Assessment of Statistical Models," Sec. 8.3 of *SPEAR Fuel
Reliability Code System: General Description*, NP-1378, Electric Power Research
Institute, Palo Alto, CA, March 1980, pp. 8-24.

QUOTES

LINFOOT (1957)

They* then define the *logarithmic index of correlations*

$$r_o = \Sigma_{ij}(p_{ij} \log p_{ij} - p_i q_i \log p_i q_i)$$

...For continuous variables x and y with joint probability density distribution $p(x,y)$ the corresponding quantity r_o is given by the equation

$$r_o = \iint \{p(x,y) \log p(x,y) - p(x)q(y) \log[p(x)q(y)]\} dx\, dy$$

$$* \quad * \quad * \quad *$$

We can now define the *informational coefficient of correlation* r_1 by the equation

$$r_1 = \sqrt{1 - e^{-2r_o}}$$

...It is zero whenever x and y are statistically independent, since then $r_o = 0$, and it is 1 whenever x and y are fully correlated, in the since that each determines the value of the other uniquely.

$$* \quad * \quad * \quad *$$

...is invariant under a change of parameterization $x' = f(x)$, $y' = g(y)$, and reduces to the classical correlation coefficient when $p(x,y)$ is normal.

> —Edward Hubert Linfoot (1905-   )
> "An Informational Measure of Correlation," *Information and Control*, 1, 1957, pp. 86, 88, 89.

KULLBACK (1959)

...we may define

$$I(2:1) = \int f_2(x) \log \frac{f_2(x)}{f_1(x)} d\lambda(x)$$

as the mean information per observation from μ_2 for discrimination in favor of H_2 against H_1.

> —Solomon Kullback (1907-   )
> *Information Theory and Statistics*, J. Wiley & Sons, Inc., NY, 1959, p. 6.

---

* Castañs Camargo, M. and Medina e Isabel, M. (1956), *Anales real soc. españ. fis. y quim. (Madrid)* Ser. A52, 117.

LEWIS (1959)

The problem in which we are interested is that of approximating one probability distribution with another.

* * * *

The closeness of approximation is then defined to be the difference between the information contained in the true distribution and the information contained in the approximating distribution about the true distribution.

$$I_{p-p'} = I_p - I_{p'}$$

$$= \sum_{0}^{2^n-1} P_j \log P_j - \sum_{0}^{2^n-1} P_j \log P_j'$$

$$= \sum_{0}^{2^n-1} P_j \log \frac{P_j}{P_j'},$$

—Philip M. Lewis, II (1931-   )
"Approximating Probability Distributions to Reduce Storage Requirements," *Information and Control*, *2*, 1959, pp. 217, 217-218.

BLASBALG and VAN BLERKOM (1962)

Let $\{X_i\}$, $i = 1, 2, \ldots, N$ be the...measurement...and let $\{P_i\}$ be the expected statistical law. An efficient and extremely important statistic for testing the validity of the law is

$$\Delta H(S) = S \sum_{i=1}^{N} \lambda_i \log \frac{\lambda_i}{P_i}$$

$$\sum_{i=1}^{N} P_i = 1$$

$$\sum_{i=1}^{N} \lambda_i = 1$$

and where $\{\lambda_i\}$ are the empirical probabilities...from S observations.

—Herman Blasbalg (1925-   ) and Richard Van Blerkom (1935-   )
"Message Compression," *IRE Trans. on Space Electronics and Telemetry*, Sept. 1962, p. 230.

THEIL (1966)

$$I(A:P) = \sum_{i=1}^{n} A_i \log \frac{A_i}{P_i}$$

Its interpretation in the language of information theory is that of the information of the indirect message $(A_1, \ldots, A_n)$ given the prior probabilities $(P_1, \ldots, P_n)$. The present interpretation is that of the degree to which the forecasts P_i are inaccurate with respect to the corresponding realizations A_i; we shall therefore describe $I(A:P)$ as the *information inaccuracy* of the P's with respect to the A's.

* * * *

It will be clear that the percentage of correct predictions cannot by itself be a sufficient measure for the evaluation of the forecasts.

* * * *

...the two measures of the quality of the forecasts (information content and frequency of correct prediction)...

> —Henri Theil (1924-   )
> *Applied Economic Forecasting*, North-Holland Pub. Co., Amsterdam, 1966, pp. 263, 287, 289.

PEARSON (1966)

Let

$$H_r = \frac{1}{2}\left[\frac{H_{X|Y}}{H_X} + \frac{H_{Y|X}}{H_Y}\right]$$

H_r will be known as the "relative uncertainty".

$$\hat{r}_{\text{ENT}} = \sqrt{1 - H_r^{\,2}}$$

> —William H. Pearson (1930-   )
> "Estimation of a Correlation Coefficient from an Uncertainty Measure," *Psychometrika*, *31*, 1966, p. 423.

CHOW and LIU (1968)

...It is well known* that the quantity

$$I(P,P_a) = \sum_x P(x) \log \frac{P(x)}{P_a(x)}$$

has the property that

$$I(P,P_a) \geq 0$$

with equality sign if and only if $P(x) \equiv P_a(x)$ for all x. Lewis** defined $I(P,P_a)$ as the measure of closeness in approximating P by P_a...

The measure...will be used as a criterion in developing a procedure of approximating an nth-order distribution by a distribution of tree dependence.

> —C.K. Chow (1928-   ) and C.N. Liu (1934-   )
> "Approximating Discrete Probability Distributions
> With Dependence Trees," *IEEE Trans. on Information
> Theory*, *IT-14*, May 1968, p. 463.

---

*   S. Kullback and R.A. Leiber, "On Information and Sufficiency," *Ann. Math. Statistics*, Vol. 22, pp. 79-86, 1951.

** P.M. Lewis, "Approximating Probability Distributions to Reduce Storage Requirement," *Information and Control*, Vol. 2, pp. 214-225, September 1959.

REFERENCES

Bishop, Yvonne M.M., Stephen E. Fienberg and Paul Holland, *Discrete Multivariate Analysis: Theory and Practice*, The MIT Press, Cambridge, MA, 1975.

Blasbalg, H. and R. Van Blerkom, "Message Compression," *IRE Trans. on Space Electronics and Telemetry*, Sept. 1962, pp. 228-238.

Brown, D.T., "A Note on Approximations to Discrete Probability Distributions," *Information and Control*, *2*, 1959, pp. 386-392.

Castañs Camargo, M., and M. Medina e Isabel, *Anales real soc españ. fis. y quim*, Ser. A52, Madrid, 1956.

Chow, C.K. and C.N. Liu, "Approximating Discrete Probability Distributions with Dependence Trees," *IEEE Trans. on Info. Theory*, *IT-14*, May 1968, pp. 462-467.

Christensen, R.A., "Crossvalidation: Minimizing the Entropy of the Future," *Info. Proc. Letters*, *4*, No. 3, Dec. 1975, pp. 73-76. [*Chapter 10 of Volume III.*]

_____, "Assessment of Statistical Models," Section 8.5 of *SPEAR Fuel Reliability Code System: General Discription*, NP-1378, Electric Power Research Inst., Palo Alto, CA, March 1980, pp. 8.15-8.24.

Conover, W.J., "Uses and Abuses of the Continuity Correction," *Biometrics*, *24*, 1968, pp. 1028.

Cramer, H., *Mathematical Methods of Statistics*, Princeton Univ. Press, Princeton, NJ, 1946.

Fano, Robert M., *Transmission of Information*, MIT Press, Camb., MA, 1961.

Fisher, R.A., *Statistical Methods for Research Workers*, Oliver & Boyd, Ltd., London, 1934.

Gauss, K.F., *Theory of Least Squares* (1803-1826), tr. by H. Trotter, Statistical Techniques Research Group, Mathematics Dept., Princeton Univ., Princeton, NJ, 1957, p. 44.

_____, *Theory of the Motion of the Heavenly Bodies Moving about the Sun in Conic Sections*, tr. by C.H. Davis, Dover Pubs., Inc., NY, 1963.

Goodman, L.A. and W.H. Kruskal, "Measures of Association for Cross-Classifications," *J. Amer. Statist. Assoc.*, *49*, 1954, pp. 732-764; *54*, 1959, pp. 123-163; *58*, 1963, pp. 310-364; and *67*, 1972, pp. 415-421.

Highleyman, W.H., "The Design and Analysis of Pattern Recognition Experiments," *Bell System Technical Journal*, *41*, 1962, pp. 723-744.

Hoel, Paul G., *Intro. to Mathematical Statistics*, J. Wiley & Sons, Inc., NY, 1954, pp. 172-175.

Ireland, C.T., H.H. Ku and S. Kullback, "Symmetry and Marginal Homogeneity of an $r \times r$ Contingency Table," *J. of The Amer. Statist. Assoc.*, *64*, March 1969, pp. 1323-1341.

Jevons, Stanley, *The Principles of Science* (1873), Dover Pubs., Inc., NY, 1958.

Klijn, N. and W.J. Zwezerignen, "Reliability and Information Contents of Weather Forecasts," Rept. 6323 of the Econometric Inst. of the Netherlands School of Economics, 1963.

Ku, H. and S. Kullback, "Approximating Discrete Probability Distributions," *IEEE Trans. of Info. Theory, IT-15,* 1969, pp. 444-447.

Kullback, S., *Information Theory and Statistics,* J. Wiley & Sons, Inc., NY, 1959.

Kullback, S. and R.A. Leiber, "On Information and Sufficiency," *Ann. Math. Statistics,* Vol. 22, pp. 79-86, 1951.

Leftwich, Preston W., Jr., "Regression Estimation of the Probability of Tropical Cyclone Recurvature," *Sixth Conf. on Probability and Statistics in Atmosphere Sciences,* Banff, Alta., Canada, Oct. 9-12, 1979, American Meteorological Society, Boston, MA, 1979, pp. 63-66.

Lewis, P.M., II, "Approximating Probability Distributions to Reduce Storage Requirements," *Information and Control, 2,* Sept. 1959, pp. 214-225.

_____, "The Characteristic Selection Problem in Recognition Systems, *IRE Trans. on Info. Theory, IT-8,* No. 2, Feb. 1962, pp. 171-178.

Lexis, W., "Ueber die Theorie der Stabilitat statistischer Reihen," Jahrb. f. nat. Ok. u. Stat. (1), *32,* (1879), p. 604; reprinted in *Abhandlungen zur Theorie der Bevolkenungs-und Moral-statistik,* Jena, 1903, p. 253. (See J.M. Keynes, *A Treatise on Probability,* Macmillan & Co., London, 1921, pp. 398-399.)

Linfoot, E.H., "AnInformational Measure of Correlation," *Information and Control, 1,* No. 1, 1957, pp. 85-89.

Metz, C.E., D.J. Goodenough and K. Rossmann, "Evaluation of Receiver Operating Characteristic Curve Data in Terms of Information Theory, with Applications in Radiography," *Radiology, 109,* Nov. 1973, pp. 297-303.

Meyer, Stuart L., *Data Analysis for Scientists and Engineers,* J. Wiley & Sons, Inc., NY, 1975, pp. 359ff.

Pearson, K., "On the Theory of Contingency and Its Relation to Association and Normal Correlation," *Draper's Co. Res. Mem. Biometric Ser. 1,* 1904. Reprinted in *Karl Pearson's Early Papers,* Cambridge Univ. Press, Cambridge, 1948.

Pearson, W.H., "Estimation of a Correlation Measure from an Uncertainty Measure," *Psychometrika, 31,* No. 3, AD462665, 1966, pp. 421-433.

Plackett, R.L., "The Continuity Correction in 2×2 Tables," *Biometrika, 5,* 1964, pp. 327-337.

Renyi, A., "On Measures of Entropy and Information," *Proceedings of the Fourth Berkeley Symposium on Mathematical Statistics and Probability,* Calif. Univ. Press, Berkeley, CA, 1961, pp. 547-561.

Rider, A., *An Intro. to Modern Statistical Methods,* J. Wiley & Sons, Inc., NY, pp. 112-115.

Theil, Henri, *Applied Economic Forecasting,* North-Holland Pub. Co., Amsterdam, 1966.

Tilanus, C.B. and H. Theil, "The Information Approach to the Evaluation of Input-Output Forecasts," Rept. 6409 of the Econometric Institute of the Netherlands School of Economics, 1964.

Yates, Frank, "Contingency Tables Involving Small Numbers and the χ^2 Test," *Journal of Roy. Statist. Soc. Suppl. 1,* 1934, pp. 217-235.

CHAPTER 17

NON-INFORMATION-THEORETIC CLUSTERING METHODS

CHAPTER 17

NON-INFORMATION-THEORETIC CLUSTERING METHODS

A. Overview

The need to classify is fundamental to analysis. We cannot even describe a problem for analysis without some form of classification. Further, the process of analysis itself often involves reclassification.

In Part III of this volume, we described how the principle of minimum entropy is used to partition the independent variable space. For purposes of contrast, and as preliminary background to one of the most serious impediments to predictive reliability, the problem of contrivedness,* we here describe some common non-information-theoretic methods for clustering observational data.

The variety of clustering methods available to researchers has grown in recent years to rival the variety of distribution functions and test-statistics used by statisticians.

These methods represent a wide range of procedures for feature selection, methods for defining clusters in terms of the features, measures of typicality and similarity, and rules for selecting clusters.

Clustering methods are sometimes grouped as "supervised" and "unsupervised." In supervised clustering, the process of partitioning the independent variable space is guided by one or more dependent variables. In unsupervised clustering, no explicit reference is made to a dependent variable. It is, however, difficult

---

* Contrivedness is discussed in Chapter 19.

to see how unsupervised clustering has any meaning other than as an arbitrary grouping of objects or events.   The more proper distinction is perhaps between clustering guided by the particular dependent variable currently under consideration, and clustering guided by a more general class of dependent variables.

It is not our purpose here to provide a comprehensive review of these methods.   A number of surveys are already available.*
Rather, we will describe, for purposes of illustration, two commonly used non-information-theoretic clustering methods: discriminant clustering, and K-nearest neighbors clustering.

---

* See, for example:
- R.R. Sokal and P.A. Sneath, *Principles of Numerical Taxonomy*, W.H. Freeman and Co., San Francisco, CA, 1963.
- G.H. Ball, "Data Analysis in the Social Sciences: What About the Details?" *Proc. of the 1965 Fall Joint Computer Conference*, *27*, Part 1, 1965, pp. 533-559.
- R.C. Tryon and D.E. Bailey, *Cluster Analysis*, McGraw-Hill, NY, 1970.
- M. Jardine and R. Sibson, *Mathematical Taxonomy*, J. Wiley & Sons, Inc., NY, 1971.
- R.M. Cormack, "A Review of Classification," *Journal of the Royal Statistical Society*, *34*, Part 3, 1971, pp. 321-353.
- M.R. Anderberg, *Cluster Analysis for Applications*, Academic Press, NY, 1973.
- R.O. Duda and P.E. Hart, *Pattern Classification and Scene Analysis*, J. Wiley & Sons, Inc., NY, 1973.
- T.Y. Young and T.W. Calvert, *Classification, Estimation and Pattern Recognition*, American Elsevier Pub. Co., NY, 1974.
- J.A. Hartigan, *Clustering Algorithms*, J. Wiley & Sons, Inc., NY, 1975.

B. Discriminant Clustering

In linear discriminant clustering, straight lines, flat surfaces, and higher-dimensional linear boundaries are drawn in feature space to separate events by outcome.   See Fig. 102.

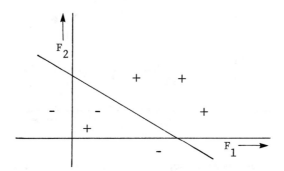

Fig. 102.   Linear Discriminant in a Two-Dimensional
            Feature Space

In piecewise linear discriminant clustering, other boundary shapes such as boxes are formed to partition the space.   See Fig. 103.

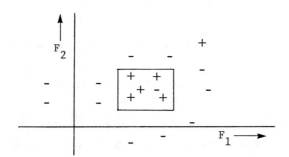

Fig. 103.   Piecewise Linear Discriminant in a
            Two-Dimensional Feature Space

Finally, in nonlinear discriminant clustering, curved boundaries
are permitted.   See Fig. 104.

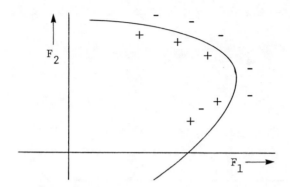

Fig. 104.   Nonlinear Discriminant in a Two-Dimensional
            Feature Space

To formulate a discriminant clustering procedure, several things
must be specified:

    o   Which variables will be used as the axes of the
        feature space?

    o   What boundary shapes will be allowed?   Linear only?
        Piecewise linear?   Polynomial?   Other?

    o   What rule will be used to determine the boundaries?

    o   How will predictions be made for each cluster, and
        how will their uncertainties be determined?

Information-theoretic methods for feature space partitioning, such
as the discrete formulation of entropy minimax, use a minimum
entropy criterion for boundary selections. Most non-information-
theoretic techniques use a form of least squared difference criterion.   Some, however, use another form of measure such as absolute difference.   In addition, it is not uncommon to introduce
weighting coefficients either for the events or the features or both.

In general, the extreme variety of the non-information-theoretic discriminant methods, together with the variety of adjustable parameters in most of the methods, give the researcher a great deal of freedom in finding a way of classifying data...perhaps too much freedom.

C.   K-Nearest Neighbors Clustering

Another form of non-information-theoretic clustering is K-nearest neighbor clustering. It goes under different names in different fields.

- o   In meteorology, one way of making weather forecasts is by using finely tuned "analogs", i.e., comparison to situations in the past in which air circulation, pressure isobars, temperature distributions, storm patterns, etc., were smaller.

- o   In real estate valuation, "comparables" are used in esti- mating the price for which a property would sell.

- o   In law, "precedents" are studied to help form an opinion about how a court would decide a case.

- o   In medicine, similar "case histories" are used to help estimate patient prognoses.

The simplest form of mathematical k-nearest neighbors (k-NN) analy- sis imagines the training sample to be displayed in an n-dimen- sional Euclidean feature space. See Fig. 105.

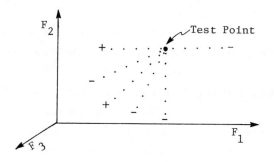

Fig. 105.   Training Sample of + and - Events Distributed in a
Three-Dimensional Feature Space (Each dotted line
indicates distance from the test point to one of the
training events.)

The test point for which we want a prediction is then located in this space. The 1-NN rule is to predict that the outcome for the test event will be the same as for that of the single nearest training event.

Many different k-NN rules can be formulated to make the prediction depend upon the test event's k-nearest neighbors. One rule is to choose the highest frequency outcome among the k neighbors. Another is to assign test point outcome probabilities according to the outcome frequencies among the k neighbors.

To formulate a k-NN rule, a number of things must be specified:

- o   What variables will be used as the axes of feature space?

- o   What distance measure will be used to determine how "far" the test point is from each training point in this space?

    --Will it be isotropic (i.e., the same in all directions)?

    --Will it be homogeneous (i.e., the same throughout the space)?

- o   How many neighbors will influence the prediction for any given event? Will it be the same regardless of how thinly or thickly the sample is populated in the vicinity of the test point?

- o   What rule will be used to transform the observed outcomes for the neighbors into the prediction for the test point? Will it be categorical or probabilistic? How will its uncertainty be determined?

The potential function formulation of entropy minimax is structurally similar to a generalized form of k-nearest neighbor analysis. All points are allowed to contribute to the outcome probabilities for the test point, the weights of the contributions being greater for the nearer points. The non-information-theoretic methods include a wide variety of procedures, ranging from the simple rule of predicting each event to have the same outcome as its nearest Euclidean-distance neighbor, to complex rules involving weighting coefficients for features and nonlinear distance metrics.

Here also, the researcher has a variety of procedures available, involving many tunable parameters, with which to analyze his data. In Chapter 19, we will consider the information price paid for this freedom. Before that, however, let us review another set of analyses techniques with an equivalent plethora of tunable parameters.

QUOTES

CORMACK (1971)

The availability of computer packages of classification techniques has led to the waste of more valuable scientific time
than any other "statistical" innovation (with the possible exception of multiple-regression techniques).

> —Richard Melville Cormack (1935-   )
> "A Review of Classification," *Journal of the Roy-
> al Statistical Society*, *34*, Part 3, 1971, p. 321.

GOWER (1971)

No doubt much "numerical taxonomic" work is logically unsound,
but it has acquainted statisticians with some new problems whose
separate identities are only now beginning to emerge. If statis-
ticians.do not like the formulations and solutions proposed, they
should do better, rather than denigrate what others have done.
Taxonomists must find it infuriating that statisticians, having
done so little to help them, laugh at their efforts. I hope
taxonomists who have real and, I think, interesting problems
find it equally funny that so much statistical work, although
logically sound, and often mathematically complicated (and surely
done for fun), has little or no relevance to practical problems.
They might prefer imperfect solutions to ill-defined problems
than perfect solutions to well-defined non-problems; at least
they can hope for improvements.

> —John C. Gower (1932-   )
> remarks on Dr. Cormack's paper, *Journal of the
> Royal Statistical Society*, *34*, Part 3, 1971, p. 365.

HARTIGAN (1975)

It may be that statistical methods are not appropriate for developing clusters, because some classification is often a pre-
requisite for statistical analyses...without a stable and appropriate classification scheme, statistical analyses are in
vain. On the other hand, clustering techniques require raw data
from some initial classification structure also, so it is doubt-
ful whether formal techniques are sufficient for organizing the
initial data gathering structure. Perhaps a mixture of informal
and formal classification techniques is required.

> —John A. Hartigan (1937-   )
> *Clustering Algorithms*, J. Wiley & Sons, Inc., NY,
> 1975, p. 8.

REFERENCES

Albano, C., W. Dunn, U. Edlund, E. Johansson, B. Norden, M. Sjöström and S. Wold, "Four Levels of Pattern Recognition," *Analytica Chimica Acta, 103,* 1958, pp. 429-443.

Anderberg, Michael R., *Cluster Analysis for Applications,* Academic Press, NY, 1973.

Andrews, Harry C., *Intro. to Mathematical Techniques in Pattern Recognition,* J. Wiley & Sons, Inc., NY, 1972.

Bartlett, M.S., *The Statistical Analysis of Spatial Pattern,* Chapman and Hall, London, 1975.

Bryan, J.G., "The Generalized Discriminant Function: Mathematical Foundation and Computational Routine," *Harvard Educ. Rev., 21,* Spring 1951, pp. 90-95.

Cole, A.J. (ed.), *Numerical Taxonomy,* Academic Press, NY, 1969.

Cormack, R.M., "A Review of Classification," *Journal of the Royal Statistical Society, 34,* Part 3, 1971, pp. 321-353.

Cover, T.M. and P.E. Hart, "Nearest Neighbor Pattern Classification," *IEEE Trans. on Info. Theory, IT-13,* No. 1, Jan. 1967, pp. 21-27.

Duda, Richard O. and Peter E. Hart, *Pattern Classification and Scene Analysis,* J. Wiley & Sons, Inc., NY, 1973.

Everitt, Brian S., *Cluster Analysis,* Halsted Press, NY, 1964.

Fisher, R.A., "The Use of Multiple Measurements in Taxonomic Problems," *Ann. Eugenics, 7,* Part II, 1936, pp. 179-188; also in *Contributions to Mathematical Statistics,* J. Wiley & Sons, Inc., NY, 1950.

Fix, E. and J.L. Hodges, Jr., *Discriminatory Analysis, Nonparametric Discrimination: Consistency Properties,* USAF School of Aviation Medicine, Randolf Field, TX, Project 21-49-004, Rept. 11, Contract AF41(128)-31, Aug. 1952.

_____, *Discriminatory Analysis, Small Sample Performance,* USAF School of Aviation Medicine, Randolf Field, TX, Project 21-49-004, Rept. 11, Contract AF41(128)-31, Aug. 1952.

Fu, K.S., *Sequential Methods in Pattern Recognition and Machine Learning,* Academic Press, NY, 1968.

Fukunaga, Keinosuke, *Intro. to Statistical Pattern Recognition,* Academic Press, NY, 1972.

Gower, J.C., remarks on Dr. Cormack's paper, *J. of the Roy. Stat. Soc., 34,* Part 3, 1971, pp. 360-365.

Hartigan, J.A., *Clustering Algorithms,* J. Wiley & Sons, Inc., NY, 1975.

Hellman, Martin E., "The Nearest Neighbor Classification Rule With a Rejection Option," *IEEE Trans. on Systems Science and Cybernetics, SSC-6,* July 1970, pp. 179-185.

Kullback, S., *Information Theory and Statistics,* J. Wiley & Sons, Inc., NY, 1959.

Mendel, J.M. and K.S. Fu (eds.), *Adaptive, Learning, and Pattern Recognition Systems, Theory and Applications,* Academic Press, NY, 1970.

Patrick, Edward A., "Decision Rules Using Local Density Estimation," Secs. 4-3 through 4-13, *Fundamentals of Pattern Recognition*, Prentice-Hall, Inc., Englewood Cliffs, NJ, 1972, pp. 217-268.

_____, "Piecewise Linear Discriminant Functions," Sec. 4-16, *Fundamentals of Pattern Recognition*, Prentice-Hall, Inc., Englewood Cliffs, NJ, 1972, pp. 275-277.

Peterson, D.W. and R.L. Mattson, "A Method of Finding Linear Discriminant Functions for a Class of Performance Criteria," *IEEE Trans. Info. Theory*, *IT-12*, 380-387, July 1966.

Rosenblatt, F., *Principle of Neurodynamics: Perceptrons and the Theory of Brain Mechanism*, Spartan Books, NY, 1962.

Sebestyen, George S., *Decision-Making Processes in Pattern Recognition*, The Macmillan Co., NY, 1962.

Sklansky, Jack (ed.), *Pattern Recognition, Intro. and Foundations*, Dowden, Hutchinson & Ross, Inc., Stroudsburg, PA, 1973.

Sonquist, J.A. and J.N. Morgan, *The Detection of Interaction Effects*, Survey Research Center, Inst. for Soc. Res., Univ. of Mich., Ann Arbor, MI, 1964.

Tatsuoka, M.M. and D.V. Tiedman, "Discriminant Analysis," *Rev. Educ. Res.*, *24*, 1954, pp. 402-420.

Tou, Julius T. (ed.), *Advances in Information Systems Science*, Vol. 3, Plenum Press, NY, 1970.

Tou, Julius T. and Rafael C. Gonzalez, *Pattern Recognition Principles*, Addison-Wesley Pub. Co., Reading, MA, 1974.

Tryon, Robert C. and Daniel E. Bailey, *Cluster Analysis*, McGraw-Hill Book Co., NY, 1970.

Yacowar, Nathan, "Probability Forecasts Using Finely-Tuned Analogs," *Fourth Conf. on Probability and Statistics in Atmospheric Science*, Tallahassee, FL, Nov. 18-21, 1975, American Meteorological Society, Boston, MA, 1975, pp. 49-50.

Young, Tzay Y. and Thomas W. Calvert, *Classification, Estimation and Pattern Recognition*, American Elsevier Pub. Co., NY, 1974.

CHAPTER 18

NON-INFORMATION-THEORETIC MULTIVARIATE ANALYSIS METHODS

CHAPTER 18

NON-INFORMATION-THEORETIC MULTIVARIATE
ANALYSIS METHODS

A. Overview

In the previous chapter, we saw that clustering techniques typically give the researcher a great deal of flexibility in fitting
a description to data. This flexibility comes in selection of
features, cluster shapes, and weighting coefficients for features
and other parameters. In this chapter, we turn to another set of
non-information-theoretic analysis methods which give the researcher an equivalent amount of freedom in describing data. These are
the multivariate methods which provide a direct mathematical estimate of the dependent variable as a function of the independent
variables, where parameters in the mathematical model are adjusted
to fit the given data.

Included among multivariate analysis methods are the following:

  o Linear regression

  o Principal components analysis

  o Discriminant analysis

  o Factor analysis procedures

  o Canonical correlation

  o Multiple partial correlation

  o Analysis of variance

  o Multiple covariance analysis

A number of publications are available describing these and other multivariate methods.* Here we will briefly describe three of the more frequently used methods: linear regression, principle components, and discriminant analysis.

B. Linear Regression

The bulk of multivariate data analyses presently conducted use linear regression, introduced by Galton in 1885. Even many modern systems analytic simulation models use linear regression to project the individual components upon which the models operate to produce their output projections.

A dependent variable Y is assumed to be approximately linearly related to a set of m features (predictors) $\{X_i\}$, i=1,...,m, formed from the observed independent variables.

Assume that we have defined the $\{X_{ij}\}$ and $\{Y_j\}$, j=1,...,n, for n data points, to have zero mean and unity standard deviation on the data. This is easily accomplished for any initial set of variables $\{X'_{ij}\}$, $\{Y'_j\}$ by the linear transformations:

$$X_{ij} = \frac{X'_{ij} - \overline{X'_i}}{s'_i}, \text{ and}$$

$$Y_j = \frac{Y'_j - \overline{Y'}}{s'},$$

*   See, for example:
  - C.R. Rao, *Linear Statistical Inference and Its Applications*, J. Wiley & Sons, Inc., NY, 1965.
  - W.W. Cooley and P.R. Lohnes, *Multivariate Data Analysis*, J. Wiley & Sons, Inc., NY, 1971.
  - G.E.P. Box and G.C. Tiao, *Bayesian Inference in Statistical Analysis*, Addison-Wesley Pub. Co., Reading, MA, 1973.
  - Y.M.M. Bishop, S.E. Fienberg and P.W. Holland, *Discrete Multivariate Analysis, Theory and Practice*, The MIT Press, Cambridge, MA, 1975.
  - W.R. Atchley and E.H. Bryant (eds.), *Multivariate Statistical Methods* (2 vols.), Dowden, Hutchinson and Ross, Inc., Stroudsburg, PA, 1975.
  - A.L. Edwards, *An Introduction to Linear Regression and Correlation*, W.H. Freeman and Co., San Francisco, CA, 1976.
  - J. Bibby and H. Toutenburg, *Prediction and Improved Estimation in Linear Models*, J. Wiley & Sons, Inc., NY, 1977.

where

$$\overline{X'_i} = \frac{1}{n} \sum_{j=1}^{n} X'_{ij} \, ,$$

$$s_i'^2 = \frac{1}{n} \sum_{j=1}^{n} (X'_{ij} - \overline{X'_i})^2 \, ,$$

$$\overline{Y'} = \frac{1}{n} \sum_{j=1}^{n} Y'_j \, , \quad \text{and}$$

$$s'^2 = \frac{1}{n} \sum_{j=1}^{n} (Y'_j - \overline{Y'})^2 \, .$$

Thus

$$\frac{1}{n} \sum_{j=1}^{n} X_{ij} = \frac{1}{n} \sum_{j=1}^{n} Y_j = 0$$

and

$$\frac{1}{n} \sum_{j=1}^{n} X^2_{ij} = \frac{1}{n} \sum_{j=1}^{n} Y^2_j = 1 \, .$$

These transformations are harmless since, taken together with $\{X'_i\}$, $\{s'_i\}$, $\overline{Y'}$ and s', they retain all the information in the original variables. They are useful because they simplify equations in the analysis.

The linear regression model expresses Y in terms of the $\{X_i\}$ by the equation

$$Y = \sum_{i=1}^{m} a_i X_i + \varepsilon \, ,$$

where $\sum_i a_i X_i$ is the regression function and ε is the error. The basis of the model is selecting a set of values for the regression coefficients $\{a_i\}$ which minimizes the squared error for a set of training events for which Y and the $\{X_i\}$ are known:

$$Y_j = \sum_{i=1}^{m} a_i X_{ij} + \varepsilon_j, \quad j=1,\ldots,n.$$

Select $\{a_i\}$ to minimize the squared error

$$\varepsilon^2 = \sum_j \varepsilon_j^2 .$$

Using the regression function we have

$$\varepsilon^2 = \sum_{j=1}^{n} (Y_j - \sum_{i-1}^{m} a_i X_{ij})^2 .$$

To find the set $\{a_i\}$ which minimizes ε^2, we perform variations with respect to the a_i's and set the result equal to zero. This gives

$$\sum_{k=1}^{m} \left[\sum_{i=1}^{m} a_i \sum_{j=1}^{n} X_{ij} X_{kj} - \sum_{j=1}^{n} Y_j X_{kj} \right] \delta a_k = 0$$

Since these variations are independent, we can set the coefficients of the δa_k individually to zero, thereby separating the expression into m equations for $k = 1, 2, \ldots, m$.

$$\sum_{i=1}^{m} a_i \sum_{j=1}^{n} X_{ij} X_{kj} - \sum_{j=1}^{n} Y_j X_{kj} = 0.$$

Define the elements

$$R_{ik} = \sum_{j=1}^{n} X_{ij} X_{kj}, \quad k=1,\ldots,m, \text{ and}$$

$$Z_k = \sum_{j=1}^{n} Y_j X_{kj}, \quad k=1,\ldots,m.$$

The matrix $\{R_{ik}\}$ contains all the intercorrelations among the predictors on the given data. The vector $\{Z_k\}$ contains all the correlations between the predictors and the dependent variable.

We can now rewrite the m equations for the regression coefficients in the form

$$\sum_{i=1}^{m} a_i R_{ik} = Z_k, \quad k=1,\ldots,m.$$

Since the R_{ik} and Z_k are known from the data, this is simply a set of m linear equations which can be solved for the m unknowns $\{a_i\}$ by standard techniques.

Now, consider the regression function for the jth data point, defined as

$$Y_{Rj} = \sum_{i=1}^{m} a_i X_{ij}.$$

Its variance over the full data set is given by

$$\sigma_R^2 = \frac{1}{n} \sum_{j=1}^{n} (Y_{Rj} - \overline{Y_R})^2.$$

After a bit of algebra, this reduces to

$$\sigma_R^2 = \sum_{k=1}^{m} a_k Z_k.$$

If we also include the error term ε_j, the full variance of the dependent variable on this data is given by

$$\sigma^2 = \frac{1}{n} \sum_{j=1}^{n} (Y_j - \overline{Y})^2.$$

Since our variables have zero mean and unity standard deviation

$$1 = \frac{1}{n} \sum_{j=1}^{n} (Y_{Rj} + \varepsilon_j)^2$$

$$= \sigma_R^2 + \frac{1}{n} \sum_{j=1}^{n} \varepsilon_j^2.$$

The variance σ_R^2 of the regression function on the given data is referred to as the fraction of the total variance σ^2 on this data which is "explained" or "accounted for" by the regression coefficients $\{a_i\}$. This is, unfortunately, loaded terminology, and the expression "modeled on the given data" may be suggested as being more accurate.

An example of an application of multiple regression is its use in fitting polynomials in x-y space to empirical data. Define the features as various powers of an independent variable $\{X, X^2, X^3, \ldots, \}$. Then the regression equation is given as

$$Y = \sum_{i=1}^{m} a_i X^i + \varepsilon.$$

This is fit to a set of data points $\{X_j, Y_j\}$

$$Y_j = \sum_{i=1}^{m} a_i X_j^i + \varepsilon_j, \quad j=1,\ldots,n.$$

by defining

$$R_{ik} = \sum_{j=1}^{n} X_j^i X_j^k, \quad \text{and}$$

$$Z_k = \sum_{j=1}^{n} Y_j X_j^k \, ,$$

and obtaining the coefficients $\{a_i\}$ by solving the set of linear equations

$$\sum_{i=1}^{m} a_i R_{ik} = Z_k \, , \quad k=1,\ldots,m.$$

The fundamental difficulty with linear regression is not its linearity, for one can define the individual features so as to absorb nonlinearities. It is, rather, its lack of a means of controlling for chance correlations considering its large number of parameters. When the number m of predictors $\{X_{ij}\}$ is not small compared to the number n of data points, regression becomes unreliable because of the increased likelihood of the regression function simply representing chance correlations in the given data. This has been recognized by many researchers.* It is less well recognized that even when m<<n, the minimum squared error criterion in itself provides no tendency to deemphasize predictors which have been contrived to fit training data but are unrepresentative of the general population. On the contrary, regression analysis tends to *emphasize* such features by assigning them high coefficients.

C. Principal Components Analysis

Principal components analysis is a form of linear regression in which the features $\{X_{ij}\}$ are transformed to new variables $\{Z_{ij}\}$ (called "principal components"). The original features $\{X_{ij}\}$ are, in general, correlated on the data. That is, for $i \neq k$,

$$\sum_{j=1}^{n} X_{ij} X_{kj} \neq 0.$$

---

* William W. Cooley and Paul R. Lohnes, *Multivariate Data Analysis*, J. Wiley & Sons, NY, 1971, pp. 53-57.

The principal components $\{Z_{ij}\}$, on the other hand, are defined so as to be uncorrelated on the data. That is,

$$\sum_{j=1}^{n} Z_{ij}Z_{kj} = \delta_{ik} = \begin{cases} 1 & \text{if } i = k \\ 0 & \text{if } i \neq k. \end{cases}$$

The advantage of transformation to principal components is that it reduces the number of nonzero elements of the R-matrix requiring interpretation from the full m^2 elements to only m elements on the diagonal. Another important advantage is that it provides a natural ranking of the components in terms of the fraction of the variance in the given data which is modeled. The disadvantage is that it increases the complexity of each of the m elements. This compli- cates the task of interpretation and runs the risk of intrafeature contrivedness.

The principal components are related to the features by a linear transformation

$$Z_{ij} = \sum_{\ell=1}^{m} V_{i\ell}X_{\ell j} .$$

Substituting this into the equation specifying that Z is uncorre- lated on the data, we obtain the following constraint on the trans- formation elements $\{V_{i\ell}\}$:

$$\sum_{j=1}^{n} \sum_{\ell=1}^{m} V_{i\ell}X_{\ell j} \sum_{\ell'=1}^{m} V_{k\ell'}X_{\ell'j} = \delta_{ik}.$$

If we define

$$R_{\ell\ell'} = \sum_{j=1}^{n} X_{\ell j} X_{\ell' j},$$

this reduces to

$$\sum_{\ell=1}^{m} \sum_{\ell'=1}^{m} V_{i\ell} R_{\ell\ell'} V_{k\ell'} = \delta_{ik}.$$

This, however, does not completely define the transformation elements. We complete the definition by specifying that V be normalized to unity sum of squares:

$$\sum_{\ell=1}^{m} V_{i\ell} V_{k\ell} = \delta_{ik}.$$

This gives us an eigenequation

$$(\underset{\sim}{R} - \lambda \underset{\sim}{I}) \underset{\sim}{V} = 0,$$

the solution of which is the first principal component.

The entire process is repeated successively in the reduced space to extract additional principal components.

D. Discriminant Analysis

Suppose we have two outcome class populations A and B displayed in a two-dimensional feature space whose axes are F_1 and F_2. See Fig. 106. Consider a related axis I. If we project the two populations A and B onto this axis, we obtain two different frequency distributions. If these two are clearly separated, then we can define line II midway between them and perpendicular to I, which discriminates between them.

Discriminant analysis formalizes this procedure using a least squared error measure separation. The mathematical formulation is written for feature spaces of arbitrary dimensionality. However, the basic principles can be most easily understood by visualizing the process in two dimensions.

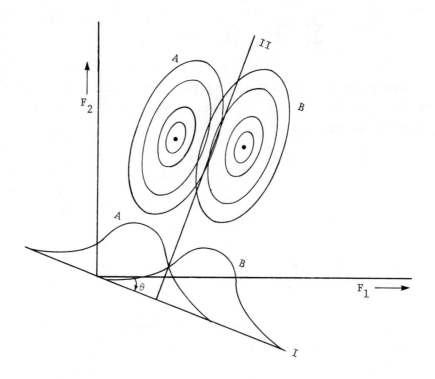

Fig. 106. Linear Function for Two-Class Discrimination

Let

$$X_{\sim ki} = \text{position of } i^{th} \text{ event in outcome class k,}$$

$$q_{\sim k} = \frac{1}{n_k} \sum_i X_{\sim ki} = \text{centroid for events of outcome class k,}$$

$$q = \frac{1}{n} \sum_{i,k} X_{ki} = \text{grand centroid for all events, and}$$

$$x_{ki} = X_{ki} - q = \begin{array}{l} \text{position, relative to grand centroid,} \\ \text{of the } i^{th} \text{ event in } k^{th} \text{ class.} \end{array}$$

Now we define the following quantities:

$$T = \sum_{k=1}^{K} \sum_{i=1}^{n_k} X_{ki} \, X_{ki}^T = \text{Total matrix,}$$

$$A = \sum_{K=1}^{K} n_k (q_k - q)(q_k - q)^T = \text{Among-groups matrix, and}$$

$$W = \sum_{K=1}^{K} \sum_{i=1}^{n_k} (X_{ki} - q_k)(X_{ki} - q_n)^T = \text{Within-groups matrix.}$$

The fundamental partition theorem of multivariate analysis of variance (MANOVA) states that

$$T = A + W.$$

We wish to find a "best" discriminatory function II. Call it y. It will be a function of our axes.

$$y = v^T x.$$

Our selection for a "best" function is that specified by a vector v for which the ratio of the among-groups score, $v^T A v$, to the within-groups score, $v^T W v$, is a maximum

$$\lambda = \max_{\{v\}} \frac{v^T A v}{v^T W v},$$

subject to the restriction

$$v^T v = 1.$$

The maximum λ is given by the eigenvector $\underset{\sim}{v}$ with maximum eigenvalue for the eigenequation

$$(\underset{\sim}{W}^{-1}\underset{\sim}{A} - \lambda\underset{\sim}{I})\underset{\sim}{v} = 0.$$

This gives us the first discriminant vector $\underset{\sim}{v}_1$ with associated eigenvalue λ_1.

We now proceed successively to extract further discriminant vectors, $\underset{\sim}{v}_i$, requiring each to be orthogonal to all the previous ones $\underset{\sim}{v}_1, \ldots, \underset{\sim}{v}_{i-1}$. The extraction continues until a zero eigenvalue of $\underset{\sim}{W}^{-1}\underset{\sim}{A}$ is encountered (or $\underset{\sim}{W}^{-1}\underset{\sim}{A}$ is fully exhausted).

The resultant set of discriminant vectors $\{\underset{\sim}{v}_i\}$ forms a test battery. The Wilks' lambda criterion for the discriminatory power of this test battery is given by

$$\Lambda = \prod_{j=1}^{n} \frac{1}{1+\lambda_j}.$$

A test-statistic used to define the significance of the discriminatory power of the remaining n-k functions after acceptance of the first k is

$$T_D = -(n - \frac{p+g}{2} - 1)\, \log \Lambda',$$

where

$$ndf = (p-k)(g-h-1)\ ,\ \text{and}$$

$$\Lambda' = \prod_{j=k+1}^{n} \frac{1}{1+\lambda_j}.$$

T_D is asymptotically chi-squared distributed with $\nu = n-k$ degrees of freedom.

QUOTES

GALTON (1885)

When Mid-Parents are taller than mediocrity, their Children tend to be shorter than they. When Mid-Parents are shorter than mediocrity, their Children tend to be taller than they.

>—Francis Galton (1822-1911)
>"Rate of Regression in Hereditary Stature," presented to the British Association for the Advancement of Science, Aberdeen, Scotland, Sept. 10, 1885.

BOX (1953)

It has frequently been suggested that a test of homogeneity of variances should be applied before making an analysis of variance test for homogeneity of means in which homogeneity of variance is assumed...when, as is usual, little is known of the parent distribution, this practice may well lead to more wrong conclusions than if the preliminary test was omitted...To make the preliminary test on variances is rather like putting to sea in a rowing boat to find out whether conditions are sufficiently calm for an ocean liner to leave port!

>—George E.P. Box (1919-   )
>"Non-Normality and Tests on Variance," *Biometrika*, *40*, 1953, p. 333.

KENDALL (1957)

...when there are collinearities in the independent variables...no reliance whatever can be put on individual coefficients in regression equations embodying all the variables.

>—Maurice George Kendall (1907-   )
>*A Course in Multivariate Analysis*, Charles Griffin & Co., London, 1957, p. 74.

NUNNALLY (1967)

The effort of science is to find a relatively small set of variables which will suffice to "explain" all other variables. A small set of variables "explains" a larger set if some combination of the smaller set correlates highly with each member of the larger set...To achieve such a small set of "explainer" variables is the essence of scientific parsimony.

>—Jum Clarence Nunnally (1924-   )
>*Psychometric Theory*, McGraw-Hill, NY, 1967, p. 151, quoted in Cooley and Lohnes, *Multivariate Data Analysis*, John Wiley & Sons, NY, 1971, p. 325.

FORSYTHE, MAY and ENGLEMAN (1971)

...For example, when there are 43 cases and 80 potential predictors...entry of all the available variables would 'explain' all the variance in the sample, but not in a meaningful way...

It is important to recognize that, whether or not the selection and reduction is done by the investigator's subjective judgment (which is, in effect, a covert equivalent of step-wise regression) or by an overt mathematical procedure, the situation poses a problem somewhat analogous to multiple t-test comparisons among means - the more variables there are to choose from, the more likely it is that any one or more of them may be erroneously declared to be (relevant) significant...

Apparently it is not sufficiently appreciated that prediction equations derived *from one particular sample* by entering a relatively large number of variables tend to perform poorly on cross-validation.

> —Alan B. Forsythe (1940-   ),
> Philip R.A. May (1920-   ) and
> Laszlo Engleman (1936-   )
> "Prediction by Multiple Regression, How Many Variables to Enter?", *J. of Psychiatric Research, 8,* 1971, pp. 119-120.

COOLEY and LOHNES (1971)

Our tendency to deemphasize the b weights stems from experience with the phenomenon of extreme fluctuation of regression weights from sample to sample when the sample size is small. Even when the sample size is moderate there is substantial fluctuation.

* * * *

We simply must plan for large samples (and good ones) if we are going to use any of the function-fitting methods and expect the functions to be taken seriously as generalizations to populations.

> —William W. Cooley (1930-   ) and
> Paul R. Lohnes (1928-   )
> *Multivariate Data Analysis,* J. Wiley & Sons, Inc. NY, 1971, pp. 55, 56.

FEINSTEIN (1972)

...regressive procedures often perform calculations of uncertain validity to provide co-efficients of uncertain meaning for numbers of unmeasurable dimension.

> —Alvan R. Feinstein (1925-   )
> "The Purposes of Prognostic Stratification," *Clin. Pharmacol. Ther., 13* (1972), pp. 285-297, reproduced in *Clinical Biostatistics,* The C.V. Mosby Co.. St. Louis, 1977, p. 393.

TOPLISS and COSTELLO (1972)

...a point which appears to be generally overlooked is the influence of the total number of variables screened for possible correlation with activity on the statistical significance of the obtained correlation...Clearly, however, the greater the number of variables tested, the greater role chance will play in the observed correlation.

*   *   *   *

For example, if $r^2 = 0.40$ is regarded as the maximum acceptable level of chance correlation then the minimum number of observations required to test five variables is about 30, for 10 variables 50 observations, for 20 variables 65 observations, and for 30 variables 85 observations.

> —John G. Topliss (1930-   ) and
> Robert J. Costello (1939-   )
> "Chance Correlations in Structure-Activity Studies
> Using Multiple Regression Analysis," *J. of Medicinal
> Chemistry, 15,* 1972, pp. 1066, 1068.

COX (1972)

Suppose now that on each individual one or more further measurements are available, say on variables $z_1, \ldots, z_p$...The z's may be functions of time. The main problem considered in this paper is that of assessing the relation between the distribution of failure time and $\underset{\sim}{z}$. This will be done in terms of a model in which the hazard is

$$(t;\underset{\sim}{z}) = \exp(\underset{\sim}{z}\underset{\sim}{\beta})\ \lambda_0(t),$$

where β is a $p \times 1$ vector of unknown parameters and $\lambda_0(t)$ is an unknown function giving the hazard function for the standard set of conditions $\underset{\sim}{z} = \underset{\sim}{0}$.

> —David R. Cox (1923-   )
> "Regression Models and Life-Tables," *J. of the
> Royal Statist. Soc., B34,* 1972, p. 189.

FORSYTHE, ENGLEMAN, JENNRICH and MAY (1973)

The usual *F* tables are of little direct use in deciding whether the "*F* to enter" of the selected variable at a step is relevant (significant) since the variable selected for inclusion is the one with the *largest "F to enter"* of all those variables not yet selected. The more variables there are to choose from, the larger the maximum "*F* to enter" we expect to observe.

> —A. Forsythe (1940-   ), L. Engleman (1936-   ),
> R. Jennrich (1932-   ) and P. May (1920-   )
> "A Stopping Rule for Variable Selection in Multiple
> Regression," *J. of the Am. Statist. Assoc., 68,*
> March 1973, p. 75.

REFERENCES

Anderson, T.W., *An Intro. to Multivariate Statistical Analysis*, J. Wiley & Sons, Inc., NY, 1958.

Bartlett, M.S., "On the Theory of Statistical Regression," *Proc. Roy. Soc. Edinburgh*, *53*, 1933, pp. 260-283.

Bibby, John and Helge Toutenburg, *Prediction and Improved Estimation in Linear Models*, J. Wiley & Sons, Inc., NY, 1977.

Box, G.E.P., "Non-Normality and Tests on Variance," *Biometrika*, *40*, 1953, pp. 318-335.

Cooley, W.W. and P.R. Lohnes, "Multivariate Data Analysis," J. Wiley & Sons, Inc., NY, 1971.

Cox, D.R., "Regression Models and Life-Tables," *J. of the Royal Statist. Soc.*, *B34*, 1972, pp. 187-202.

Feinstein, Alvan R., *Clinical Biostatistics*, The C.V. Mosby Co., St. Louis, MO, 1977.

Fisher, Ronald A., "The General Sampling Distribution of the Multiple Correlation Coefficient," *Proc. of the Royal Society*, Ser. A, *121*, 1928, pp. 654-673.

_____, "The Use of Multiple Measurements in Taxonomic Problems," *Ann. Eugenics*, *7*, 1936, pp. 179-189.

Forsythe, A.B., P.R.A. May and L. Englemen, "Prediction by Multiple Regression, How Many Variables to Enter?" *J. Psychiat. Res.*, *8*, 1971, pp. 119-126.

Forsythe, A.B., L. Engleman, R. Jennrich, and P.R.A. May, "A Stopping Rule for Variable Selection in Multiple Regression," *J.Am. Stat. Assoc.*, *68*, 1973, pp. 75-77.

Galton, Francis, "Rate of Regression in Hereditary Stature," presented to the British Assoc. for the Advancement of Science, Aberdeen, Scotland, Sept. 10, 1885.

Hotelling, Harold, "The Generalization of Student's Ratio," *Ann. Math. Stat.*, *2*, (3), 1931, pp. 360-378.

_____, "Analysis of a Complex of Statistical Variables into Principal Components," *Jour. of Educational Psychology*, *24*, 1933, pp. 417-441, 498-520.

Kendall, M.G., *A Course in Multivariate Analysis*, Charles Griffin & Co., London, 1957.

Krishnaiah, P.R., "Simultaneous Test Procedures under General MANOVA Models," *Multivariate Analysis, Vol. II*, ed. by P.R. Krishnaiah, Academic Press, NY, 1964, pp. 121-143.

Lachenbruch, P.A. and R.M. Mickey, "Estimation of Error Rates in Discriminant Analysis," *Technometrics*, *10*, 1968, pp. 1-11.

Leontiades, K., "Computationally Practical Entropy Minimax Rotations, Applications to the Iris Data and Comparison to Other Methods," Bio-Medical Engr. Program, Carnegie-Mellon Univ., Pittsburgh, PA, 1976.

Morgan, J.N. and J.A. Sonquist, "Problems in the Analysis of Survey Data," *J. Am. Statist. Assoc.*, *58*, 1963, pp. 415-435.

Nunnally, J.C., *Psychometric Theory*, McGraw-Hill, NY, 1967.

Rao, C. Radhakrishna, *Linear Statistical Inference and Its Applications*, J. Wiley & Sons, Inc., NY, 1965.

Reichert, T., "Patterns of Overuse of Health Care Facilities—A Comparison of Methods," *Proc. of the 1973 Intl. Conf. on Cybernetics and Society*, 73 CHO 799-7-SMC, IEEE Systems, Man and Cybernetics Society, Nov. 5-7, 1973, Boston, MA, pp. 138-329.

Schonbein, William R., "Analysis of Decisions and Information in Patient Management," *Current Concepts in Radiology*, *2*, The C.V. Mosby Co., St. Louis, MO, 1975, pp. 31-58.

Topliss, J. and R. Costello, "Chance Correlations in Structure-Activity Studies Using Multiple Regression Analysis," *J. Med. Chem.*, *15*, 1972, pp. 1066-1068.

Tryon, Robert C. and Daniel E. Bailey, *Cluster Analysis*, McGraw-Hill Book Co., NY, 1970.

Wilks, S.S., "Certain Generalizations in the Analysis of Variance," *Biometrika*, *24*, 1932, pp. 471-474.

CHAPTER 19

CHANCE CORRELATIONS AND CONTRIVEDNESS

CHAPTER 19

CHANCE CORRELATIONS AND CONTRIVEDNESS

A.   Contrivedness in Fitting Training Data

One of the more remarkable endeavors in the search for patterns in nature was Eddington's *Fundamental Theory*. He began by assuming three constants as fundamental: the velocity of light, the Rydberg constant, and the Faraday constant. He then proceeded to construct a theory which gave numerical values to Planck's constant, the electron mass, the proton mass, the fine-structure constant, and a number of other physical quantities in very close agreement to measurement data available at the time...suspiciously close:

> [One possible explanation] that would occur to
> many physicists is that Eddington's theory is
> artificial throughout, and that by skillful
> juggling with numbers he has produced a forced
> agreement.

> —Harold Jeffreys
> *Theory of Probability*, Oxford Univ.
> Press, London, 1961, p. 311.

In Chapter 13, we saw how an entropy minimizing computer algorithm can be given considerable flexibility to fit training data. Suppose we allowed the pattern formation algorithm to construct cells of arbitrary shape.

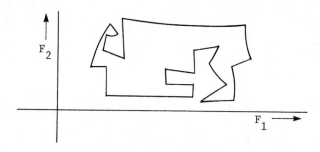

Fig. 107.   Complex Pattern in Feature Space

The algorithm could then select a cell with a very intricate boundary, such as the one shown in Fig. 107. The algorithm could fit irregular data arbitrarily well if we permitted it this much freedom. It might be objected that it requires a complex description to specify such a pattern in a language of two features and linear segments. The extra complexity of the nooks and crannies enables us to extract more information about the dependent variable out of the independent variables. However, this is at the cost of spending additional independent variable information specifying the complexities.

We could try to disguise an obviously contrived partition by rede-fining the features and using a very "simple" partition in the new space. This happens whenever we make a transformation which happens to sort the training data very well. What we have done is simply shift the complexity from the patterns relating the features to the definitions of the features themselves.

As we shall see below, there is a limit to our ability to gain addition-al information with increasingly complex boundaries. Beyond that point we actually lose information. We refer to excessively com-plex boundaries as *contrived*.

Contrivedness appears in two forms:

Among-features contrivedness

> This is the most commonly recognized form. It arises when complex combinations of many features are used to match a relatively limited number of training data points. What are matched may be chance correlations which just happen to have arisen in the particular sample, but are unrepresentative of the population as a whole. Here the contrivedness arises from among-features complexity of the structure of the model.

Within-features contrivedness

> Another form of contrivedness occurs when individual features have complex definitions which just happen to allow the par-ticular training sample to be matched with little error using few features, but these features would not provide a good match for the population as a whole. Here the contrivedness arises from the complexity of individual feature definitions.

Figs. 108 and 109 illustrate these two types of contrivedness. Among-features contrivedness can be further subdivided as contrivedness in the number of features used to form the patterns, and contrivedness in the intricacy of the logic interlinking them.

The symptom of both among-features and within-features contrivedness is a good fit to the training sample but a poor fit to the general population.

Our extensive array of feature extraction tools, including formation of linear and nonlinear combinations, Fourier and K-L analyses, gives us the ability to form patterns in feature space which, were they transformed back to the original independent variables space, would be seen to have very odd shapes. Upon closer inspection we would notice that these odd shapes happen to be arranged with nooks and crannies devised to do a good job separating the training data by outcome class.

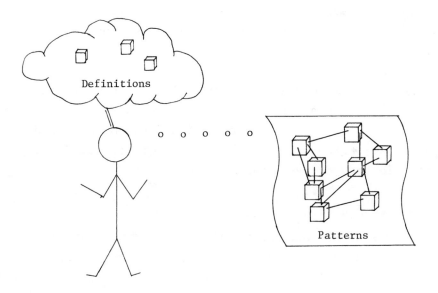

Fig. 108.   Among-Features Contrivedness

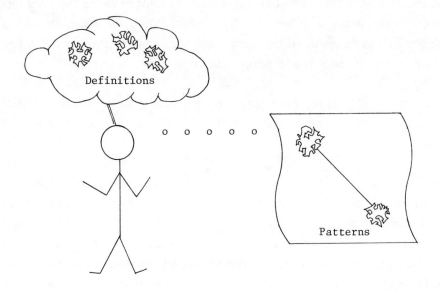

Fig. 109. Within-Features Contrivedness

We might be suspicious that these little oddities in the shapes did not just happen by accident; but, rather, they were purposely put there specifically to fit the training data. We would, of course, be right.

We have caught our pattern seeker committing the sin of contrivedness. It has used too many degrees of freedom in formulating its model of the training data. With that much freedom, it could have fit random data. Why should we believe such a model has the slightest bit of relevancy to new events?

B. Pattern Specification Entropy

In exploring possible means of avoiding contrivedness, we might try introducing an explicit bias into our analysis toward patterns which can be specified with less information. As an example, assume that all cells have a common fixed shape (e.g., m-dimensional rectangles).

Then we can introduce such a bias by minimizing the joint entropy, S(O and C), rather than the conditional entropy, S(O given C). These two are related by

$$S(O \text{ and } C) = S(O \text{ given } C) + S(C),$$

where

$$S(C) = -\sum_{i=1}^{L} P(C_i) \log P(C_i)$$

is the variable portion of the pattern specification entropy for this case. L is the number of feature space cells, and $P(C_i)$ is the probability of falling in the ith cell. In cases where each cell is equally probable, S(C) is simply

$$S(C) = \log L.$$

This has the desired biasing effect since it puts a premium on "simpler" partitions, in this case those with smaller numbers of cells.

However, in assessing the possible utility of such a procedure, we need to ask a question: Why should we not be willing to pay any price, however high, in pattern specification information in order to gain additional information about the dependent variable?

Our ultimate goal is not to maximize information about the dependent variable, given the available data, at minimum specification cost. It is, rather, purely to maximize information about the dependent variable given the available data. Only if reducing specification detail is relevant to increasing this latter information is such reduction important to our goal. It has no independent significance to our analysis. Thus, for example, we are not necessarily willing to trade an increase in S(O given C) for a larger decrease in S(C). We will take greater than the minimum S(O given C) only when this improves our ability to predict the dependent variable. To determine when such circumstances arise, we turn to the topic of chance correlations.

C.   Entropy of Chance Correlations

Consider a situation in which we have a sample of data comprising several independent variables plus one dependent variable.   See Table 40.

Table 40.   Sample Data in Original Order

	Independent Variables				Dependent Variable	
Event #	IV1	IV2	...	IVn	Event #	DV
1	8.6	16.4		8.5	1	4
2	5.3	3.2		-1.2	2	3
3	4.1	-5.8		-4.3	3	2
4	6.2	4.1		6.4	4	1
5	5.3	16.3		9.8	5	1
'	'	'		'	'	'
'	'	'		'	'	'
'	'	'		'	'	'

Suppose that we compute the entropy of each feature space partition Π_j in a set of partitions $\Pi_\nu = \{\Pi_j\}_\nu$ defined with ν degrees of freedom.   The lowest such entropy we will call S_{min}.

Now suppose that we perform a number of random shuffles of the ordering of the events in the dependent variable data column.   For each permutation i, we find the feature space partition with lowest entropy $S_{min}(i)$.   See Table 41.

Table 41.   Sample Data with Dependent
Variable in ith Shuffled Order

	Independent Variables				Shuffled Dependent Variable	
Event #	IV1	IV2	...	IVn	Event #	DV
1	8.6	16.4		8.5	1	2
2	5.3	3.2		-1.2	2	1
3	4.1	-5.8		-4.3	3	4
4	6.2	4.1		6.4	4	1
5	5.3	16.3		9.8	5	3
'	'	'		'	'	'
'	'	'		'	'	'
'	'	'		'	'	'

Averaging these results, we compute the expected value $S^c(\nu)$ of the minimum entropy on a random permutation of the dependent variables:

$$S^c(\nu) = \sum_i S_{min}(i) \; Prob(i^{th} \text{ permutation}).$$

This quantity is called the entropy of chance correlations associated with the ν degrees of freedom.*

Suppose that, when using ν degrees of freedom in forming patterns, we obtain a minimum entropy of S_{min}. If

$$S_{min} \approx S^c(\nu) \; ,$$

we have not found any better a set of patterns in the actual data than we could have expected to find in random data with the same statistical properties but a randomized outcome.

Suppose, however, that

$$S_{min} \ll S^c(\nu) \; .$$

In this case we have, indeed, done significantly better than chance.

Our objective, therefore, is to find a partition Π_j for which the difference $S_j - S^c(\nu)$ is a (generally negative) minimum, where $S^c(\nu)$ is the chance correlation entropy for the number of degrees of freedom associated with the partition Π_j. The negative of this difference, $R_j(\nu) = S^c(\nu) - S_j$, is a measure of the extent of reality of the phenomena to us as viewed using the partition Π_j with ν degrees of freedom. The unreality $U_j = -R_j$ is the entropy exchange between randomness and the data (where the conceptual vacuum "randomness" is, of course, relative to our information).

Fig. 110 shows four general shapes for the dependence of the chance correlation entropy $S^c(\nu)$ upon the number of degrees of freedom ν.

---

* R.A. Christensen, "Entropy Minimax, A Non-Bayesian Approach to Probability Estimation From Empirical Data," *Proc. of the 1973 Intl. Conf. on Cybernetics and Society*, IEEE Systems, Man and Cybernetics Society, Nov. 5-7, 1973, Boston, MA, 73 CHO 799-7-SMC, p. 323 (*Volume III*, p. 113).

In each case, the curve starts out at the whole unpartitioned space entropy (zero degrees of freedom), given by

$$S^c(\nu) = -\sum_k P(O_k) \log P(O_k).$$

As $\nu \to \infty$, $S^c(\nu)$ approaches a limit in which each observed data point can be isolated into a separate cell in the partition.*

Fig. 110(d) shows the case in which $S^c(\nu)$ reaches its minimum at $\nu = 0$, while in 110(a) it approaches its minimum at $\nu \to \infty$. Figs. 110(b) and (c) are intermediate cases in which the minimum is reached at finite ν.

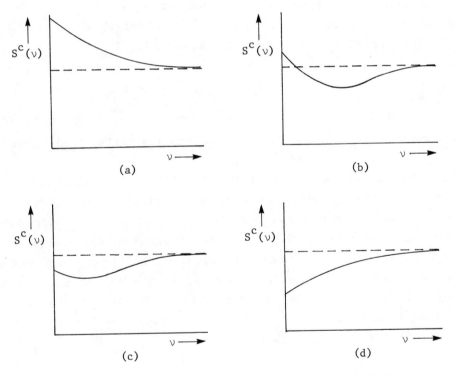

(a)

(b)

(c)

(d)

Fig. 110.   General Shapes of Chance Correlation Entropy
As Function of Number of Degrees of Freedom

---

* In special cases, data points may happen to line up in certain ways so that they cannot be separated even with infinite ν.

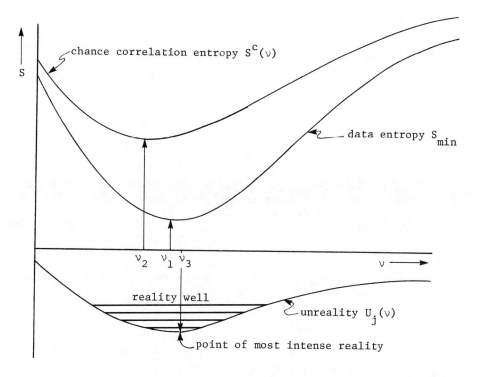

Fig. 111. Discounting Data Entropy S_{min} by Chance
Correlation Entropy $S^c(\nu)$

Fig. 111 illustrates the effect of discounting for chance correlation
entropy. The undiscounted entropy is a minimum S_{min} for a pattern set
with ν_1 degrees of freedom. This is the pattern set we would choose
if we were not considering $S^c(\nu)$. However, if we base our predictions on this pattern set, we run the risk of their being unreliable due to contrivedness.

The chance correlation curve $S^c(\nu)$ has a minimum at ν_2. Using this
to discount S_{min}, we arrive at a chosen pattern set based on ν_3 degrees
of freedom. Whether ν_1 is greater or less than ν_2, ν_3 will be farther from ν_2 than ν_1. The dip in the $U_j(\nu) = S_j - S^c(\nu)$ curve forms
a "reality well", the bottom of which is the point at which the reality
of the phenomenon is the most intense to the observer.

QUOTES

POINCARÉ (1902)

Without generalisation, prediction is impossible. The circumstances under which one has operated will never again be reproduced simultaneously. The fact observed will never be repeated.
All that can be affirmed is that under analogous circumstances
an analogous fact will be produced.

*   *   *   *

It is clear that any fact can be generalised in an infinite number of ways, and it is a question of choice. The choice can
only be guided by considerations of simplicity. Let us take
the most ordinary case, that of interpolation. We draw a continuous line as regularly as possible between the points given
by observation. Why do we avoid angular points and inflexions
that are too sharp? Why do we not make our curve describe the
most capricious zigzags? It is because we know beforehand, or
think we know, that the law we have to express cannot be so complicated as all that.

—Henri Poincaré (1854-1912)
Science and Hypothesis (1902), tr. by W.J.G.,
Dover Pubs., Inc., NY, 1952, pp. 142, 146.

YULE (1926)

If we took samples of...observations at random from a record in
which the correlation for the entire aggregate was zero, would
there be any appreciable chance of our getting such a correlation...merely by the chances of sampling?...

...to imply that the observed correlation *is* only a fluctuation
of sampling, whatever the ordinary formula for the standard
error may seem to imply: we are arguing that the result given
by the ordinary formula is not merely wrong, but very badly
wrong.

*   *   *   *

...some time-series are conjunct series with conjunct differences, and...when we take samples from two such series the distribution of correlations between them is U-shaped—we tend to
get high positive or high negative correlations between the
samples, without any regard to the true value of the correlation
between the series that would be given by long experience over
an indefinitely extended time.

—G. Udny Yule (1871-1951)
"Why Do We Sometimes Get Nonsense-Correlations Be-
tween Time-Series? A Study in Sampling and the
Nature of Time-Series?" *Journal of the Royal Statis.
Soc., 89*, Jan. 1926, pp. 4, 39.

ALLAIS (1966)

Investigators in the field of pattern recognition have shown that the performance of a recognition system sometimes deteriorates when additional measurements are included in the pattern analysis. This strange phenomenon is explained theoretically for the related problem of normal prediction, in which the parameters must be estimated.

A practical method is proposed for making the predictor more accurate by reducing the number of measurements. The effectiveness of this technique is shown by experiments with artificial data...in weather forecasting and character recognition.

* * * *

Although no rigorous justification exists for these procedures, experiments show them to be markedly superior to unselective prediction when the number of measurements is the same order of magnitude as the number of learning samples.

* * * *

...the fractional increase in error due to using a learning sample was found to be approximately $p/(N-p-2)$, where p is the number of measurements.

> —David Charles Allais (1933-   )
> "The Problem of Too Many Measurements in Pattern Recognition and Prediction," *IEEE Intl. Convention Record*, Part 7 (Discrimination, Measurements), March 21-25, 1966, pp. 124, 126, 128.

HUGHES (1968)

With a fixed design pattern sample, recognition accuracy can first increase as the number of measurements made on a pattern increases, but decay with measurement complexity higher than some optimal value...

The penalty extracted for the generality of the analysis is the use of the mean accuracy itself as a recognizer optimality criterion.

* * * *

...the existence of an *optimal measurement complexity* n_{opt} which maximizes the mean accuracy for any given $m < \infty$. Examples are $n_{opt} = 17$ for $m = 500$ and $n_{opt} = 23$ for $m = 1000$...With m fixed, the accuracy first begins to rise with n... It ultimately must fall back as $n/m \to \infty$ because the precision of the probability estimates monotonically degrades...as $\sigma(s/m_1)/p(x|c_1) \propto \sqrt{n/m}$.

> —Gordon F. Hughes (1924-   )
> "On the Mean Accuracy of Statistical Pattern Recognizers," *IEEE Trans. on Information Theory*, *IT-14*, Jan. 1968, pp. 55, 59.

FOLEY (1972)

The deleterious effects of inadequate sample size have been
discussed in the past...All of these results pertain to the
expected performance of a classifier on future test samples
when the classifier has been designed using a design set of
finite size.

*   *   *   *

Although it is common knowledge that the error rate on the design set is a biased estimate of the true error rate of the
classifier, the amount of bias as a function of sample size per
class and feature size has been an open question. In this paper,
the design-set error rate for a two-class problem with multivariate normal distributions is derived as a function of the
sample size per class (N) and dimensionality (L)...It is demonstrated that the design-set error rate is an extremely biased
estimate of either the Bayes or test-set error rate if the
ratio of samples per class to dimensions (N/L) is less than
three. Also the variance of the design-set error rate is approximated by a function that is bounded by $1/8N$.

—Donald Foley (1916-   )
"Considerations of Sample and Feature Size," *IEEE
Trans. on Info. Theory*, IT-18, Sept. 1972, p. 618.

BISHOP, FIENBERG and HOLLAND (1975)

When several equally sensible models provide a reasonable fit to
a set of data, the scientist usually chooses the simplest such
model. What do we mean by simplest? The simplest model might
be one that (i) has the fewest parameters and (ii) has the least
complicated parametric structure. It is difficult, if not impossible, to work with criteria (i) and (ii) simultaneously.

...Many authors have attempted to define the concept of simplicity precisely (see, for example, Kemeny [1953])*, but we know
of no particular definition that seems natural for every problem.

*   *   *   *

---

* Kemeny, J.G., "The Use of Simplicity in Induction," *Phil. Rev.*,
62, 391-408 (1953).

BISHOP, FIENBERG and HOLLAND (1975, cont.)

*Suppose we have two models for predicting the cell frequencies
in a table of counts, both of which are compatible with the
observed data, one model being a special case of and having
fewer parameters than the other. Then the "overall variability"
of the estimates from the simpler model about the "true" values
for the cells is smaller than the "overall variability" for the
model with more parameters requiring estimation.* We have no
general proof of this theorem, although it is specifically true
for many regression problems, and we believe it is true in a
rather general way.

> —Yvonne M.M. Bishop (1925-   ),
> Stephen E. Fienberg (1942-   ) and
> Paul W. Holland (1940-   )
> *Discrete Multivariate Analysis: Theory and
> Practice,* The MIT Press, Camb., MA, 1975,
> pp. 312-313.

LEE, McNEER, STARMER, HARRIS and ROSATI (1980)

In today's studies of therapeutic effects in a complex chronic
disease, investigators generally accept the importance of complete, carefully collected baseline data, a comparable control
group, and physiologic rationale for treatment effects. Other
points such as the likelihood of chance differences achieving
statistical significance when multiple subgroups are analyzed
and re-analyzed, or the influence on outcome comparisons of the
combined effect of several relatively small differences in
baseline prognostic variables may not be so obvious. These
points are often ignored or not mentioned in the presentation of
results of clinical research.

> —Kerry L. Lee (1941-   ), J. Frederick McNeer
> (1946-   ), C. Frank Starmer (1941-   ),
> Phillip J. Harris (1945-   ), and Robert A.
> Rosati (1941-   )
> "Clinical Judgment and Statistics," *Circulation,*
> *61,* March 1980, p. 513.

REFERENCES

Allais, D.C., "The Selection of Measurements for Prediction," Rept. SEL-64-115 (TR No. 6103-9), Stanford Electronics Laboratories, Stanford, CA, Nov. 1964.

_____, "The Problem of Too Many Measurements in Pattern Recognition and Prediction," *IEEE Intl. Convention Record*, Part 7 (Discrimination, Measurements), March 21-25, 1966, pp. 124-130.

Bishop, Y., S. Fienberg and P. Holland, *Discrete Multivariate Analysis: Theory and Practice*, The MIT Press, Cambridge, MA, 1975.

Box, G.E.P, and G.M. Jenkins, *Time-Series Analysis: Forecasting and Control*, Holden Day, San Francisco, CA, 1976, p. 17.

Christensen, R.A., "General Solution to Analogical Screening Problems," section F of "A General Approach to Pattern Discovery," Tech. Rept. No. 20, Computer Center, Univ. of Calif., Berkeley, CA, June 29, 1967 (revised Nov. 15, 1967). [*Chapter 1 of Volume III, pp. 21-23.*]

_____, "Seminar on Entropy Minimax Method of Pattern Discovery and Probability Determination," presented at Carnegie-Mellon Univ., and MIT, Feb. and March, 1971, Tech. Rept. No. 40.3.75, Carnegie-Mellon Univ., Pittsburgh, PA, April 7, 1971. [*Chapter 3 of Volume III, pp. 53-56.*]

_____, "Entropy Minimax Method of Pattern Discovery and Probability Determination," Arthur D. Little, Inc., Camb. MA, March 7, 1972. [*Chapter 4 of Volume III, pp. 87-90.*]

_____, "Entropy Minimax, A Non-Bayesian Approach to Probability Estimation from Empirical Data," *Proc. of the 1973 Intl. Conf. on Cybernetics and Society*, IEEE Systems, Man and Cybernetics Society, Nov. 5-7, 1973, Boston, MA, 73 CHO 799-7-SMC, pp. 321-325. [*Chapter 5 of Volume III, pp. 113.*]

_____, "Contrived Patterns, Trying to Avoid Them Without Trying Too Hard," Carnegie-Mellon Univ., Pittsburgh, PA, Tech. Rept. No. 40.12-75, June 19, 1975. [*Chapter 9 of Volume III.*]

_____, "Crossvalidation: Minimizing the Entropy of the Future," *Info. Proc. Letters, 4*, No. 3, Dec. 1975, pp. 73-76. [*Chapter 10 of Volume III.*]

_____, "Statistical Tables for the Entropy Decrement Distribution," E.E. Dept., Univ. of Maine, Orono, ME, Nov. 20, 1978 (revised Dec. 30, 1979). [*Chapter 11 of Volume III.*]

Christensen, R.A. and R. Eilbert, "Estimating Chance Correlation Likelihood for Hazard Axis Analysis," *Fuel Rod Mechanical Performance Modeling, Task 3: Fuel Rod Modeling and Decision Analysis*, Entropy Limited, Lincoln, FRMPM33-2, Aug. - Oct. 1979, pp. 7.12-7.47 and FRMPM34-1, Nov. 1979 - Jan. 1980, pp. 32-39.

Cover, T.M., "Geometrical and Statistical Properties of Systems of Linear Inequalities with Applications in Pattern Recognition," *IEEE Trans. Elec. Comp., EC-14*, 1965, pp. 326-334.

Davis, R.E., "Predictability of Sea Surface Temperature and Sea Level Pressure Anomalies over the North Pacific Ocean," *J. Phys. Oceanography, 6*, 1976, pp. 249-266.

Davis, R.E., Techniques for Statistical Analyses and Prediction of Geophysical Fluid Systems," *Geophys. Astrophys. Fluid Dynamics, 8,* 1977, pp. 245-277.

Dempster, A.P., "An Overview of Multivariate Data Analysis," *J. Multivariate Anal., 1,* 1971, pp. 316-346.

Eddington, Arthur, *Fundamental Theory,* Camb. Univ. Press, London, 1948.

Eilbert, R.F. and R.A. Christensen, "Contrivedness, the Boundary between Pattern Recognition and Numerology," *Pattern Recognition* (to be published).

Foley, D.H., "Considerations of Sample and Feature Size," *IEEE Trans. Info. Theory, IT-18,* 1972, pp. 618-626.

Forsythe, A.B., P.R.A. May and L. Engleman, "Prediction by Multiple Regression, How Many Variables to Enter?", *J. Psychiat. Res., 8,* 1971, pp. 119-126.

Forsythe, A.B., L. Engleman, R. Jennrich and P.R.A. May, "A Stopping Rule for Variable Selection in Multiple Regression," *J. Amer. Stat. Assoc., 68,* 1973, pp. 75-77.

Groner, G.F., "Statistical Analysis of Adaptive Linear Classifiers," Rept. SEL-64-026 (JR No. 6761-1), Stanford Electronics Laboratory, Stanford, CA, April 1964.

Hills, M., "Allocation Rules and Their Error Rate," *J. Royal Stat. Soc., 28,* 1968, pp. 1-31.

Hughes, G.F., "On the Mean Accuracy of Statistical Pattern Recognizers," *IEEE Trans. Info. Theory. IT-14,* Jan. 1968, pp. 55-63.

Kanal, L.N. and B. Chandrasekaran, "On Dimensionality and Sample Size in Statistical Pattern Classifications," *Proc. Nat'l Electronics Conf., 24,* 1968, pp. 2-7, also *Pattern Recognition, 3,* Oct. 1971, pp. 225-234.

Krishnaiah, P.R., "Simultaneous Test Procedures under General MANOVA Models," *Multivariate Analysis, Vol. II,* P.R. Krishnaiah (ed.), Academic Press, NY, 1964, pp. 121-143.

Lachenbruch, P.A. and R.M. Mickey, "Estimation of Error Rates in Discriminant Analyses," *Technometrics, 10,* 1968, p. 1-11.

Lee, Kerry L., J. Frederick McNeer, C. Frank Starmer, Philip J. Harris and Robert A. Rosati, "Clinical Judgment and Statistics," *Circulation, 61,* March 1980, pp. 508-515.

Poincaré, Henri, *Science and Hypothesis* (1902), tr. by W.J.G., Dover Pubs., Inc., NY, 1952.

Topliss, J. and R. Costello, "Chance Correlations in Structure-Activity Studies Using Multiple Regression Analysis," *J. Med. Chem., 15,* 1972, pp. 1066-1068.

Yule, G.U., "Why Do We Sometimes Get Nonsense Correlations Between Time-Series? A Study in Sampling and the Nature of Time-Series," *J. Royal Stat. Soc., 89,* 1976, pp. 1-69.

CHAPTER 20

INFORMATION-THEORETIC CROSSVALIDATION
AND WEIGHT NORMALIZATION

CHAPTER 20

INFORMATION-THEORETIC CROSSVALIDATION AND WEIGHT NORMALIZATION

A. Training and Test Samples

In the previous three chapters, we saw that non-information-theoretic methods use numerous tunable parameters, often 10, 20, 30 or more, to fit data, and that this seriously degrades the reliability of their results. In this chapter, we see how parametric entropy mini-max analysis fits data with a single parameter, the weight normal-ization, maximizing predictive reliability.

Assume that our data base contains information on a total of 1,000 events, and imagine that we have formulated predictive patterns based on all 1,000 of these events. How do we convince ourselves that we have not contrived a set of patterns that happens to fit the data but has no predictive value? We might wait to see how well the patterns perform in predicting outcomes for new events. However, this could be a very time-consuming process.

We can short-circuit the wait period by setting aside a portion of our original data for the purpose of crossvalidation.

There are a number of crossvalidation protocols that can be used. The leave-one-out protocol involves training on 999 events and predicting the one left out. This procedure is repeated 1,000 times, leaving a different one out each time, and accumulating statistics on prediction versus observation. These statistics provide as thorough an assessment as can be obtained for the predictive reli-ability of patterns found by the pattern discovery process. The 1,000 sets of patterns found will have only minor differences, since for each pair, there will be 998 common training events. It will matter little which of the 1,000 pattern sets is used for future predictions.

This is a good approach if one can afford the computer time for generating 1,000 pattern sets. To reduce computer time, one can use a
leave-n-out protocol. For example, one can train on 990 events and
predict the remaining 10. Repeating this 100 times with a different
10 left out each time, one will sequence through the entire sample.

To minimize computer time, one can use a leave-half-out protocol.
Suppose we divide the sample into two halves: A and B. We train on
A and assess predictions on B; then train on B and test on A. This
saves considerably on computer time. But we pay a two-fold price:

o The assessment results now depend upon which events
happen to be lumped together.

o The patterns based on the respective halves can be quite
different from each other and from a set based on the
whole data set.

To avoid introducing bias into the results depending upon how the
events happen to be sorted between A and B, the separation should
be random relative to all information we have concerning any possible connection between the dependent and independent variables.
Assume that we measure the extent of similarity of two pattern sets in
terms of the difference between their performance results on their
test data. We do not weight different errors differently. We are
only interested in comparing overall error rates. With random separation, there is greater likelihood that the assessment results will
be similar for the two halves than that they will be very different.

What do we do if the error rates are quite different? If this happens, we cannot trust either as an estimate of the reliability of
the pattern discovery process. One remedy is to make another random separation and try again. This can be repeated till we obtain
reasonable closeness of error rates. This reduces our risk to the
unlikely occurrence of nearly equal error rates but unrepresentative patterns. Even if several passes are necessary, it still saves
considerably on computer time compared to the leave-one-out protocol.

The expected number of passes can be reduced by constraining the random separator. For example, we can require that the dependent variable distribution be similar for the two halves, A and B. We can also require that the independent variable distributions be similar. However, we must be careful not to impose any informational link between the dependent and independent variables by such constraints.

B.   Two-Stage Crossvalidation

The central purpose of crossvalidation is to guard against contrivedness. The use of test samples, whether new before-the-fact events or historical events set aside for crossvalidation purposes, provides a mechanism for measuring contrivedness. Reducing contrivedness can then be used as a criterion for parameter estimation or prediction system selection.

The safest contrivedness assessment test data are obviously before-the-fact events. This provides an absolute guarantee that the persons involved in the variables definition and feature selection processes do not contaminate the features with contrivedness from knowledge of test sample outcomes.

When before-the-fact crossvalidation is not feasible or practical, retrospective crossvalidation can be used. This involves taking past data and separating out a portion for test purposes. In most cases, it is difficult to ensure that knowledge of the outcomes for retrospective test cases does not somehow, directly or indirectly, even unconsciously, influence some aspects of the pattern discovery process. Because of the much greater convenience of retrospective crossvalidation over prospective crossvalidation, it is very useful to guide feature selection and to set the weight normalization.

However, true before-the-fact predictive success is the ultimate validation of the predictive capability of a set of patterns. Extensive descriptive freedom confers no special advantage in making predictions for unknown events as it does in matching past data.

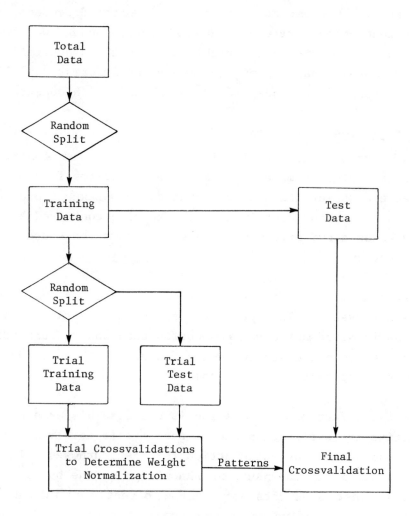

Fig. 112.   Two-Stage Crossvalidation

Fig. 112 shows a two-stage crossvalidation scheme. The total available data are first randomly split into training and test sets. The training set is further split into trial training and trial test sets. Trial crossvalidations are used to determine one parameter, the weight normalization. The patterns found using this normalization are then crossvalidated against the test set.

C. Weight Normalization Determination

The difficulty of calculating $S^c(\nu)$ in the previous chapter can be avoided by an alternative treatment of contrivedness. Instead of treating contrivedness as a reference against which we compare the partition entropies, we treat it as a reference against which to compare the data counts.

In Chapter 8, we derived the expression for the probability of outcome k given by

$$P_k = \frac{x_k + w_k}{n + w} ,$$

where

$$w_k = t_k + a_k .$$

The quantity a_k is the effective virtual event count for background data with the kth outcome. If the total weight w is large, we need large data counts x_k to overcome the weight ratios w_k/w and push the probability toward the frequency limit x_k/n. The value of w thus sets the scale for sample size.

Assume we have randomly split the data into two portions: a training portion and a test portion. We set w at an arbitrary value, say w = 10, and find the minimum entropy partition for the training data. We use this partition and its associated conditional probabilities to make predictions for the test data. We then compute the accuracy of the prediction on the test data as the entropy decrement:

$$\Delta S(w) = \sum_k F_k(w) \ \log \ [F_k(w)/P_k(w)] .$$

Next, we change w to a new value, say w = 100, and repeat the process. We will obtain a new entropy $\Delta S(100)$. Note that F_k, as well as P_k, is a function of w. This is because the minimum entropy feature space partition depends upon w.

After performing a sequence of such computations, we identify a value of w for which $\Delta S(w)$ is a minimum.

Now we repeat the whole process for a new random test sample. By this means, we finally arrive at an expected value for the w producing minimum $\Delta S(w)$. We call this the background weight normalization.

We make our predictions based on the patterns producing minimum entropy with this value for w. Large values for w occur when many features are used and the data is quite erratic, necessitating strong measures to avoid contrivedness. The large values cause the partition to have fewer cells, each more highly populated. They shrink the individual cell probabilities more severely toward the overall population averages.

Small values for w occur when few features are used and the data are highly consistent. The small values cause the partitions to become more finely structured into smaller cells, with the cell probabilities more nearly approximating the cell frequencies.

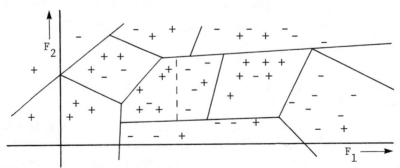

Fig. 113.   Midway in the Pattern Search Process Where a Tentative
Boundary (dotted line) Is Being Assessed

Fig. 113 illustrates an intermediate stage in a pattern generation and assessment process. The question being analyzed is whether or not the introduction of an additional boundary segment, shown by the dotted line, lowers the entropy of the sample.

Three sets of probabilities are relevant to this question:

$P(O_k$ given C) $=$ P(outcome k given event in undivided cell)

$P(O_k$ given $C_L)$ $=$ P(outcome k given event in left portion of divided cell)

$P(O_k$ given $C_R)$ $=$ P(outcome k given event in right portion of divided cell)

The portion of the whole space entropy under investigation is given for the two cases by:

Undivided case

$$-P(C) \sum_k P(O_k \text{ given C}) \log P(O_k \text{ given C})$$

Divided case

$$-P(C_L) \sum_k P(O_k \text{ given } C_L) \log P(O_k \text{ given } C_L)$$

$$- P(C_R) \sum_k P(O_k \text{ given } C_R) \log P(O_k \text{ given } C_R)$$

The probabilities are related to the data counts as follows:

$$P(O_k \text{ given C}) = \frac{n_{k,L} + n_{k,R} + w_k}{n_L + n_R + w} ,$$

$$P(O_k \text{ given } C_L) = \frac{n_{k,L} + w_k}{n_L + w} , \text{ and}$$

$$P(O_k \text{ given } C_R) = \frac{n_{k,R} + w_k}{n_R + w} ,$$

where

$$n_{k,L} = \text{number of real events in left portion with } k^{th} \text{ outcome,}$$

$$n_{k,R} = \text{number of real events in right portion with } k^{th} \text{ outcome,}$$

$$n_L = \sum_k n_{k,L} \text{ , and}$$

$$n_R = \sum_k n_{k,R}.$$

If we happen to have well-defined local background information such as the output from a well-conceived theoretical model, we can use it to provide the weights w_k. Otherwise, the weight ratios may be estimated by using the entire space as the background for the local portion under examination.   This gives

$$\frac{w_k}{w} = \text{fraction of events throughout the entire space with } k^{th} \text{ outcome.}$$

At this point there is one undetermined parameter, the weight normalization w.

Variation of w will affect which pattern boundaries are selected by the minimum entropy principle. With lower w, there will be a preference toward many smaller patterns, each containing fewer events but with greater outcome purity. With higher w, the preference will lean toward fewer and larger patterns sacrificing purity for increased statistical reliability.

Increasing w thus acts as a brake to avoid contrivedness.   If the situation is such that fine-screened predictions hold up under crossvalidation, then this brake can be relaxed. The normalization w can be lowered and more particularized patterns can be used.

It is therefore appropriate to use crossvalidation as the process by which we estimate the weight normalization.* To review this process,

---

* R.A. Christensen, "Crossvalidation: Minimizing the Entropy of the Future," *Information Processing Letters, 4,* Dec. 1975, p. 74.

we find the minimum entropy partition of the entire training sample for a variety of values of the normalization. Then we assess each pattern set by crossvalidation on the test sample. The normalization is fixed at that value for which the patterns crossvalidate best.

This increase by one of the number of degrees of freedom used must, of course, be accounted for when computing the significance levels for the crossvalidation results. It will generally have only a small effect, considering the complexity of typical applications.

This parametric approach replaces use of

$$P_k = \frac{x_k + t_k}{\sum_j (x_j + t_j)}$$

and minimization of

$$S(O \text{ given } C)\bigg|_{\substack{real \\ data}} - \hat{S}(O \text{ given } C)\bigg|_{\substack{random \\ data}} ,$$

with use of

$$P_k = \frac{x_k + w_k}{\sum_j (x_j + w_j)}$$

and minimization of

$$S(O \text{ given } C)\bigg|_{\substack{real \\ data}} .$$

The effect of the contrived entropy $\hat{S}(O \text{ given } C)\big|_{\substack{random \\ data}}$ upon the patterns is absorbed into the background via crossvalidation normalization of the weights, $\{w_i\}$.

We obtain an informationally equivalent partition $C = \{C_j\}$ by selecting the parameter $W = \sum_i w_i$ so as to minimize the expected value of the information disagreement

$$\Delta S = \sum_k F_k \log F_k/P_k,$$

where

F_k = observed outcome frequency for random test sample, and

P_k = predicted outcome probability for that test sample based on minimum entropy patterns in the training sample.

D.   Variable Resolution Normalizations

The weight normalization plays the role of a parameter controlling the resolution of the pattern seeking algorithm.   See Fig. 114.   It determines the level of detail with which we view the data.

Imagine a range of resolutions tagged by the index i.   Define

n   = number of events in training sample,

n_i = number of training sample events in i^{th} resolution pattern containing point in question, and

x_{ik} = number of the n_i events in the kth outcome class.

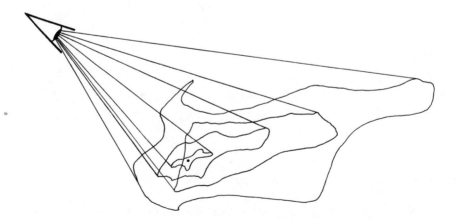

Fig. 114.   Variations of the Weight Normalization Focuses the Resolution of the Feature Space Partition on Finer or Coarser Patterns Enclosing the Test Point

We can now define a generalized linear shrinkage outcome class probability estimator for the point in question as

$$\bar{P}_k = \sum_i \alpha_i \frac{x_{ik}}{n_i} .$$

The generalized maximum entropy estimator is

$$\hat{P}_k = \frac{\sum_i \beta_i x_{ik}}{\sum_i \beta_i n_i} .$$

The coefficients $\{\alpha_i\}$ or $\{\beta_i\}$ are determined by minimizing an error measure with respect to the test sample $\{m_k\}$, where m_k is the number of test sample events with the kth outcome in a region enclosing the test point, and $m = \Sigma m_k$. The minimization is performed subject to the constraint

$$\sum_k P_k = I ,$$

where $I < 1$ if there are irrelevancies, $I > 1$ if there are ambiguities, $I = 1$ if there are neither irrelevancies nor ambiguities, and may be either greater or less than 1 if there are both.

For linear shrinkage estimation, a least squared error measure is mathematically convenient:

$$\epsilon^2 = \sum_k \left(\frac{m_k}{m} - \bar{P}_k \right)^2 .$$

We set $\delta\epsilon^2 = 0$, and include the constraint $\sum_k \bar{P}_k = I$ via a Lagrange multiplier. This leads to a set of linear equations in the $\{\alpha_i\}$.

Minimization of Pearson's T ,

$$T = \sum_i \frac{(m_k - m\bar{P}_k)^2}{m\bar{P}_k} ,$$

is another alternative, but more involved.

For maximum entropy estimation, we select the $\{\beta_i\}$ so as to minimize the entropy decrement

$$\Delta S = \sum_k \frac{m_k}{m} \log \left(\frac{m_k/m}{\hat{P}_k}\right) .$$

Note that, in each case (ε^2, T or ΔS), the error measure must be
accumulated over all regions of feature space for which predictions
are desired.   Thus there is a double summation involved.

E.   Why Bother with a Test Sample?

From a purely logical perspective it would appear that we really
have but one grand sample.   The division between test and training
portions is arbitrary.   We use the training portion to estimate
our probabilities and to guide us in drawing pattern boundaries.
We use the test portion to estimate one parameter: the background
weight.   Why not just use the entire sample to make all the estimates and boundary decisions?   What difference could it possibly
make whether we make the hypothesis before or after we see the
test data?

If this were so, then a key aspect of the modern scientific method
would be useless, namely experimental verification.   When a scientist constructs a new theory to fit data after-the-fact, it may be
interesting, but it is not particularly impressive.   However, when
Gell-Mann and Ne'eman constructed their new theory which success-
fully predicted the existence of the omega-particle prior to its

experimental discovery, the confidence level of the eight-fold way among physicists was significantly raised. Why was this? Is it just a psychological phenomenon of being impressed by successful forecasts because accurate predictions have decision-making utility? Is it just the principle of surprisal in play: complex new theories in general have a low probability of making accurate predictions; so when we find one that does we are impressed (surprised)? Or is there something more fundamental at stake with respect to the probable truth of the verified hypothesis?

Imagine a World Scientific Research Council. An important research question is at stake: how to predict whether or not a person will have cancer. The Council randomly divides the data into two samples--half in a training sample, half in a test sample.

Copies of the training half are sent to research workers throughout the world. The test half is kept secret. Each researcher analyzes the training data, formulates theories to explain them, and submits his or her best theory to the WSRC.

WSRC then tests each theory by comparing its predictions for the test data to the actual test outcomes.

Because of variations in schooling, experience, personality, etc., of the different research workers, WSRC can well expect to receive a wide variety of different theories. This will be especially so on a research question with respect to which there are still many unknowns. How can WSRC guard against the possibility that some researcher will formulate a theory which by chance alone does well on its particular test data, but which is really a poor theory for the general population? Has not WSRC, in fact, loaded the die in favor of this possibility by taking the test and training samples from the same grand sample, thereby ensuring that they will tend to have similar statistical properties regardless of whether the grand sample is really representative of the population at large?

502 ENTROPY MINIMAX SOURCEBOOK

Why does WSRC go to all this trouble to obtain a theory which does well on both the training and test halves? Presumably, each research worker will construct a theory which does a more or less good job matching the training data. Then WSRC is picking among them for the one which does the best job on the test data. Would it not be simpler just to give out both samples in the first place?

In fact, would not WSRC be likely to obtain, by a full disclosure approach, a theory which does a better overall job on the training and test data?

If so, by what rationale can WSRC prefer a closed-test theory which matches the test data only moderately over an open-test theory which matches it better?

WSRC may wish to overcome the objections to the randomization by pointing out that it was very careful in drawing the grand sample from the general population. It may also argue that it could well have decided that time was an irrelevant or, at least, insignificant factor and issued as the training sample all data it had collected as of the start of the contest. It subsequently used, as the test sample, all data it collected between then and the assessment time. The "unrepresentativeness possibility" is a criticism of the sampling procedure, not of the training/test randomization.

Suppose we tabulate how well the theories do according to some scoring system adopted by WSRC. See Table 42. (We assume individual sample scores must be between 0 and 100, higher being better.)

Table 42. WSRC Scores

Theory	Test Score	Training Score	Total Score
A	50	70	120
B	45	45	90
C	45	40	85
D	40	85	125
E	35	75	110
F	30	50	80
G	30	65	95
H	25	90	115

The theories are ranked in the order in which they scored on the test data. So theory A wins under the original rules. But is it the best theory? Why should we stake people's lives on it over theories D or H, for example?

WSRC might reason as follows: The theories generally score lower on the test half than on the training half (C is an exception). This reflects contrivedness at work in the theory formation process. So we are completely justified in ranking them solely by how well they score on the test half.

Suppose that the research worker who submitted theory A had also considered a theory A' but rejected it because its score on the training half was only 65. Further, suppose that WSRC had asked all the research workers to append their complete notebooks to their submittals. Upon reviewing these notes, WSRC uncovers theory A'. To its surprise, it discovers that A' scores 65 on the test data also, for a total score of 130! Why should WSRC not select this theory over the original A?

By opening up the notebooks, WSRC is opening up the possible examination of many hundreds or thousands of additional theories. It is putting itself in the place of a research worker, assessing numerous possibilities (though using an enlarged training sample). Thus, it becomes prone to the very contrivedness which it was attempting to avoid.

It is too late for WSRC to pretend it never heard of A'. Does it have grounds for suspecting A' to be contrived? It seems difficult to see how, since the test data was not available to the researcher who formulated A'.

If not, then WSRC must accept A' over A. Assuming this line of reasoning to be correct, if it has not already done so, WSRC should open everybody's research notes on the chance that there may be an even better theory. In fact, WSRC should itself search for better theories using the whole data, regardless of any possible problem of contrivedness. If this is so, successful after-the-fact fitting is as good as successful before-the-fact predicting.

Since we are not willing to accept this last conclusion (for reasons stated in Chapter 19), there must be a flaw in the above reasoning. For example, perhaps there are grounds for suspecting A' to be contrived.

It is not necessary for the creator of A' to have purposely made it contrived. The contrivedness enters when WSRC uses up degrees of freedom by opening the research notes to explore a larger class of theories.

Suppose that we conduct a two-stage crossvalidation. On the trial training data we formulate our patterns and in the trial crossvalidation we set the weight normalization. Suppose that on final crossvalidation we are not quite satisfied with the results, but notice that if we make a slight adjustment to the patterns or the normalization (or both) we can improve the overall results to our satisfaction. It is important to recognize that the new results are unvalidated precisely to the extent of this "adjustment". We need new independent data to perform this validation. No amount of smoothing or other analysis can eliminate the need for crossvalidation at every stage. Data fitting without validation is precisely that, data fitting *without validation*, and has no greater predictive reliability. When someone presents you with a "predictive model", always ask on what basis the *final* adjustment was made, and how much validation was performed afterwards on data not known at the time of this adjustment.

F.   Separating Causal From Non-Causal Correlations

Methods of generating predictive models can be divided between observational analysis and experimental analysis. Observational analysis involves fitting data after-the-fact. It is the only type of analysis possible when the data are historically settled (as generally is so in geology), when it is impractical to attempt experiments (as often is the case in economics), or when an experiment is unethical (as may be the case in psychology).

Crossvalidation analysis of observational data permits one to probabilistically separate chance correlations from population representative patterns. It does not, however, permit one to distinguish between situations in which outcome O is *caused* by condition C, and situations in which O and C share a common cause C'.

This can be regarded as a sort of meta-contrivedness problem. In the observational samples (training and test), it may have just so happened that the hidden cause C' was generally the reason C occurred and thus the "true" reason that O occurred. Thus, although the presence of C has information indicating the likelihood of O occurring under total circumstances in which C and C' are similar to the observational samples, we do not know what would happen if C occurred without C' (assuming such to be possible).

Experimental analysis is used to provide information for establishing probable causation as distinct from probable correlation. In the natural setting, some of the conditions $\{C_i\}$ arise from the uncontrolled causes C', concerning which we may or may not have complete information. In the laboratory setting, the conditions $\{C_i\}$ are experimentally controlled. By appropriately manipulating them, we structure the sample to enhance the probabilistic separation between direct correlation of O to C and indirect correlation via an unknown common cause. We make it highly unlikely that the unknown cause will have just such a pattern of occurrences as to produce our observed results.

Causality* is a fractal relationship. The length of a causal chain depends upon the fineness of our analytic yardstick. If we turn up the resolution, we may find that our controlling C in the laboratory in fact causes an intermediate effect O' which in turn causes O. O' may not be caused by C in non-laboratory circumstances.

The extent to which this resolution may be enhanced by more refined analysis is a question of entropy minimization. We treat the description of C and all other surrounding circumstances as forming the feature space. How finely we subdivide feature space depends upon the amount and variability of our observational data, as sorted by the minimum entropy partition of this space.

---

* It has been asserted that both the sufficiency and the necessity definitions of causality suffer a logical defect, viz: anything satisfying a minimal independence requirement causes any effect. See Ernest Sosa, *Causation and Conditionals*, Oxford Univ. Press, London, 1975, p. 2. However, there is an assumption of causality buried in the posited "minimal" independence requirement.

QUOTES

GOOD (1965)

...Given a sample of r successes in N trials, the final Type II expectation and variance of p are

$$\mathcal{E}_{II}(p|r,s) = \frac{\alpha + r + 1}{\alpha + \beta + N + 2}$$

and

$$\text{var}_{II}(p|r,s) = \frac{(\alpha + r + 1)(2\alpha + 2\beta + r + 1)}{(\alpha + \beta + N + 2)(\alpha + \beta + N + 3)}$$

The parameters α and β can be determined by equating the guessed initial Type II expectation and variance to

$$\left. \begin{array}{l} \mathcal{E}_{II}(p|0,0) = \dfrac{\alpha + 1}{\alpha + \beta + 2} \\[3mm] \text{var}_{II}(p|0,0) = \dfrac{(\alpha + 1)(2\alpha + 2\beta + 1)}{(\alpha + \beta + 2)(\alpha + \beta + 3)} \end{array} \right\}$$

* * * *

Jeffreys and Wrinch therefore suggested that a nonzero initial Type II probability should be associated with the end points of the unit interval, in order to provide a rationale for scientific induction.

* * * *

Suppose that we decide, for one reason or another, by model-building, experience, discussion, and intuition...

* * * *

The device of imaginary results helps you to decide on an initial Type II distribution before sampling...

> —Irving John Good (1916-   )
> *The Estimation of Probabilities, An Essay on Modern Bayesian Methods,* MIT Press, Cambridge, MA, 1965, pp. 17, 18, 19, 20.

KANAL and CHANDRASEKARAN (1971)

If m is the total number of samples, take all possible partitions of size 1 for the test set and m-1 for the design set. This results in successively omitting one sample in the design procedure.

> —Laveen Kanal (1931-   ) and
> B. Chandrasekaran (1942-   )
> "On Dimensionality and Sample Size in Statistical Pattern Recognition," *Pattern Recognition, 3,* 1971, p. 226.

REFERENCES

Christensen, R.A., "Crossvalidation: Minimizing the Entropy of the Future," *Information Processing Letters, 4,* Dec. 1975, pp. 73-76. [*Chapter 10 of Volume III.*]

_____, "Statistical Tables for the Entropy Decrement Distribution," E.E. Dept., Univ. of Maine, Orono, ME, Nov. 20, 1978 (revised Dec. 30, 1979). [*Chapter 11 of Volume III.*]

Good, I.J., *Probability and the Weighing of Evidence,* Hafner Pub. Co., NY, 1950.

_____, *The Estimation of Probabilities: An Essay on Modern Bayesian Methods,* MIT Press, Cambridge, MA, 1965.

Highleyman, W.H., "The Design and Analysis of Pattern Recognition Experiments," *Bell System Tech. Journ., 41,* March 1962, pp. 723-744.

Jeffreys, Harold, *Theory of Probability,* Oxford at the Clarendon Press, London, 1961.

Kanal, Laveen and B. Chandrasekaran, "On Dimensionality and Sample Size in Statistical Pattern Recognition," *Pattern Recognition, 3,* 1971, pp. 225-234.

Mandelbrot, B., *Fractals: Form, Chance and Dimension*, W.H. Freeman & Co., San Francisco, 1977.

Oldberg, S., "New Code Development Activities," *Planning Support Document for the EPRI Light Water Reactor Fuel Performance Program,* prepared by J.T.A. Roberts, F.E. Gelhaus, H. Ochen, N. Hoppe, S.T. Oldberg, G.R. Thomas and D. Franklin, EPRI NP-1024-SR, Electric Power Research Inst., Palo Alto, CA, Feb. 1979, pp. 2.32-2.39.

_____, "SPEAR Methodology," *SPEAR Fuel Performance Reliability Code System: General Description,* ed. by R. Christensen, EPRI NP-1378, Electric Power Research Institute, Palo Alto, CA, March 1980, pp. 2.1-2.17.

Perks, Wilfred, "Some Observations on Inverse Probability Including a New Indifference Rule," *J. Inst. Actuar., 73,* 1974, pp. 285-312.

Skyrms, Brian, *Causal Necessity, A Pragmatic Investigation of the Necessity of Laws,* Yale Univ. Press, New Haven, CT, 1980.

Wrinch, D. and H. Jeffreys, "On Certain Fundamental Principles of Scientific Inquiry," *Phil. Mag., 42,* 1920, pp. 369-390 and *45,* 1923, pp. 368-374.

PART V

FOUNDATIONS OF ENTROPY MINIMAX

CHAPTER 21

COUNTERFACTUALS AND THE MEANING OF PREDICTION

CHAPTER 21

COUNTERFACTUALS AND THE MEANING OF PREDICTION

A.   Counterfactuals

Counterfactuals is the study of "what would be the case if."   It typically arises in conceptual explorations such as the following:

- o   A scientist is planning an experiment.   He runs through a series of thought-experiments (gedankenexperimente) in a process of formulating hypotheses he may wish to test.

- o   A scientist is trying to explain an unexpected result in an experiment he has just conducted.   He explores a variety of counterfactuals for hypothetically similar situations in seeking a possible explanation and a means of testing the possible explanation.

- o   A teacher is demonstrating certain laws or principles to students.   He poses a number of hypothetical "what if" situations to illustrate his points, and to test the students' understanding.

- o   Two individuals are negotiating a contract under which they will each want to be able to rely on the actions of the other.   They exchange a number of "what if" scenarios in the process of reaching agreement.

- o   A salesman is attempting to use reason to persuade a client to make a specific purchase.   He describes possible outcomes if the client purchases one item versus another or nothing at all.

- o   A military planner is establishing strategic policy.   He reviews a number of alternative scenarios.

- o   A lawyer is attempting to establish legal responsibility for a factual situation.   He analyzes related counter-factual situations to establish the causal ties in the factual case.

- o   A philosopher is studying the meaning of cause-effect, induction, and related concepts.   He uses the analysis of counterfactuals as a framework in which to conduct this study.

We will examine a physics problem as illustrative of the study of counterfactuals. Consider the following statements (see Fig. 115):

Law L: After any observation of a particle with rest mass m_1 and velocity v_1 colliding with another particle with rest mass m_2 which is stationary ($v_2 = 0$) in the observer's pre-collision coordinate system under conditions when gravitational effects can be neglected, it will be observed that the two particles recoil with velocities v_1' and v_2' ($v_2' > 0$) in the observer's post-collision coordinate system, obeying the relationship:

$$m_1 v_1' + m_2 v_2' = m_1 v_1.$$

Cause C: A particle with rest mass M_1 and velocity V_1 is observed to collide with another particle with rest mass M_2 which is stationary ($V_2 = 0$) in the observer's pre-collision coordinate system.

Condition D: Conditions are such that gravitational effects can be neglected.

Effect E: Afterward it is observed that the two particles recoil with velocities V_1' and V_2' ($V_2' > 0$), and these velocities obey the relationship:

$$M_1 V_1' + M_2 V_2' = M_1 V_1$$

in the observer's post-collision coordinate system.

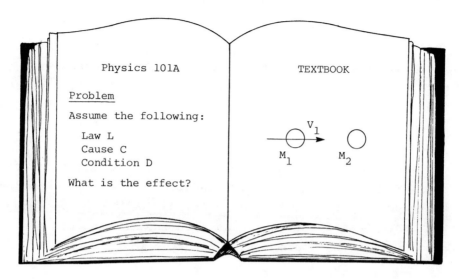

Physics 101A

TEXTBOOK

Problem

Assume the following:

   Law L
   Cause C
   Condition D

What is the effect?

V_1

M_1 M_2

Fig. 115.   Factual World.   Condition D Holds but neither Cause C nor
Effect E are Real.   Rather, Cause C is Simply Posed as an
Hypothetical Example in a Physics Textbook.

We assume that suitable measurement apparatus is used for making
the observational determinations of the number and identity of the
particles, and their masses and velocities, both before and after
collision, and for establishing the negligibility or non-negligi-
bility of gravitational effects.

We further assume the observer's post-collision *memory* of the pre-
collision state to have precision and accuracy consistent with
physical limitations on communication of information.

The counterfactual problem is: Suppose the *real* facts are that condition D is so but that *neither* cause C *nor* effect E is observed.
What can we conclude about *would have been* the case if cause C *had
been* observed?

The factual and counterfactual worlds may differ in various ways
with respect to cause C.   For example, one or more of the quantities M_1, V_1, M_2, V_2 may be different in the two worlds.   The particles may miss and fail to collide in the factual world.   As part
of the statement of the problem, we assume our state of knowledge
of the factual world to have an accuracy consistent with physical

measurement limitations. Our state of knowledge of the make-be-
lieve counterfactual world is assumed to be constructed with a
precision equivalent to the factual accuracy. We will explore examples of factual worlds (without cause C) which differ from the
posited counterfactual world (with cause C) in various ways.

Listed below are some possible answers to the counterfactual problem. We assume that law L has been induced from a given experience
base by a procedure such as entropy minimax. We further assume that
the experience base for L is the same in the factual and the counterfactual worlds, but permit variations in L, C, D, and E.

Answer #1:   Effect E would have been observed to occur, under con-
             dition D, with law L not violated.

    Rationale: This is the usual answer. It is in accord with
                the actual condition D and the assumed law L.
                See Fig. 116.

Fig. 116. Counterfactual World of Type #1. Effect E
Is Observed: $M_1 V_1' + M_2 V_2' = M_1 V_1$

Answer #2:   Effect E would not have been observed to occur, al-
             though condition D would be so; rather law L would
             have been violated.

    Rationale: This is the answer which is properly entertained
                as a possibility whenever some aspect of cause C
                goes *beyond* the range of the experience base upon
                which the law L is based. For example, law L may
                be based on experience with massive particles
                while the posited cause C may involve a massless
                particle $M_1 = 0$. In another example, law L may be
                based on experience involving only massive particles

with velocities small compared to the velocity of light, while the posited cause C may involve a velocity V_1 beyond the range of this experience. In such cases we are prepared to question the *extrapolation* of the law to the posited cause C.

Answer #2 may also be a possibility when some aspect of cause C is *between* instances in the experience base for law L--examples include specific heats as a function of temperature, shattering of glass as a function of sound frequency, and nuclear cross-sections as a function of energy, which can have very sharp resonances which may not appear in the finite sample even when the sample includes points on both sides of the resonance. So as a matter of logic, we are prepared to question the *interpolation* of the law to the posited cause. (Suppose, for example, that within a narrow range of velocities, $V_0 < V_1 < V_0 + \Delta V$, not observed in the finite experience base for L, the two particles will coalesce and form a single new particle, thereby contradicting the explicit wording of the result stated in law L.) See Fig. 117.

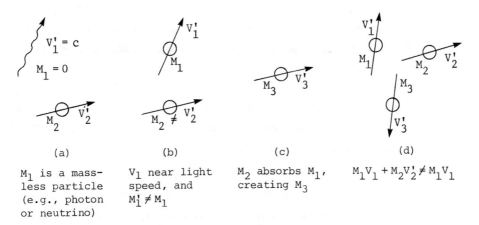

(a)	(b)	(c)	(d)

M_1 is a massless particle (e.g., photon or neutrino)

V_1 near light speed, and $M_1' \neq M_1$

M_2 absorbs M_1, creating M_3

$M_1 V_1 + M_2 V_2' \neq M_1 V_1$

Fig. 117. Counterfactual Worlds of Type #2. Effect C Not Observed: $M_2 V_1' + M_2 V_2' \neq M_1 V_1$. Condition D Holds, but Law L is Violated.

Finally, answer #2 may even be possible when all
aspects of the posited cause C exactly match one
or more examples in our experience base for law
L. It may be that the conditions stated in law
L are actually *insufficient* to determine the result,
but that the finite experience base we have for
the law just does not happen to include an exception. There may exist some unstated condition
that happened to be the same for the events in the
experience base, but is now different. For example, all the collisions in the experience base
may have involved only two particles, while there
may be a third particle involved in the particular
case under discussion, since this possibility is
not excluded by condition D. (Thus, it is not
even necessary for identical causes and conditions
in the experience base to always have yielded the
same effects. This will be the case with "statistical" laws.)

Alternatively, the effect may be "fundamentally"
indeterminant, as in the no-hidden-variables interpretation of quantum effects. For example, one
or the other of the two particles is radioactive
and spontaneously decomposes into two particles
during the collision in question, though this never happened in the experience base for L.

Answer #3:   Effect E would not have been observed to occur, although
law L would not have been violated; rather, we would
have observed condition D not to be so.

Rationale: This answer should be considered as a possibility
when the particular circumstances together with
the particular elements of effect E may be inconsistent with condition D. It may be that condition

D and cause C are interlinked (physically or logically) such that when cause C occurs, condition D
does not. For example, if the velocity V_1 specified by cause C in the counterfactual world is
sufficiently close to the velocity of light, then
the mass of the particle may be so amplified according to the principle of special relativity
that gravitational effects are no longer negligible.   (This would be a physical interlinking. Examples of logical interlinking can be constructed
by giving tricky definitions of the cause and the
condition.)   In the factual world we did not observe this high velocity and the gravitational effects were negligible.

An even simpler example is a case in which a very
massive third particle M_3 is moving toward M_2. If
the counterfactual situation is that V_2 is sufficiently slow to permit M_3's gravitational field to
be non-negligible by the time of the collision,
then it will interfere with our observing effect E.
See Fig. 118.

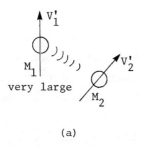

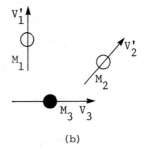

(a)

Counterfactual V_1 near light speed,
causing M_1 to be so large its gravitational attraction on M_2 is non-
negligible

(b)

Counterfactual V_1 so slow it gives
time for the approaching massive
M_3's gravitational field to become
non-negligible

Fig. 118.   Counterfactual Worlds of Type #3.   Effect E Not Observed:
$M_1 V_1' + M_2 V_2' \neq M_1 V_1$.   Law L Not Violated, but Condition D Fails.

Answer #4:   Effect E would not have been observed to occur, and yet
condition D would still be so, and law L would not have
been violated.

Rationale: One of the first things to suspect in this case
is that an intervening additional condition T,
absent in the factual world but present in the
counterfactual world, may prevent effect E from
occurring. For example, there may be a force operating upon the particles that is so weak it is not
detectable in the factual world in which V_1 is of
"ordinary" magnitude, but would result in our failure to observe E in a counterfactual world in which
V_1 is very small. Equivalently, think of a third
particle approaching the original pair and interfering with the collision if and only if V_1 is
slow enough to delay the collision sufficiently.
Thus, effect E may not be observed, despite the
validity of law L, when the velocity is so small
that the time to the collision exceeds the natural
period of stability of the environment. This can
be a very small duration in the case of short-
lived particles.

Another possibility is that the intervening condition T has distributed the observer's coordinate
system, so that the post-collision coordinate system measurements have nothing to do with measurements in the pre-collision coordinate system.

If no intervening condition can be found, another
possible explanation may be uncovered by an error
analysis. We can only know the aspects of cause
C and effect E to accuracies consistent with the
accuracies of our measurement apparatus. Thus, we

might be sufficiently in error on some aspect of cause C to contradict the literal statement of effect E. (Suppose, for example, we measure the resulting velocities V_1' and V_1' very accurately, but that there is significant inaccuracy in our measurement of the original velocity V_1.) See Fig. 119.

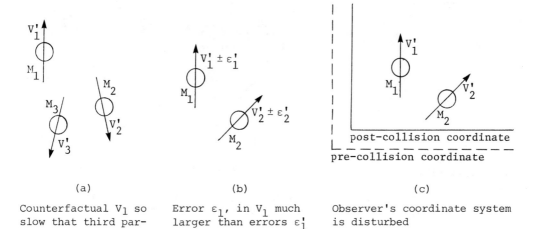

| (a) | (b) | (c) |
| Counterfactual V_1 so slow that third particle has time to intervene | Error ε_1, in V_1 much larger than errors ε_1' and ε_2' in V_1' and V_2' | Observer's coordinate system is disturbed |

Fig. 119. Counterfactual Worlds of Type #4. Effect E not Observed: $M_1(V_1' \pm \varepsilon_1') + M_2(V_2' \pm \varepsilon_1') \neq M_1(V_1 \pm \varepsilon_1)$. Yet Condition D Holds and Law L not Violated.

Answer #5:   Effect E would have been observed to occur, with law L not violated, but condition D would not have been so.

Rationale: As illustrated in answer #3 above, once we shift from what *is* to what *might have been* we entertain the possibility of different conditions. However, we might be faced with a situation in which the stated condition is *not necessary* for the production of effect E; or a situation in which the new conditions are *equivalent* to the original condition D in yielding effect E from cause C. See Fig. 120.

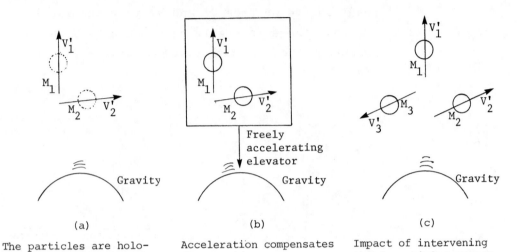

(a) The particles are holo-graphic images of distant "real" particles

(b) Acceleration compensates for gravity

(c) Impact of intervening third particles compensates for gravity

Fig. 120.   Counterfactual Worlds of Type #5. Effect E Observed:
$M_1V_1' + M_2V_2' = M_1V_1$.   L Not Violated, but Condition D Fails.

A third possibility is that effect E might result in a counterfactual case from a second condition T in addition to condition D while T happened to be absent from the factual case. An approaching third particle may just overcome the gravitational effects if the counterfactual velocity V_1 is exactly right.

Answer #6:   Effect E would have been observed to occur, under condition D, yet law L would be violated.

Rationale: Consider the case of near-light velocities. Just because law L is violated does not mean that effect E cannot result. There may be an entirely separate cause, law, condition or combination to bring about E. For example, the intruding third particle is just right for the two original particles to end up with velocities V_1' and V_2' in accord with a three-body interaction law L'.   See Fig. 121.

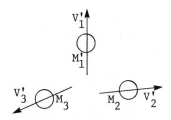

$M_1' \neq M_1$ because counterfactual V_1 close to
light speed, but intervening third particle
just compensates to produce effect E anyway

Fig. 121.   Counterfactual Worlds of Type #6.   Effect E Observed:
$M_1 V_1' + M_2 V_2' \neq M_1 V_1$.   Condition D Holds, but Law L violated.

Since E occurs in this case, we might ask in what sense the literal interpretation of the law as stated *is* violated.   Law L states "After *any* particle..."   Thus, effect E is now an instance of *this* law.   Of course, we need further information in addition to the mere singular observation of E to push us one way or the other on this issue.

It may appear that we have now exhausted the logical possibilities:
    E: occurs or does not occur
    L: violated or not violated
    D: so or not so

However, there are other dimensions to the counterfactual problem, as the following answers illustrate:

Answer #7:   Effect E would have been observed to occur, under condition D, with Law L not violated, but E would not have been due to cause C under condition D and law L.

Rationale: Effect E might result from some different cause,
condition, law, or combination. Examples include
those mentioned in answers #5 and #6 above. In answer
#1 above, we implicitly assumed E to result from C
under D and L. Here we explicitly distinguish
cases in which this is not so. See Fig. 122.

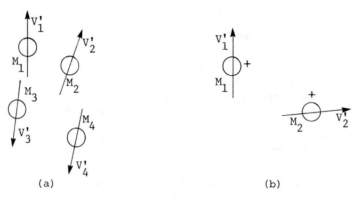

(a)   (b)

Other particles happen M_1 and M_2 never collide. They are
to intervene just so as   electrically charged with the same
to produce the effect.   sign and happen to repel each other
                              just so as to produce the effect.

Fig. 122. Counterfactual Worlds of Type #7. Effect E Occurs.
$M_1V_1' + M_1V_2' \neq M_1V_1$. Condition D Holds and Law L Not
Violated. But E Not *due to* C under D and L.

Answer #8: Effect E would be neither observed to occur nor observed
to not occur, although condition D would be so, and law L
would not be violated.

Rationale: A well known example is the double slit experiment.
See Fig. 123. A low intensity particle source (e.g.,
photons, electrons, etc.) is located in a box with
particle-absorbing walls, one of which has two slits,
A and B. Outside the box, facing the wall, is a
particle-sensitive screen which scintillates each
time a particle strikes it, revealing when and where
the particle struck.

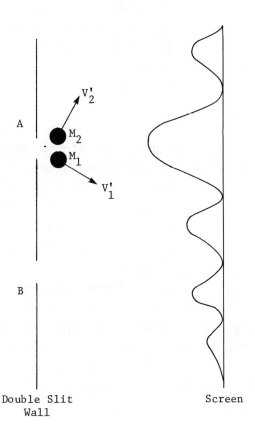

particle
source

Double Slit   Screen
Wall

Fig. 123.   Counterfactual World of Type #8. Violation of law of
            the excluded middle. Effect E neither occurs nor fails
            to occur. In the factual world M_2 does not obstruct the
            slit and we observe M_1 reaching the screen as part of an
            interference pattern statistics for probability density
            of particles at the screen. Obstructing the slit A with
            M_2 in the counterfactual world, we contradict factual
            physics by assuming M_1 to be observed at A by M_2's re-
            coil, and we also contradict factual physics by assuming
            it not to be observed at A.

In the factual world, both slits are unobstructed. Consider the scintillation of a single particle M_1 at a particular location on the screen. We ask the counterfactual question: If mass M_2 had obstructed slit A (cause C), would we have observed the presence of M_1 at slit A via the recoil of M_2 (effect E)?

Suppose we assume that in the counterfactual world we either *would* or *would not* have observed effect E. This assumption can be shown* to lead to an erroneous conclusion, despite the fact that, if we actually do obstruct slit A, we will either observe M_1 at this position or not. The physics of the factual world is inconsistent with either counterfactual observation.

It has been an accepted principle of physics that information cannot be transmitted at a velocity exceeding that of light in vacuum, on the supposition that energy is the carrier of information. However, if form is the carrier, then information transmittal is not so limited.

What would we accept as proof that one particle could, for example, signal its position to another with information traveling outside the forward light cone? Wouldn't the second particle's being able to predict, i.e., exhibit behavior contingent upon, the position of the first qualify as such a proof? What would we accept as proof that one ensemble of particles could similarly signal its positional distribution to another? Wouldn't the second ensemble's being able to predict the distribution of the first qualify?

---

* For a detailed discussion, see Josef M. Jauch, *Foundations of Quantum Mechanics*, Addison-Wesley Publ. Co., Reading, MA, 1968, pp. 112-114.

In a sense, Einstein violated his own principle of separability
when he wrote the equations of relativity as generalizations about
the motion of material bodies. For now we, as ensembles of particles, need only the addition of memory to have knowledge of events
outside the light cone extending from our present position in
space-time.* What gave him reason to believe in the general validity of these equations?

B.   The Meaning of Prediction

The prediction problem is to estimate, based on available data,
the frequency that a statement about the physical world will be
found to be (or to have been) the case.

- o *a statement:*   A proposition.

- o *the physical world:*   This puts us in the realm of
     scientific inquiry. We are speaking of the actual
     world. We are not speaking of a hypothetical world
     following a different course while obeying the same
     "laws", or of a hypothetical world following the same
     course (up to a point) but obeying different "laws".

- o *about the:* Included, but not exclusively, are state-
     ments that certain events will happen, that certain
     matter bears a certain spatial/temporal relation to
     other matter. We generally err to some extent in our
     perception, memory, or understanding of the physical
     world. We do not demand absolute accuracy of the
     statement in order for it to qualify as being "about
     the" physical world.

---

*   Consider, for example, the following situation:

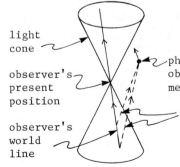

light
cone

photon given off by collision about which
observer's   observer presently knows through use of
present   memory and Einstein's equations
position

world lines of two particles observer
observer's   remembers viewing in the past
world
line

o *the case:* True, i.e., correspondent with observation.

o *to be (or to have been):* The subject of the proposition may be in the future; it may be in the past; it may be in the immediate present; it may have a duration which spans from the past to the future.

o *found:* We only refer to observables. They may be directly or indirectly observed, i.e., they may be deduced from another (directly or indirectly) observed event. The "finding" may be the observing or the deducing or both.

o *will be:* The finding is not present. This is the predictive aspect.

o *the frequency that:* Only frequency is being predicted. However, values of 0 and 1 are logically admissible.

o *to estimate:* We do not assume it is possible to determine the probability with certainty. Thus, there may be an uncertainty associated with the probability.

o *data:* Consist of propositions and the associated probabilities of truth. Include, but are not necessarily limited to, propositions describing sense perceptions and memories of sense perceptions.

o *available:* Propositions present for aiding solution of the problem, and deductions therefrom.

o *based on:* The prediction would, in general, be different if the available data were different.

In the philosophy of science, the question of how we make and justify predictions is studied as the problem of induction. Considering the vast number of predictive successes of modern science, one would expect inductive reasoning to be fairly well understood. To the contrary, this is far from true. In the next chapter, we examine this methodological skeleton in science's closet.

QUOTES

ARISTOTLE (c. 335-323 B.C.)

Everything, we say, that undergoes alteration is altered by
sensible causes, and there is alteration only in things that
are said to be essentially affected by sensible things.

> —Aristotle (384-322 B.C.)
> *Physics*, Book VII, Ch. 2, (c. 335-323 B.C.), tr.
> by R. Hardie and R. Gaye, in *The Basic Works
> of Aristotle*, ed. by R. McKeon, Random House,
> NY, 1941, p. 345.

MINKOWSKI (1908)

Lorentz called the t' combination of x and t the local time of
the electron in uniform motion, and applied a physical construction of this concept for the better understanding of the hypothesis of contraction. But the credit of first recognizing clearly
that the time of the one electron is just as good as that of the
other, that is to say, that t and t' are to be treated identi-
cally, belongs to A. Einstein.* Thus time, as a concept unequi-
vocally determined by phenomena, was first deposed from its high
seat. Neither Einstein nor Lorentz made any attack on the concept of space...Nevertheless, this further step is indispensable
for the true understanding...only the four-dimensional world in
space and time is given by phenomena...

> —Hermann Minkowski (1864-1909)
> "Space and Time," address delivered at the 80th
> Assembly of German Natural Scientists and Physi-
> cians, Cologne, 21 September 1908, in *The Prin-
> ciple of Relativity*, tr. by W. Perrett and
> G. Jeffery, pp. 82-83.

EINSTEIN (1916)

What is the reason for this difference in the two bodies? No
answer can be admitted as epistemologically satisfactory unless
the reason given is an *observable fact of experience*. The law of
causality has not the significance of a statement as to the world
of experience, except when *observable facts* ultimately appear
as causes and effects.

> —Albert Einstein (1879-1955)
> "The Foundation of the General Theory of Relativ-
> ity," *Annalen der Physik, 49*, 1916, in *The Prin-
> ciple of Relativity*, tr. by W. Perrett and
> G. Jeffery, pp. 112-113.

---

* A. Einstein, Ann. d. Phys., 17, 1905, p. 891, Jahrb. d.
Radioaktivität und Elektronik, 4, 1907, p. 411.

PLANCK (1923)

In all physical science there is no more fundamental difference
than that between reversible and irreversible processes.

* * * *

There can be no doubt now, in the mind of the physicist who has
associated himself with inductive methods, that matter is constituted of atoms, heat is movement of molecules, and conduction
of heat, like all other irreversible phenomena, obeys, not dynam-
ical, but statistical laws, namely, the laws of probability.

> —Max Planck (1858-1947)
> *A Survey of Physics* (1923), tr. by R. Jones and
> D. Williams, reprinted as *A Survey of Physical
> Theory*, Dover Pubs., Inc., NY, 1960, pp. 61, 63.

WIENER (1949)

If we are dealing with any field of science where long-time
observations are possible and where experiments can be repeated,
it is desirable that our operations not be tied to any specific
origin of time. If a certain experiment started at ten o'clock
today would give a certain distribution of results by twelve
o'clock, then we must expect that if this experiment is carried
out under similar conditions at ten o'clock tomorrow, by twelve
o'clock we should get the same distribution of results. Without
at least an approximate repeatability of experiments, no comparisons of results at different times are possible, and there can
be no science. *That is, the operators which come into consideration are invariant under a shift in the origin of time.*

> —Norbert Wiener (1894-1964)
> *Extrapolation, Interpolation, and Smoothing of
> Stationary Time-Series, With Engineering Appli-
> cations*, J. Wiley & Sons, Inc., NY, 1949, p. 11.

REICHENBACH (1956)

It is an empirical fact that in all branch systems the entropy
increases in the same direction. For this empirical reason, the
convention of defining positive time through growing entropy is
inseparable from accepting causality as the general method of
explanation.

> —Hans Reichenbach (1891-1953)
> *The Direction of Time*, Univ. of Calif. Press,
> Berkeley, CA, 1956, p. 154.

BRILLOUIN (1961)

When Einstein formulated the principle of relativity, or when
de Broglie invented wave mechanics, these thinkers really
created new processes of scientific prediction. They supplied
humanity with information up to then unknown. From those remarks, we can draw the following suggestion: thought creates
negative entropy.

> —Leon Brillouin (1889-1969)
> "Thermodynamics, Statistics and Information,"
> *Amer. J. Phys.*, *29*, 1961, p. 326.

MISNER, THORNE and WHEELER (1973)

By *"time-oriented"* one means that at each event in spacetime a
distinct choice has been made as to which light cone is the
future cone and which is the past, and moreover that this choice
is continuous from event to event throughout spacetime.

* * * *

...black holes, which form by stellar collapse, and which collide,
coalesce, accrete matter, and generally wreak havoc in their
immediate vicinities. The surfaces of all black holes ("future
horizons") separate the external universe, which can send signals
out to ϑ^+ [future null infinity], from the black-hole interiors,
which cannot.

* * * *

*Hawking assumes that spacetime is "future asymptotically pre-
dictable." In essence this means that spacetime possesses no
"naked singularities" - i.e., no singularities visible from*
ϑ^+. (Naked singularities could influence the evolution of the
external universe; and, therefore, unless one knew the laws of
physics governing singularities — which one does not — they
would prevent one from predicting the future in the external
universe.)

> —Charles W. Misner (1932-   ), Kip S. Thorne (1940-   )
> and John Archibald Wheeler (1911-   )
> *Gravitation*, W.H. Freeman & Co., San Francisco,
> CA, 1973, pp. 922, 923, 933.

RYAN and SHEPLEY (1975)

...ask: If our universe is everything, what need is there to
study other universes? And again the answer is very straightforward. We must simplify...the real universe in a mathematically
tractable structure...we study aspects of the real universe and
possible aspects of the real universe. It is in fact often instructive to study features which we know not to be present in
the real universe.

> —Michael P. Ryan, Jr. (1943-   ) and
> Lawrence C. Shepley (1939-   )
> *Homogeneous Relativistic Cosmologies*, Princeton
> Univ. Press, Princeton, NJ, 1975, p. 3.

REFERENCES

Aristotle, *Physics*, tr. by R. Hardie and R. Gaye, in *The Basic Works of Aristotle*, ed. by R. McKeon, Random House, NY, 1941.

Brillouin, Leon, "Thermodynamics, Statistics, and Information," *Amer. J. Phys.*, *29*, 1961, pp. 326-327.

Cooper, Leon N., and Deborah van Vechten, "On the Interpretation of Measurement Within the Quantum Theory," *Am. Journal of Physics*, *37*, Dec. 1969, pp. 1212-1220.

Einstein, Albert, "The Foundation of the General Theory of Relativity," *Annalen der Physik, 49,* 1916, in *The Principle of Relativity*, tr. by W. Perrett and G. Jeffery, Dover Pubs., Inc., NY, 1952.

Glymour, Clark, *Theory and Evidence*, Princeton Univ. Press, Princeton, NJ, 1980.

Griffiths, A. Phillips, ed., *Knowledge and Belief*, Oxford University Press, London, 1967.

Minkowski, H., "Space and Time," address delivered at the 80th Assembly of German Natural Scientists and Physicians, Cologne, 21 Sept. 1908, in *The Principle of Relativity*, tr. by W. Perrett and G. Jeffery, pp. 75-91.

Misner, Charles W., Kip S. Thorne and John Archibald Wheeler, *Gravitation*, W.H. Freeman & Co., San Francisco, CA, 1973.

Planck, Max, *A Survey of Physics* (1923), tr. by R. Jones and D. Williams, reprinted as *A Survey of Physical Theory*, Dover Pubs., Inc., NY, 1960.

Reichenbach, Hans, *The Direction of Time,* Univ. of Calif. Press, Berkeley, CA, 1956.

Ryan, Michael P. and Lawrence C. Shepley, *Homogeneous Relativistic Cosmologies,* Princeton Univ. Press, Princeton, NJ, 1975.

Sosa, Ernest, ed., *Causality and Conditionals,* Oxford University Press, London, 1975.

Wiener, Norbert, *Extrapolation, Interpolation, and Smoothing of Stationary Time-Series, With Engineering Applications,* J. Wiley & Sons, Inc., NY, 1949.

CHAPTER 22

THE PROBLEM OF INDUCTION
AND ITS JUSTIFICATION

CHAPTER 22

THE PROBLEM OF INDUCTION AND ITS JUSTIFICATION

A.   Hume's Dilemma

Philosophical inquiry into inductive reasoning asks two questions:

1. What procedures do we use to assign truth values (categor-
    ical or probabilistic) to more general propositions about
    the physical world, given truth values for less general
    propositions?

2. Can we justify (in some sense) belief in the truth values
    for the more general propositions assigned by such proce-
    dures?

The first question has been approached both from the viewpoint of
historical/psychological study of procedures which have been or are
being used, and from the viewpoint of logical/mathematical investigation into procedures that might be proposed.

Inquiry into the second question has floundered over an inability
to reach agreement on what it means to "justify" a belief.   The
sharp contrast between the ease with which we *do* inductive reasoning
in everyday life and the tenacious difficulty of *justifying* it under
the high magnification of the philosopher's microscopic analysis
makes one suspect that the intended subject matter slips somehow
out of view when the magnification is turned up.

Consider a scientist using a Geiger counter to monitor an unknown
box which has been discovered to emit radioactivity.   See Fig. 124.
He has been monitoring this box for ten minutes.   The intensity of
its radioactivity has been such that the average frequency of at least
one Geiger counter click during a 0.1 second interval has been 0.5.

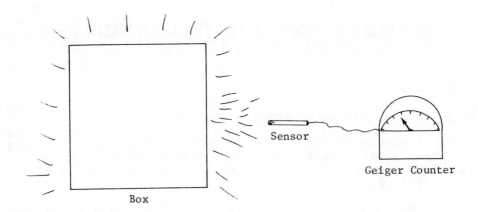

Fig. 124.   Will the Box Continue to Give Off Radioactivity?

In the ten minutes of observation, there have been 6,000 of these
0.1 second intervals.   We are concerned with the future radioactivity of the box.

Our less general proposition is: "During a past 0.1-second interval there was at least one click."   Its observed (frequency) truth
value is 0.5.

A more general proposition is: "During future interval, there will
be at least one click."   What procedure do we use to induce its
(probabilistic) truth value?   How do we justify this induction procedure?

An often used induction procedure sets the future probability equal
to the past frequency.

Suppose we classify ways of justifying such a procedure into one
or the other of the following two categories:

   o   Experience-based justifications

   o   Non-experience-based justifications

Hume's dilemma invokes the law of the excluded middle, that these are our only alternatives, and asserts that they both lead to dead ends. This is an extension of Berkeley's skepticism concerning reality of matter to a skepticism concerning reality of cause-effect relations, and an extension to probabilistic cause-effect relations of earlier Greek and Indian skepticism concerning categorical cause-effect relations.

Experience-based justifications fail. They assert, for example, that past frequencies will be good estimates of future frequencies in the future, because the frequency of past frequencies being good estimates of future frequencies has been high in the past. But we have no basis for this assertion. It is entirely circular. It attempts to use one induction procedure to justify another induction procedure.   In the example, it even attempts to use the very procedure for which we seek a justification to justify itself! (See Fig. 125.)

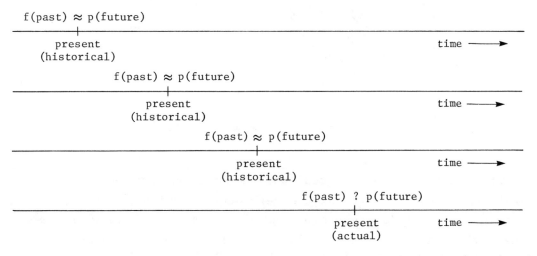

Fig. 125. Will the Future Be Like the Past Again?

Non-experience-based justifications also fail. Regardless of the chain of logic we use to reason from such a justification to the induction procedure, the justification must in the last analysis

contain the induction procedure itself as an assumption. This is
similar to the situation in which the axioms of geometry implicitly
contain all of the propositions of geometry.

Thus the induction procedure must either be a tautology or an axiom
assumed without justification. In either case, a single counterexample will suffice to destroy it. We have no way of ruling out
the counterexample. This is so even for probabilistic induction,
for then we must contend with counterexamples in which *the probabilities* are in error. This reasoning may be extended to probabilities
of probabilities, ad infinitum.

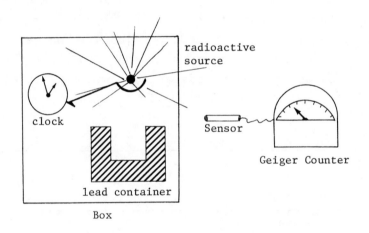

Fig. 126.   What Can We Assume about What Is
inside the Radioactive Box?

Counterexamples abound in real-world induction problems. Consider
again the scientist with his Geiger counter monitoring the radioactive box. It is conceivable that, unknown to him, there is a timing device in the box that will cause the radioactive source to be
dropped into a lead container one minute from now, and thereby shield
the radioactivity from his sensor. See Fig. 126.

Thus the non-experience-based justification is destroyed by the conceptual possibilities of experience.

Hume's conclusion was that induction is fundamentally unjustifiable. It is simply a linguistically conditioned psychological process by which we happen to acquire beliefs in general propositions that go beyond our actual experience. No amount of reasoning can, however, provide it any justification. Ultimately, all statements about the physical world, such as

"The sun will rise tomorrow,"

or

"Newton's laws will continue to hold tomorrow under such and such conditions,"

are mere opinions. No amount of reason or experience can justify any degree, however small, of belief in them. (See, e.g., Fig. 127.)

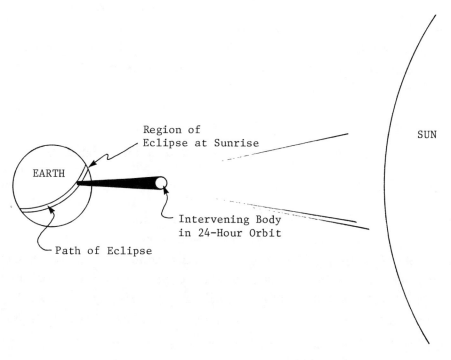

Fig. 127.   What Is the Necessity of
the Sun Rising Tomorrow?

B.   The Futile Search for Necessary Non-Tautological
     Propositions about the Physical World

Kant led the abortive attack upon Hume's dilemma from the non-
experiential flank. He drew deterministic implications from the
success of Newtonian mechanics, and maintained that there are
necessary truths about the physical world for which there are no
counterexamples. From these truths, he asserted, we may deduce,
by pure reason, the behavior (if not the existence) of things
physical.

A fine distinction can be made as to whether a counterexample must
be physically possible or need only be conceptually possible to
destroy an induction axiom. An even finer distinction can be made
between counterexamples possible in the factual world and those
only possible in a counterfactual world. But all this proved un-
necessary to Kant's undoing. Eventually real-life counterexamples
arose to challenge even the most "obvious" of Kant's necessary
truths. (See Fig. 128.) For instance:

   o   The straight line between two points is the shortest.*
       Not so, near strong gravitational fields.

   o   In all changes in the corporeal world the quantity of
       matter remains unchanged.** Not so, when matter changes
       into energy.

One has the option, of course, of following Kant's philosophy by
continually shifting one's list of "necessary" truths to new
broader theories (e.g., "enlarged" definitions of "straightness",
"matter", etc.). However, the experience of being proved wrong in
asserting something as "self-evident" about the physical world
made a deep impression. We are now much more cautious. We do not
assert that we know any "self-evident" matters of fact.

---

* I. Kant, *Prolegomena* (1783), Univ. of Manchester Press, Manchester, 1953,
p. 20.

** *Ibid.*, p. 22.

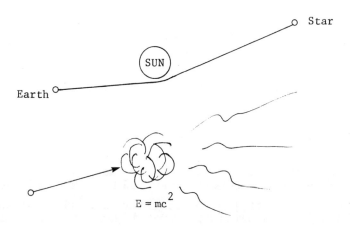

Fig. 128.   Does Any Knowable Logical Necessity
Pertain to Physical Reality?

The situation is even clearer in engineering, economics, medicine, psychology, and sociology than it is in physics.   In these fields we are so far removed from the fundamental (if there be such) laws of the physical world that counterexamples are even easier to find.

C.   Failure of the Principle of Insufficient Reason to Solve
     Hume's Dilemma

Developments in probability theory led some to think that it provided a way around Hume's dilemma. This notion stemmed from a misreading of Hume.   They assumed that Hume only argued against the justifiability of procedures for arriving at categorical assertions about the physical world.   This would leave open the justifiability of procedures for arriving at probabilistic assertions.

The most frequently used procedure of this kind is the principle of insufficient reason: If there is insufficient reason to give two probabilities different values, then make their values equal.

Applications of this principle have led to a rich variety of developments in probability theory, statistics, statistical mechanics, thermodynamics, and other fields.   However, despite these practical successes, it is recognized that mere use of probabilities does not resolve Hume's dilemma.   The dilemma is just as devastating in probabilistic approaches as it is in categorical approaches.   The example of the Geiger counter at the beginning of this chapter illustrates one such failure.

Questions left unanswered by the principle of insufficient reason include:

o What constitutes "reason"?

o How does one distinguish a "sufficient" from an "insufficient" reason? For example, how large a sample of past observations can one possess and still have "insufficient reason"?

o How does one assign probabilities when there is sufficient evidence in the form of an observed sample to overcome the principle?

o How does one assign probabilities when there are other forms of information? How much other information can one have and still have "insufficient reason"? What does one use for the probabilities when this is exceeded?

o How does one reconcile the contradictory probabilities obtained by applying the principle to different ways of describing the events?

Yet if the principle of insufficient reason is really so devoid of substantive content, how do we explain its apparent success in a wide variety of applications? The answer is that it was *not* the principle of insufficient reason that produced these successes. See Fig. 129.

Fig. 129. If Induction Has No Justification, Has Its Success Just Been Luck?

D.   Failure of Attempts to Counter Hume's Dilemma with a Frequency
     Limit Hypothesis

One of the most significant of the early contributions to probability theory was Bernoulli's proof in 1713* of the law of large
numbers.

Assume that, for a defined event, a particular outcome has a fixed
probability p. Let n be the number of instances of the event, and
x be the number of occurrences of the particular outcome.   Let P
be the probability that the frequency $\frac{x}{n}$ will differ from p by less
than an amount ε.   The law of large numbers is that P approaches unity
as n increases without bound, for any positive ε however small.
That is:

$$\lim_{n \to \infty} P\left(\left|\frac{x}{n} - p\right| < \varepsilon\right) = 1 \qquad \text{for any } \varepsilon > 0 .$$

The rate of convergence can be estimated from the approximation

$$P\left(\left|\frac{x}{n} - p\right| < \varepsilon\right) \approx \sqrt{\frac{n}{2\pi p(1-p)}} \int_{-\varepsilon}^{\varepsilon} e^{-\frac{nz^2}{2p(1-p)}} \, dz$$

$$= \text{erf}(\alpha) ,$$

where the error function is given by

$$\text{erf}(\alpha) = \frac{2}{\sqrt{\pi}} \int_0^{\alpha} e^{-t^2} \, dt$$

$$= \frac{2}{\sqrt{\pi}} e^{-\alpha^2} \sum_{k=0}^{\infty} \frac{2^k \alpha^{2k+1}}{1 \cdot 3 \cdot 5 \cdots (2k+1)} ,$$

_____

* James (Jacob) Bernoulli, *Ars Conjectandi* (1713); see also Abraham De Moivre,
Miscellanea Analytica (1733).

and the argument is

$$\alpha = \sqrt{\frac{n}{2p(1-p)}}\ \varepsilon\ .$$

Thus, if we know p we can compute the rate of convergence of P to unity for any assumed ε as $n \to \infty$.

From the form of the argument, we see that convergence is faster for larger values of $1/p(1-p)$. It is slowest when $p = \frac{1}{2}$.* For this case, we can set

$$\alpha = \sqrt{2n}\ \varepsilon$$

to estimate a lower limit for P. See Table 43. For example, if we toss a perfectly unbiased ($p = \frac{1}{2}$) coin 5,000 times, the probability of observing a frequency in the range 0.49 to 0.51 is 0.84; while the probability for this range with 10,000 tosses is 0.95.

---

* This special position of $p = \frac{1}{2}$ in the law of large numbers derives from an assumed uniform distribution over probability and the binomial analyses into two outcome options, occurrence and nonoccurrence, in much the same way as $p = \frac{1}{2}$ is picked out by the maximum entropy principle when we analyze the event in terms of two outcome possibilities. Were we to set up a multinomial analysis for the law of large numbers on the basis of m outcome possibilities, with probabilities $p_1, p_2, \ldots, p_m$, convergence would depend upon $\dfrac{1}{p_1 p_2 \cdots p_m}$ and be slowest at $p_1 = p_2 = \ldots = p_m = \dfrac{1}{m}$.

Table 43. Values of P at p = ½ for Various Values of ε and n

n \ ε	0.4	0.2	0.1	0.05	0.02	0.01	0.005
1	0.58	0.31	0.16	0.08	0.03	0.02	0.01
5	.93	.63	.35	.17	.07	.03	.02
10	.99	.79	.47	.25	.10	.05	.02
50	>.99	>.99	.84	.52	.22	.11	.06
100	"	"	.95	.68	.31	.16	.08
500	"	"	>.99	.97	.63	.35	.17
1000	"	"	"	>.99	.79	.47	.25
5000	"	"	"	"	>.99	.84	.52
10000	"	"	"	"	"	.95	.68

Despite the fact that this probability approaches unity, it cannot be shown that the frequency $\frac{x}{n}$ will approach arbitrarily close to the probability p as the sample size n is increased without bound.

$$\lim_{n \to \infty} \frac{x}{n} = p \ ?$$

This requires the existence of a limit to the series x/n as n→∞. Since such a limit does not necessarily exist, the frequency limit hypothesis, applied to any sequence of real-world observations, is an unjustified assumption.

In fact, there are numerous counterexamples to this frequency limit hypothesis, all consistent with random sampling with fixed probability p. Indeed, an infinite number of counterexamples can be generated.* The law of large numbers gives us no way of knowing

---

* Harold Jeffreys, *Theory of Probability* (1939), Oxford University Press, London, 3rd ed., 1961, pp. 63.

whether or not any real-world series is such a counterexample, or even the probability that it has a limit.

What the law of large numbers does show is referred to as *convergence in probability* of x/n to p. The frequency limit hypothesis, on the other hand, refers to *mathematical convergence*.

In an attempted answer to criticism of the frequency limit hypothesis, Reichenbach argued "What do we have to lose?" The components of his argument run as follows:

o   Although the limit does not, in general, necessarily exist, it *might* exist for a specific sequence of observations under investigation.

o   If it does not exist, then the series x/n supplies us with no data about the underlying p. In this case, the frequency estimate is certainly no worse than any other.

o   If it does exist and if p is already fairly close to it and will get progressively closer, then we can use the series x/n to make increasingly accurate estimates of p as n increases.

The technical flaw in this reasoning is that it assumes its conclusion. The fundamental flaw is that it confuses reality with our model of reality. Frequency is an observable measure of reality. Probability is a parameter in a model. There is no necessary reason why this model parameter should stay constant as n increases, i.e., as we accumulate more data.

It makes no sense to say that an event outcome *has* a probability independent of our concept of the event. Probability is a measure of our information about the event, not a property of the event itself. More properly, it is a measure of our information about ensembles of events that we choose to lump together under a common designation as "similar".

A coin is tossed. It either will or will not turn up heads. When
we model this situation by saying that the probability of heads is
one-half, we are talking about our state of mind and not about any
measurable property of the coin toss independent of our conceptual
categories. If, alternatively, we say that the heads probability
is one-half for all tosses in a specific set of one hundred "similar" coin tosses, the situation is no different except that we
are referring to our model of an ensemble of events rather than of
a single event.

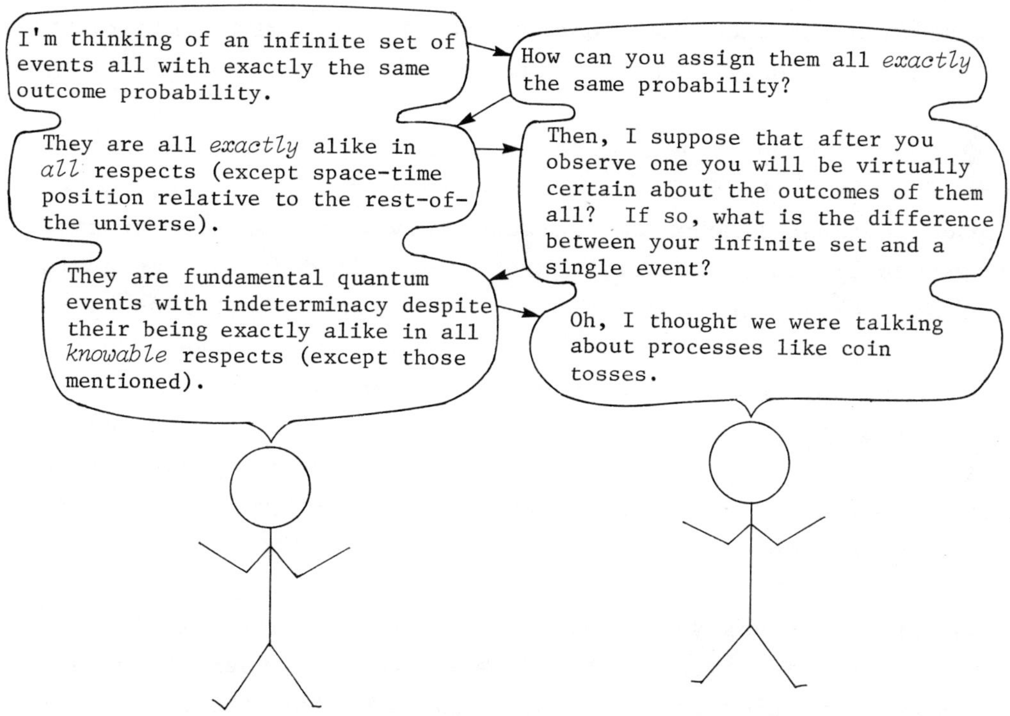

Fig. 130. How Can Two Different Events Have Exactly
the Same Outcome Probabilities?

E. Irrelevance of Bayes' Theorem to the Problem of Induction
    and Its Justification

Some authors have maintained that Bayes' inverse probability theorem,

$$P(O_k \text{ given } C_i) = \frac{P(C_i \text{ given } O_k) \, P(O_k)}{\sum_j P(C_i \text{ given } O_j) \, P(O_j)} \, ,$$

provides a proper basis for a theory of inductive reasoning. Bayesian
analysis is, however, a purely deductive process relating one set of
conditional probabilities under assumed circumstances to other sets
of conditional and unconditional probabilities. It is entirely silent on both major issues of induction.

First, Bayesian analysis does not address the problem of the extraction of probability estimates from raw observations. To use Bayes'
theorem, we must somehow have already solved this problem to obtain
the probabilities on the right-hand side. It does not matter in
Bayesian analyses whether these probabilities are obtained by assuming equiprobabilities, by equating them with observed frequencies, by entropy maximization, or by some other procedure. They simply assert that, if these probabilities are available and "consistent", then it logically follows that the inverse probabilities
can be computed. This begs the basic inductive question of how
to determine probabilities from observational data.

Second, Bayesian analysis avoids the problem of how best to construct and partition the feature space in which observations are
analyzed and predictions are made. It assumes that the problem of
defining the set of conditions for the conditional probabilities
has somehow already been solved. Thus, it is silent on the key
issues underlying the justification of inductive reasoning.

Suppose, for example, we wish to estimate P(D given C) where

        D = "person dies before age 60"

        C = "person is you".

According to Bayes' theorem:

$$P(D \text{ given } C) = \frac{P(C \text{ given } D)P(D)}{P(C \text{ given } D)P(D) + P(C \text{ given non-}D)P(\text{non-}D)} \cdot$$

Thus we need to know

P(C given D) = Probability that "person is you" given that "person dies before age 60",

P(C given non-D) = Probability that "person is you" given that "person does not die before age 60", and

P(D) = Probability that "person dies before age 60".

How do we, for example, assign a numerical value to P(D)? We could assume equiprobability. Or we could use the frequency in observational death statistics. Or we could use some other estimator. But this is not the issue. The point is that Bayes' theorem has nothing to say about how we form the probability. So when we are faced with small samples or special constraints, we need an induction procedure to guide us in arriving at P(D). The same problem is encountered in assigning values to P(C given D) and P(C given non-D).

The second point is that Bayes' theorem is silent on the justification problem. It may be better, for example, to decompose condition C into a set of sub-conditions:

$$C = X_1 \text{ and } X_2 \text{ and } X_3 \text{ and } \ldots,$$

where

$$X_1 = \text{age } 30$$
$$X_2 = \text{sex male}$$
$$\vdots$$
$$X_n = \ldots$$

Such a decomposition may enable us to obtain an estimate of the desired conditional probability which uses more of the available

information by first computing a more general conditional probability such as:

$$P(D \text{ given } C'),$$

where

$$C' = X_2 \text{ and } X_5.$$

Bayes' theorem has nothing to say about whether conditioning the
probability on C or C' extracts more information out of our available data.

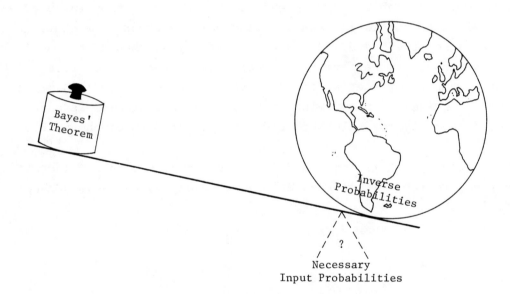

Fig. 131.   Is It Possible to Determine Any Probability
without Arbitrarily Assuming Another?

QUOTES

DEMOCRITUS (c. 420 B.C.)

In fact we do not know anything infallibly, but only that which changes according to the condition of our body and of the [influences] that reach and impinge upon it.

> —Democritus of Abdera (c. 460-370 B.C.)
> "Fragments" (Thrace, c. 420 B.C.), *Source Book in Ancient Philosophy*, ed. by C. Bakewell, Charles Scribner's Sons, Inc., NY, 1907, p. 59.

CARNEADES (c. 159 B.C.)

There is absolutely no criterion for truth. For reason, senses, ideas, or whatever else may exist are all deceptive.

Even if such criterion were at hand, it could not stand apart from the feelings which sense impressions produce. It is the faculty of feeling that distinguishes the living creature from inanimate things. By means of that feeling, the living creature becomes perceptive of both itself and the external world. There is no sensation or perception of anything unless the sense is irritated, agitated, or perturbed. When an object is indicated, then the senses become irritated and somewhat disturbed. It is impossible that there be an unperturbed presentation of external things.

The subject is more or less persuaded by the image it perceives. The strength of that persuasion depends on the disposition of the subject and or the degree of irritation produced by the image. It is not the distinctness of the image that constitutes its credibility.

The only way we can ever obtain certitude is by the difficult process of examination. We cannot be satisfied with evidence that is incomplete and only probable. Our certitude is always a precarious one. Science relies on probability, not on certitude.

> —Carneades of Cyrene (c. 214-129 B.C.)
> "The Fallacy of the Criterion of Truth" (Africa, c. 159 B.C.), *Treasury of Philosophy*, ed. by D.D. Runes, Philosophical Library, NY, 1955, pp. 224-225.

SEXTUS EMPIRICUS (c. 200 A.D.)

The proof, then, by which proof is established will not be evident and agreed...and being thus in dispute and non-evident it
will need another proof, and this again a third, and so on *ad infinitum*.

* * * *

It is also easy, I consider, to set aside the method of induction.
For, when they propose to establish the universal from the particulars by means of induction, they will effect this by a review either of all or of some of the particular instances. But
if they review some, the induction will be insecure, since some
of the particulars omitted in the induction may contravene the
universal; while if they are to review all, they will be toiling
at the impossible, since the particulars are infinite and indefinite. Thus on both grounds, as I think, the consequence is
that induction is invalidated.

> —Sextus Empiricus (160-210)
> *Outlines of Pyrrhonism*, Chapter II, (c. 200 A.D.),
> tr. by R.G. Bury, reproduced in *Sourcebook in
> Ancient Philosophy*, Charles Bakewell, ed., Charles
> Scribner's Sons, Inc., NY, 1907, pp. 362, 363-364.

JAYARĀŚI BHAṬṬA (7th Century A.D.)

A relation subsisting between one pair of terms cannot serve as
the ground of inference for another [pair], for that would be an
undue extension...for it is an essential characteristic of a
contact that it gives rise to the knowledge of an object with
reference only to a particular time and place.

...the perception of smoke could not give rise to the inference
of fire, because the relation [of invariability] subsisting between them is...unknown...

If from the cognition of that which is related to fire [smoke]
you argue to a cognition of the relation [between the two], then
there is the fallacy of undue assumption...

* * * *

From this it follows also that there is no possibility of understanding the relation of cause and effect.

> —Jayarāśi Bhaṭṭa (7th Century A.D.)
> *Refutation of Inference*, Ch. VII (India, 7th Centu-
> ry A.D.) tr. by S. Shastri and S. Saksena; rev. by
> S. Chatterjee, from *Tattvopaplavasiṁha*, ed. by P.
> Sanghavi and R. Parikh, Gaekwad's Oriental Insti-
> tute, 1940), reproduced in *A Source Book in Indian
> Philosophy*, ed. by S. Radhakrishnan and C. Moore,
> Princeton Univ. Press, Princeton, NJ, 1957, pp. 238.
> 242.

AQUINAS (c. 1256)

If, therefore, truth is principally in the mind, the judgment of truth is according to the estimation of the mind; and thus will return the error of the philosophers of antiquity, who said that all that which any one believes, is true, and that two contradictories are true at the same time; which is absurd.

Moreover, if truth is principally in the understanding, it is necessary that something which pertains to the understanding of truth be posited in the definition of truth. But Augustine in his book of *Soliloquies*, book II, chapters 4 and 5, condemns definitions of such sort as the following, *That is true which is as it is seen*, because according to this, that would not be true which is not seen, which is obviously false in the case of the deeply hidden stones which are in the bowels of the earth.

> —Thomas Aquinas (1225-1274)
> "The Disputed Questions of Truth," in *Opera Omnia* (Italy, c. 1256), ed. by S. Frette, Vives, Paris, 1875, Vol. XIV, pp. 315-341, from *Selections from Medieval Philosophers*, *II*, R. McKeon, ed., Charles Scribner's Sons, Inc., NY, 1930, pp. 170-171.

BACON (1270)

For if a man who has never seen fire should prove by sufficient argument that fire burns and that it injures things and destroys them, the mind of one hearing it would never be satisfied by that nor would a hearer avoid fire until he had put a hand or a combustible object into the fire that he might prove by experience what argument had taught. But once he has had experience of combustion, his mind is made sure and rests in the brightness of truth. Therefore, argumentation does not suffice but experience does.

> —Roger Bacon (1215-1294)
> *Opus Majus* (Paris, 1270), *The Opus Majus of Roger Bacon*, ed. by J.H. Bridges, Williams and Norgate, London, 1900, Vol. II, p. 167, from *Selections from Medieval Philosophers*, *II*, R. McKeon, ed., Charles Scribner's Sons, Inc., NY, 1930, p. 73.

DUNS SCOTUS (1302)

Sometimes, however, we experience a principle in such a way that
it is impossible to discover by further division any self- evident
principle from which it could be derived. Instead we must be
satisfied with a principle whose terms are known by experience
to be frequently united, for example, that a certain species of
herb is hot. Neither do we find any other prior means of demonstrating just why this attribute belongs to this particular subject, but must content ourselves with this as a first principle
known from experience. Now even though the uncertainty and
fallibility in such a case may be removed by the proposition
"What occurs in most instances by means of a cause that is not
free is the natural effect of such a cause", still this is the
very lowest degree of scientific knowledge - and perhaps we have
here no knowledge of the actual union of the terms but only a
knowledge of what is apt to be the case. For if an attribute is
an absolute entity other than the subject, it could be separated
from its subject without involving any contradiction. Hence, the
person whose knowledge is based on experience would not know
whether such a thing is actually so or not, but only that by its
nature is it apt to be so.

> —John Duns Scotus (1270-1308)
> "Concerning Human Knowledge" (Oxford, 1302), *Duns
> Scotus Philosophical Writings*, tr. and ed. by
> A. Wolter, T. Nelson & Sons, Ltd., London, 1963,
> pp. 110-111.

OCKHAM (1315-1319)

...a plurality must not be asserted without necessity...

*   *   *   *

And if you say that if one sees the sun and later enters a dark
place, it appears to one that one sees the sun in the same place
and of the same magnitude: therefore, the sight of the sun
remains when it is itself absent; and for the same reason it
would remain if it were itself non-existent; I reply that the
sight of the sun does not remain, but there remains some quality,
namely the light impressed on the eye, and that quality is seen;
and if the understanding forms such a proposition as this, *the
light is seen in the same place*, etc. and assents to it, it is
deceived because of that impressed quality which is seen.

> —William of Ockham (c. 1300-1349)
> *The Seven Quodlibeta* (Oxford, 1315-1319), 2nd
> edition (Strasbourg, 1491), *Selections from
> Medieval Philosophers*, II, ed. and tr. by R.
> McKeon, Ch. Scribner's Sons, NY, 1930, pp. 368-369,
> 374.

MĀDHAVA ĀCĀRYA (14th century A.D.)

Now this invariable connection must be a relation destitute of any condition accepted or disputed; and this connection does not possess its power of causing inference by virtue of its existence, as the eye, &c., are the cause of perception, but by virtue of its being known. What then is the means of this connection's being known?

We will first show that it is not perception. Now perception is held to be of two kinds, external and internal [i.e., as produced by the external senses, or by the inner sense, mind]. The former is not the required means; for although it is possible that the actual contact of the senses and the object will produce the knowledge of the particular object thus brought in contact, yet as there can never be such contact in the case of the past or the future, the universal proposition which was to embrace the invariable connection of the middle and major terms in every case becomes impossible to be known. Nor may you maintain that this knowledge of the universal proposition has the general class as its object, because, if so, there might arise a doubt as to the existence of the invariable connection in this particular case [as, for instance, in this particular smoke as implying fire].

Nor is internal perception the means, since you cannot establish that the mind has any power to act independently towards an external object, since all allow that it is dependent on the external senses, as has been said by one of the logicians, "The eye, &c. have their objects as described; but mind externally is dependent on the others."

Nor can inference be the means of the knowledge of the universal proposition, since in the case of this inference we should also require another inference to establish it, and so on, and hence would arise the fallacy of an *ad infinitum* retrogression.

—Mādhava Ācārya (14th Century A.D.)
Sarvadarśanasaṅgraha (India, 14th Century), tr. by
E. Cowell and A. Gough, Kegan Paul, Trench, Trubner
& Co., Ltd., London, 1904, p. 6-7.

NEWTON (1686)

RULE I

We are to admit no more causes of natural things than such as are both true and sufficient to explain their appearances.

To this purpose the philosophers say that Nature does nothing in vain, and more is in vain when less will serve; for Nature is pleased with simplicity, and affects not the pomp of superfluous causes.

RULE II

Therefore to the same natural effects we must, as far as possible, assign the same causes...

RULE III

The qualities of bodies, which admit neither intensification nor remission of degrees, and which are found to belong to all bodies within the reach of our experiments, are to be esteemed the universal qualities of all bodies whatsoever.

*   *   *   *

RULE IV

In experimental philosophy we are to look upon propositions inferred by general induction from phenomena as accurately or very nearly true, notwithstanding any contrary hypotheses that may be imagined, till such time as other phenomena occur, by which they may either be made more accurate, or liable to exceptions.

—Isaac Newton (1642-1727)
System of the World (Book 3 of *Principia*, London, 1686) *Sir Isaac Newton's Mathematical Principles of Natural Philosophy and His System of the World,* Vol. 2, tr. by A. Motte, (1729) ed. by F. Cajori, Univ. of Calif. Press, Berkeley, CA, 1962, pp. 398, 400.

BERKELEY (1709)

Upon the whole, I think we may fairly conclude that the proper
objects of Vision constitute the Universal Language of Nature;
whereby we are instructed how to regulate our actions, in order
to attain those things that are necessary to the preservation and
well-being of our bodies, as also to avoid whatever may be hurt-
ful and destructive of them. It is by their information that
we are principally guided in all the transactions and concerns
of life. And the manner wherein they signify and mark out unto
us the objects which are at a distance is the same with that of
languages and signs of human appointment; which do not suggest
the things signified by any likeness or identity of nature, but
only by an habitual connexion that experience has made us to
observe between them.

> — George Berkeley (1685-1753)
> *An Essay Towards A New Theory of Vision* (Dublin,
> 1709), reproduced in *Berkeley: Essay, Principles,
> Dialogues, with Selections from Other Writings*,
> ed. by M. Calkins, Charles Scribner's Sons, NY,
> 1929, p. 92.

BERKELEY (1710)

But, though it were possible that solid, figured, moveable sub-
stances.may exist without the mind, corresponding to the ideas
we have of bodies, yet how is it possible for us to know this?
Either we must know it by Sense or by Reason. As for our senses,
by them we have the knowledge only of our sensations, ideas, or
those things that are immediately perceived by sense, call them
what you will: but they do not inform us that things exist
without the mind, or unperceived, like to those which are perceived. This the materialists themselves acknowledge. --It
remains therefore, that if we have any knowledge at all of external things, it must be by reason inferring their existence
from what is immediately perceived by sense. But [I do not see]
what reason can induce us to believe the existence of bodies
without the mind, from what we perceive, since the very patrons
of Matter themselves do not pretend there is any necessary connexion betwixt them and our ideas? I say it is granted on all
hands (and what happens in dreams, frensies, and the like, puts
it beyond dispute) that it is possible we might be affected with
all the ideas we have now, though no bodies existed without resembling them. Hence it is evident the supposition of external
bodies is not necessary for the producing our ideas; since it is
granted they are produced sometimes, and might possibly be produced always, in the same order we see them in at present, without their concurrence.

* .* * *

BERKELEY (1710, cont.)

Now the set rules, or established methods, wherein the Mind we depend on excites in us the ideas of Sense, are called *the laws of nature;* and these we learn by experience, which teaches us that such and such ideas are attended with such and such other ideas in the ordinary course of things.

This gives us a sort of foresight, which enables us to regulate our actions for the benefit of life. And without this we should be eternally at a loss: we could not know how to act anything that might procure us the least pleasure, or remove the least pain of sense. That food nourishes, sleep refreshes, and fire warm us; that to sow in the seed-time is the way to reap in the harvest; and in general that to obtain such or such ends, such or such means are conducive -- all this we know, not by discovering any *necessary connexion* between our ideas, but only by the observation of the *settled laws* of nature; without which we should be all in uncertainty and confusion, and a grown man no more know how to manage himself in the affairs of life than an infant just born.

> —George Berkeley (1685-1753)
> *A Treatise Concerning the Principles of Human Knowledge* (Dublin, 1710), reproduced in *Berkeley: Essay, Principles, Dialogues, with Selections from Other Writings,* ed. by M. Calkins, Charles Scribner's Sons, NY, 1929, pp. 133-134, 140.

BERKELEY (1713)

Hyl. You say you believe your senses; and seem to applaud yourself that in this you agree with the vulgar. According to you, therefore, the true nature of a thing is discovered by the senses. If so, whence comes that disagreement? Why is not the same figure, and other sensible qualities, perceived all manner of ways? and why should we use a microscope the better to discover the true nature of a body, if it were discoverable to the naked eye?

Phil. Strictly speaking, Hylas, we do not see the same object that we feel; neither is the same object perceived by the microscope which was by the naked eye. But, in case every variation was thought sufficient to constitute a new kind or individual, the endless number or confusion of names would render language impracticable. Therefore, to avoid this, as well as other inconveniences which are obvious upon a little thought, men combine together several ideas, apprehended by divers senses, or by the same sense at different times, or in different circumstances, but observed, however, to have some connexion in nature, either with respect to co-existence or succession; all which they refer to one name, and consider as one thing. Hence it follows that

BERKELEY (1713, cont.)

when I examine, by my other senses, a thing I have seen, it is
not in order to understand better the same object which I had
perceived by sight, the object of one sense not being perceived
by the other senses. And when I look through a microscope, it
is not that I may perceive more clearly what I perceived already with my bare eyes; the object perceived by the glass being
quite different from the former. But, in both cases, my aim is
only to know what ideas are connected together; and the more a
man knows of the connexion of ideas, the more he is said to know
of the nature of things. What therefore, if our ideas are variable; what if our senses are not in all circumstances affected
by the same appearances? It will not thence follow they are not
to be trusted; or that they are inconsistent either with themselves
or anything else: except it be with your preconceived notion of
(I know not what) one single, unchanged, unperceivable, real
Nature, marked by each name. Which prejudice seems to have
taken its rise from not rightly understanding the common language
of men, speaking of several distinct ideas as united into one
thing by the mind. And, indeed, there is cause to suspect
several erroneous conceits of the philosophers are owing to the
same original: while they began to build their schemes not so
much on notions as on words, which were framed by the vulgar,
merely for conveniency and dispatch in the common actions of
life, without any regard to speculation.

Hyl. Methinks I apprehend your meaning.

Phil. It is your opinion the ideas we perceive by our senses are
not real things, but images or copies of them. Our knowledge,
therefore, is no farther real than as our ideas are the true
representations of those *originals*. But, as they supposed originals are in themselves unknown, it is impossible to know how
far our ideas resemble them; or whether they resemble them at
all. We cannot, therefore, be sure we have any real knowledge.
Farther, as our ideas are perpetually varied, without any change
in the supposed real things, it necessarily follows they cannot
all be true copies of them: or, if some are and others are not,
it is impossible to distinguish the former from the latter. And
this plunges us yet deeper in uncertainty. Again, when we consider the point, we cannot conceive how any idea, or anything
like an idea, should have an absolute existence out of a mind:
nor consequently, according to you, how there should be any real
thing in nature.

> —George Berkeley (1685-1753)
> *Three Dialogues Between Hylas and Philonous*
> (London, 1713), reproduced in *Berkeley: Essay,
> Principles, Dialogues, with Selections from Other
> Writings,* ed. by M. Calkins, Charles Scribner's
> Sons, NY, 1929, pp. 319-321.

HUME (1739)

A very material question has been started concerning *abstract* or *general* ideas, *whether they be general or particular in the mind's conception of them*. A[1] great philosopher has disputed the receiv'd opinion in this particular, and has asserted, that all general ideas are nothing but particular ones, annexed to a certain term, which gives them a more extensive signification, and makes them recall upon occasion other individuals, which are similar to them...I look upon this to be one of the greatest and most valuable discoveries that has been made of late years in the republic of letters...

* * * *

'Tis the same case with *identity* and *causation*. Two objects, tho' perfectly resembling each other, and even appearing in the same place at different times, may be numerically different: And as the power, by which one object produces another, is never discoverable merely from their idea, 'tis evident *cause* and *effect* are relations, of which we receive information from experience, and not from any abstract reasoning or reflexion. There is no single phenomenon, even the most simple, which can be accounted for from the qualities of the objects, as they appear to us; or which we cou'd foresee without the help of our memory and experience.

* * * *

The idea, then, of causation must be deriv'd from some *relation* among objects; and that relation we must now endeavour to discover. I find in the first place, that whatever objects are consider'd as causes or effects, are *contiguous;* and that nothing can operate in a time or place, which is ever so little remov'd from those of its existence.

* * * *

The second relation I shall observe as essential to causes and effects, is not so universally acknowledg'd but is liable to some controversy. 'Tis that of PRIORITY of time in the cause before the effect.

* * * *

Having thus discover'd or suppos'd the two relations of *contiguity* and *succession* to be essential to causes and effects, I find I am stopt short, and can proceed no farther in considering any single instance of cause and effect.

* * * *

---

[1] Dr. *Berkeley*.

HUME (1739, cont.)

When we consider these objects with the utmost attention, we find only that the one body approaches the other; and that the motion of it precedes that of the other, but without any sensible interval. 'Tis in vain to rack ourselves with *farther* thought and reflexion upon this subject. We can go no *farther* in considering this particular instance. Shou'd any one leave this instance, and pretend to define a cause, by saying it is something productive of another, 'tis evident he wou'd say nothing. For what does he mean by *production?* Can he give any definition of it, that will not be the same with that of causation? If he can; I desire it may be produc'd. If he cannot; he here runs in a circle, and gives a synonimous term instead of a definition.

* * * *

Your appeal to past experience decides nothing in the present case; and at the utmost can only prove, that that very object, which produc'd any other, was at the very instant endow'd with such a power; but can never prove, that the same power must continue in the same object or collection of sensible qualities; much less, that a like power is always conjoin'd with like sensible qualities. Shou'd it be said, that we have experience, that the same power continues united with the same object, and that like objects are endow'd with like powers, I wou'd renew my question, *why from this experience we form any conclusion beyond those past instances, of which we have had experience.* If you answer this question in the same manner as the preceding, your answer gives still occasion to a new question of the same kind, even *in infinitum;* which clearly proves, that the foregoing reason had no just foundation.

* * * *

We are next to consider what effect a superior combination of chances can have upon the mind, and after what manner it influences our judgment and opinion. Here we may repeat all the same arguments we employ'd in examining that belief, which arises from causes; and may prove after the same manner, that a superior number of chances produces our assent neither by *demonstration* nor *probability.*

* * * *

HUME (1739, cont.)

Shou'd it be said, that tho' in an opposition of chances 'tis
impossible to determine with *certainty*, on which side the event
will fall, yet we can pronounce with certainty, that 'tis more
likely and probable, 'twill be on that side where there is a
superior number of chances, than where there is an inferior:
Shou'd this be said, I wou'd ask, what is here meant by *likelihood* and *probability*? The likelihood and probability of chances
is a superior number of equal chances; and consequently when we
say 'tis likely the event will fall on the side, which is superior,
rather than on the inferior, we do no more than affirm, that
where there is a superior number of chances there is actually a
superior, and where there is an inferior there is an inferior;
which are identical propositions, and of no consequence. The
question is, by what means a superior number of equal chances
operates upon the mind, and produces belief or assent; since
it appears, that 'tis neither by arguments deriv'd from demonstration, nor from probability.

*   *   *   *

But 'tis evident, in the first place, that the repetition of
like objects in like relations of succession and contiguity
discovers nothing new in any one of them; since we can draw no
inference from it, nor make it a subject either of our demonstrative or probable reasonings...

*   *   *   *

I am sensible, that of all the paradoxes, which I have had, or
shall hereafter have occasion to advance in the course of this
treatise, the present one is the most violent, and that 'tis
merely by dint of solid proof and reasoning I can ever hope it
will have admission, and overcome the inveterate prejudices of
mankind.

> —David Hume (1711-1776)
> *A Treatise of Human Nature* (London, 1739),
> L.A. Selby-Bigge, ed., Oxford at the Clarendon
> Press, London, 1888, pp. 17, 69-70, 75, 75-76,
> 77, 91, 126, 127, 163, 166.

HUME (1748)

Though there be no such thing as *Chance* in the world; our ignorance of the real cause of any event has the same influence on the understanding, and begets a like species of belief or opinion.

> —David Hume (1711-1776)
> "Of Probability," *Enquiries Concerning the Human Understanding* (London, 1748), L.A. Selby-Bigge, ed., Oxford at the Clarendon Press, London, 1902, p. 56.

The first time a man saw the communication of motion by impulse, as by the shock of two billiard balls, he could not pronounce that the one event was *connected:* but only that it was *conjoined* with the other. After he has observed several instances of this nature, he then pronounces them to be *connected*. What alteration has happened to give rise to this new idea of *connexion?* Nothing but that he now *feels* these events to be *connected* in his imagination, and can readily foretell the existence of one from the appearance of the other. When we say, therefore, that one object is connected with another, we mean only that they have acquired a connexion in our thought...

> —David Hume (1711-1776)
> "Of the Idea of Necessary Connexion," *Enquiries Concerning the Human Understanding* (London, 1748), L.A. Selby-Bigge, ed., Oxford at the Clarendon Press, London, 1902, pp. 75-76.

KANT (1781)

Space is not a conception which has been derived from outward experiences. For, in order that certain sensations may relate to something without me (that is, to something which occupies a different part of space from that in which I am); in like manner, in order that I may represent them not merely as without of and near to each other, but also in separate places, the representation of space must already exist as a foundation.

* * * *

Time is not an empirical conception. For neither coexistence nor succession would be perceived by us, if the representation of time did not exist as a foundation *a priori*. Without this presupposition we could not represent to ourselves that things exist together at one and the same time, or at different times, that is, contemporaneously, or in succession.

* * * *

Substances (in the world of phenomena) are the substratum of all determinations of time. The beginning of some, and the ceasing to be of other substances, would utterly do away with the only condition of the empirical unity of time; and in that case phenomena would relate to two different times, in which, side by side, existence would pass; which is absurd.

* * * *

"All alteration (succession) of phenomena is merely change;" for the changes of substance are not origin or extinction, because the conception of change presupposes the same subject as existing with two opposite determinations, and consequently as permanent.

* * * *

But the conception which carries with it a necessity of synthetical unity, can be none other than a pure conception of the understanding which does not lie in mere perception; and in this case it is the conception of the *relation of cause and effect,* the former of which determines the latter in time, as its necessary consequence, and not as something which relate[s] to a preceding point of time; from a given time, on the other hand, there is always a necessary progression to the determined succeeding time.

> —Immanuel Kant (1724-1804)
> *Critique of Pure Reason* (Königsberg, 1781), tr. by J. Meiklejohn, Willey Book Co., NY, 1943, pp. 23, 28, 127, 128, 128-129.

KANT (1783)

...it was David Hume's remark that first, many years ago, inter-rupted my dogmatic slumber...

So I tried first whether Hume's objection could not be repre-sented universally, and I soon found that the concept of the connection of cause and effect is by no means the only one by which connections between things are thought *a priori* by the understanding; indeed that metaphysics consists of nothing else whatever. I tried to make certain of the number of these con-cepts, and when I had succeeded in doing this in the way I wished, namely from a single principle, I proceeded to the deduction of them. I was now assured that they are not, as Hume had feared, deduced from experience, but have their origin in pure under-standing.

* * * *

Thus neither outer experience, which provides the source of physics proper, nor inner experience, which provides the basis for empirical psychology, will be the ground of metaphysics. Metaphysics is thus knowledge *a priori*, or out of pure under-standing and pure reason.

* * * *

Geometry is grounded on the pure intuition of space. Arithmetic forms its own concepts of numbers by successive addition of units in time; and pure mechanics especially can only form its concepts of motion by means of the representation of time. But both repre-sentations are merely intuitions; for if everything empirical, namely what belongs to sensation, is taken away from the empirical intuitions of bodies and their changes (motion), space and time are still left.

* * * *

...there are also several things in it which are not wholly pure and independent of sources in experience: such as the concept of *motion*, of *impenetrability* (on which the empirical concept of matter rests), of *inertia* and others, which prevent it from being called quite pure natural science...

...*that substance remains* and is permanent, that *everything that happens is* always previously *determined* according to constant laws *by a cause*, etc. These really are universal laws of nature which subsist wholly *a priori*. Thus there is in fact pure natural science...

—Immanuel Kant (1724-1804)
Prolegomena to Any Future Metaphysics That Will Be Able to Present Itself as a Science (Königsberg, 1783), tr. by P. Lucas, Manchester Univ. Press, Manchester, 1953, pp. 9-10, 15, 39, 53.

DE MORGAN (1838)

No finite experience whatsoever can justify us in saying that the future shall coincide with the past in all time to come, or that there is any probability for such a conclusion.

—Augustus De Morgan (1806-1871)
"An Essay on Probabilities and on Their Application to Life Contingencies and Insurance Offices,"
Cabinet Cyclopedia, ed. by Dr. D. Lardner, Longman
and Co., London, 1838, p. 128.

MAXWELL (1877)

Absolute, true and mathematical Time is conceived by Newton as flowing at a constant rate, unaffected by the speed or slowness of the motions of material things.

* * * *

Absolute space is conceived as remaining always similar to itself and immovable. The arrangement of the parts of space can no more be altered than the order of the portions of time. To conceive them to move from their places is to conceive a place to move away from itself.

...We cannot describe the time of an event except by reference to some other event, or the place of a body except by reference to some other body. All our knowledge, both of time and place, is essentially relative.

* * * *

"The difference between one event and another does not depend on the mere difference of the times or the places at which they oc-cur, but only on differences in the nature, configuration, or motion of the bodies concerned."

It follows from this that if an event has occurred at a given time and place it is possible for an event exactly similar to occur at any other time and place.

—James Clerk Maxwell (1831-1879)
Matter and Motion (1877), Dover Pubs., Inc., NY,
1952, pp. 11, 12, 13.

FREUD (1899)

...the shortest path to the fulfilment of the wish is a path leading direct from the excitation produced by the need to a complete cathexis of the perception. Nothing prevents us from assuming that there was a primitive state of the psychical ap-paratus in which this path was actually traversed, that is in which wishing ended in hallucinating.

—Sigmund Freud (1856-1939)
The Interpretation of Dreams (1899), ed. and tr. by
J. Strachey, Avon Books, Discus Edition, NY, 1965,
p. 605.

JAMES (1907)

Locke, Hume, Berkeley, Kant, Hegel, have all been utterly sterile
so far as shedding any light on the details of nature goes, and
I can think of no invention or discovery that can be directly
traced to anything in their peculiar thought...

*   *   *   *

Sensations are forced upon us, coming we know not whence. Over
their nature, order, and quantity we have as good as no control.
They are neither true nor false; they simple *are*. It is only
what we say about them, only the names we give them, our theories
of their source and nature and remote relations, that may be or
not.

> —William James (1824-1910)
> *Pragmatism: A New Name For Some Old Ways of Think-*
> *ing* (1907), Harvard Univ. Press, Cambridge, MA, 1978,
> pp. 91-92, 117.

EINSTEIN (1921)

The only justification for our concepts and system of concepts
is that they serve to represent the complex of our experiences;
beyond this they have no legitimacy. I am convinced that the
philosophers have had a harmful effect upon the progress of
scientific thinking in removing certain fundamental concepts
from the domain of empiricism, where they are under our control,
to the intangible heights of the *a priori*. For even if it should
appear that the universe of ideas cannot be deduced from experience by logical means, but is, in a sense, a creation of the
human mind, without which no science is possible, nevertheless
this universe of ideas is just as little independent of the
nature of our experiences as clothes are of the form of the
human body. This is particularly true of our concepts of time
and space, which physicists have been obliged by the facts to
bring down from the Olympus of the *a priori* in order to adjust
them and put them in a serviceable condition.

> —Albert Einstein (1879-1955)
> "Space and Time in Pre-Relativity Physics," *The*
> *Meaning of Relativity: Four Lectures Delivered*
> *at Princeton University, May 1921*, tr. by E. Adams,
> Princeton Univ. Press, Princeton, NJ, 1922, p. 2.

WHORF (1940)

> We are thus introduced to a new principle of relativity, which holds that all observers are not led by the same physical evidence to the same picture of the universe, unless their linguistic backgrounds are similar, or can in some way be calibrated.

* * * *

> What surprises most is to find that various grand generalizations of the Western world, such as time, velocity, and matter, are not essential to the construction of a consistent picture of the universe. The psychic experiences that we class under these headings are, of course, not destroyed; rather, categories derived from other kinds of experiences take over the rulership of the cosmology and seem to function just as well. Hopi may be called a timeless language. It recognizes psychological time, which is much like Bergson's "duration", but this "time" is quite unlike the mathematical time, T, used by our physicists.

> —Benjamin Lee Whorf (1897-1941)
> "Science and Linguistics," *The Technology Review*, *42*, April 1940, pp. 231, 247.

> ...automatic, involuntary patterns of language are not the same for all men but are specific for each language...

> From this fact proceeds what I have called the "linguistic relativity principle," which means, in informal terms, that users of markedly different grammars are pointed by their grammars toward different types of observations and different evaluations of externally similar acts of observation, and hence are not equivalent as observers but must arrive at somewhat different views of the world.

> —Benjamin Lee Whorf (1897-1941)
> "Linguistics as an Exact Science," *The Technology Review*, *43*, Dec. 1940, p. 61.

MATURANA, LETTVIN, McCULLOCH and PITTS (1960)

> How do frogs recognize the universals, *prey* and *enemy*?

> —Humberto R. Maturana (1928-   ), Jerome Y. Lettvin (1920-   ), W.S. McCulloch (1868-1969) and Walter H. Pitts (1923-1969)
> "Anatomy and Physiology of Vision in the Frog (*Rana pipiens*)," *The Journal of General Physiology*, *43*, Supp., July 1960, p. 129.

JUNG (1961)

Thus a word or an image is symbolic when it implies something
more than its obvious and immediate meaning. It has a wider
"unconscious" aspect that is never precisely defined or fully
explained. Nor can one hope to define or explain it. As the
mind explores the symbol, it is led to ideas that lie beyond the
grasp of reason...

Because there are innumerable things beyond the range of human
understanding, we constantly use symbolic terms to represent
concepts that we cannot define or fully comprehend. This is one
reason why all religions employ symbolic language or images.
But this conscious use of symbols is only one aspect of a psychological fact of great importance: Man also produces symbols unconsciously and spontaneously, in the form of dreams.

It is not easy to grasp this point. But the point must be
grasped if we are to know more about the ways in which the human
mind works. Man, as we realize if we reflect for a moment, never
perceives anything fully or comprehends anything completely. He
can see, hear, touch and taste; but how far he sees, how well he
hears, what his touch tells him, and what he tastes depend upon
the number and quality of his senses. These limit his perception
of the world around him. By using scientific instruments he can
partly compensate for the deficiencies of his senses. For example, he can extend the range of his vision by binoculars or of
his hearing by electrical amplification. But the most elaborate
apparatus cannot do more than bring distant or small objects
within range of his eyes, or make faint sounds more audible. No
matter what instruments he uses, at some point he reaches the
edge of certainty beyond which conscious knowledge cannot pass.

> —Carl G. Jung (1875-1961)
> "Approaching the Unconscious," (1961), *Man
> and His Symbols*, Dell Pub. Co., NY, 1968, p. 4.

ERICKSON, ROSSI and ROSSI (1976)

In most experiences of trance some observer ego is present,
quietly taking in the scene; the patient is quietly watching
what is happening within*. It is this observing ego that gives
a detached, impersonal and objective quality to much of the conscious ideation in trance. The objective quality of this ideation makes it particularly useful in psychotherapy. As long as
this observer ego is present, however, many patients will insist
they are not hypnotized; they equate the observer function with
being conscious in the normal sense of the word.

> —Milton H. Erickson (1901-   ),Ernest L. Rossi
> (1933-   ) and Sheila I. Rossi (1940-   )
> *Hypnotic Realities, The Induction of Clinical Hyp-
> nosis and Forms of Indirect Suggestion*, Irvington
> Pub., Inc., NY, 1976, p. 301.

---

* Gill, M. and M. Brenman, *Hypnosis and Related States*, Intl. Univ.
Press, NY, 1959.

REFERENCES

Aquinas, Thomas, "The Disputed Question of Truth" (c. 1256), *Opera Omnia*, ed. by S. Frette, Vives, Paris, 1875, Vol. XIV, from *Selections from Medieval Philosophers*, II, R. McKeon, ed., Charles Scribner's Sons, Inc., NY, 1930.

Ayer, Alfred J., *The Foundations of Empirical Knowledge*, Macmillan & Co., Ltd., London, 1961.

_____, *Language, Truth and Logic* (1936), Dover Pubs., Inc., NY, 1946.

_____, *The Problem of Knowledge*, Penguin Books Ltd., Harmondsworth, 1956.

_____, *Probability and Evidence*, Columbia Univ. Press, NY, 1972.

Bacon, Roger, *Opus Majus* (1270), *The Opus Majus of Roger Bacon*, ed. by J.H. Bridges, Williams and Norgate, London, 1900, from *Selections from Medieval Philosophers*, II, R. McKeon, ed., Charles Scribner's Sons, Inc., NY, 1930.

Barker, Stephen F., *Induction and Hypothesis*, Cornell Univ. Press, Ithaca, NY, 1957.

Berkeley, George, *An Essay Towards a New Theory of Vision* (Dublin, 1709), reproduced in *Berkeley: Essay, Principles, Dialogues, with Selections from Other Writings*, ed. by M. Calkins, Charles Scribner's Sons, NY, 1929.

_____, *A Treatise Concerning the Principle of Human Knowledge* (Dublin, 1710), reproduced in *Berkeley: Essay, Principles, Dialogues, with Selections from Other Writings*, ed. by M. Calkins, Charles Scribner's Sons, Inc., NY, 1929.

_____, *Three Dialogues Between Hylas and Philonous* (London, 1913), reproduced in *Berkeley: Essay, Principles, Dialogues, with Selections from Other Writings*, ed. by M. Calkins, Charles Scribner's Sons, NY, 1929.

Box, George E.P. and George C. Tiao, *Bayesian Inference in Statistical Analysis*, Addison-Wesley Pub. Co., Reading, MA, 1973.

Braithwaite, R.B., *Scientific Explanation, A Study of the Function of Theory, Probability and Law in Science* (1946), Camb. Univ. Press, Cambridge, 1975.

Bunge, Miro, *The Myth of Simplicity, Problems of Scientific Philosophy*, Prentice-Hall, Englewood Cliffs, NJ, 1963.

Carneades of Cyrene, "The Fallacy of the Criterion of Truth" (c. 159 B.C.), *Treasury of Philosophy*, ed. by D.D. Runes, Philosophical Library, NY, 1955.

Christensen, R.A., "Truth and Probability," Harvard Univ., Cambridge, MA, winter 1960-1961. [*Chapter 1 of Volume II.*]

_____, "Induction and the Evolution of Language," Physics Dept., Univ. of Calif., Berkeley, CA, July 19, 1963. [*Chapter 7 of Volume II.*]

_____, *Foundations of Inductive Reasoning*, Berkeley, CA, 1964.

Cohen, Morris R. and Ernest Nagel, *An Intro. to Logic and Scientific Method*, Harcourt, Brace & Co., NY, 1934.

Day, John Patrick, *Inductive Probability*, The Humanities Press, NY, 1961.

Democritus of Abdera, "Fragments" (c. 420 B.C.), *Sourcebook in Ancient Philosophy*, ed. by C. Bakewell, Charles Scribner's Sons, Inc., NY, 1907.

De Morgan, Augustus, "An Essay on Probabilities and Their Applications to Life Contingencies and Insurance Offices," *Cabinet Cyclopedia*, Dr. Dionysius Lardner, ed., Longman and Co., London, 1838.

Duns Scotus, John, "Concerning Human Knowledge" (1302), *Duns Scotus Philosophical Writings*, tr. and ed. by A. Wolter, T. Nelson & Sons, Ltd., London, 1963.

Einstein, Albert, "Space and Time in Pre-Relativity Physics," *The Meaning of Relativity: Four Lectures Delivered at Princeton University, May, 1921*, tr. by E. Adams, Princeton Univ. Press, Princeton, NJ, 1922.

Erickson, Milton H., Ernest L. Rossi and Sheila I. Rossi, *Hypnotic Realities, The Induction of Clinical Hypnosis and Forms of Indirect Suggestion*, Irvington Pubs., Inc., NY, 1976.

Freud, Sigmund, *The Interpretation of Dreams* (1899), ed. and tr. by J. Strachey, Avon Books, Discus Edition, NY, 1965.

Harrod, Roy, *Foundations of Inductive Logic*, Harcourt, Brace & Co., NY, 1956.

Hume, David, *A Treatise of Human Nature* (London, 1739), L.A. Selby-Bigge, ed., Oxford at the Clarendon Press, London, 1888.

_____, *Enquiries Concerning the Human Understanding* (London, 1748), L.A. Selby-Bigge, ed., Oxford at the Clarendon Press, London, 1902.

James, William, *Pragmatism: A New Name For Some Old Ways of Thinking* (1907), Harvard Univ. Press, Cambridge, MA, 1978.

Jayarāśi Bhaṭṭa, "Refutation of Inference" (7th century A.D.), *A Sourcebook in Indian Philosophy*, ed. by S. Radhakrishnan and C. Moore, Princeton Univ. Press, Princeton, NJ, 1957.

Jeffreys, Harold, *Theory of Probability*, Oxford at the Clarendon Press, London, 1961.

Jung, Carl G. (ed.), *Man and His Symbols*, Dell Pub. Co., NY, 1968.

Kahl, Russell (ed.), *Studies in Explanation*, Prentice-Hall, Englewood Cliffs, NJ, 1963.

Kant, Immanuel, *Critique of Pure Reason* (1781), tr. by J. Meiklejohn, Willey Book Co., NY, 1943.

_____, *Prolegomena to any Future Metaphysics That Will Be Able to Present Itself as a Science* (1783), Manchester Univ. Press, Manchester, 1953.

Kaplan, Abraham, *The Conduct of Inquiry*, Chandler Pub. Co., San Francisco, CA, 1964.

Katz, Jerrold J., *The Problem of Induction and Its Solution*, The Univ. of Chicago Press, Chicago, IL, 1962.

Kemeny, John G., *A Philosopher Looks at Science*, D. Van Nostrand Co., NY, 1959.

Kneale, William, *Probability and Induction*, Oxford at the Clarendon Press, London, 1949.

Körner, Stephen, *Conceptual Thinking*, Dover Pubs., Inc., NY, 1959.

Kyburg, Henry E., *Probability and the Logic of Rational Belief*, Wesleyan Univ. Press, Middletown, CT, 1961.

Lewis, Clarence Irving, *Mind and the World-Order, Outline of a Theory of Knowledge* (1929), Dover Pubs., Inc., NY, 1952.

Mādhava Ācārya, *Sarvadarśanasaṁgraha* (14th century A.D.), tr. by E. Cowell & A. Gough, Kegan Paul, Trench, Trubner & Co., London, 1904.

Maturana, H.R., J.Y. Lettvin, W.S. McCulloch and W.H. Pitts, "Anatomy and Physiology of Vision in the Frog (*Rana pipiens*)," *The Journal of General Physiology, 43*, Supp., July 1960, pp. 129-175.

Maxwell, James Clerk, *Matter and Motion* (1877), Dover Pubs., Inc., NY, 1952.

Newton, Isaac, *Sir Isaac Newton's Mathematical Principles of Natural Philosophy and His System of the World*, Vol. 2, tr. by A. Motte, ed. by F. Cajori, Univ. of Calif. Press, Berkeley, CA, 1962.

Ockham, William of, *The Seven Quodlibeta* (Oxford, 1315-1319), 2nd ed., (Strasbourg, 1491), *Selections from Medieval Philosophers, II*, ed. and tr. by R. McKeon, Charles Scribner's Sons, NY, 1930.

Popper, Karl R., *Conjectures and Refutations*, Basic Books, NY, 1962.

Potterer, Ray H., *Postulates and Implications*, Philosophical Library, NY, 1955.

Reichenbach, Hans, *Experience and Predictions*, Univ. of Chicago Press, Chicago, IL, 1938.

_____, *The Rise of Scientific Philosophy*, Univ. of Calif. Press, Berkeley, CA, 1951.

Russell, Bertrand, *An Inquiry Into Meaning and Truth*, George Allen & Unwin Ltd., London, 1940.

_____, *Human Knowledge, Its Scope and Limitations*, Simon and Schuster, NY, 1948.

Salmon, Wesley C., "Statistical Explanation," *Nature and Function of Scientific Theories*, ed. by R. Coloday, Univ. of Pittsburgh Press, Pittsburgh, PA, 1970, pp. 173-231.

Santayana, George, *Scepticism and Animal Faith* (1923), Dover Pubs., Inc., NY, 1955.

Sextus Empiricus, *Outlines of Pyrrhonism* (c. 200 A.D.), tr. by R.G. Bury, in *Sourcebook in Ancient Philosophy*, ed. by C. Bakewell, Charles Scribner & Sons, Inc., 1907.

Swinburne, Richard (ed.), *The Justification of Induction*, Oxford Univ. Press, London, 1974.

Unger, Peter, *Ignorance, A Case for Skepticism*, Oxford Univ. Press, London, 1975.

Whorf, Benjamin Lee, "Science and Linguistics," *The Technology Review, 42*, April 1940, pp. 229-231, 247-248.

_____, "Linguistics as an Exact Science," *The Technology Review, 43*, Dec. 1940, pp. 61-63, 80-83.

Wisdom, John, *Paradox and Discovery*, Philosophica Librum, NY, 1965.

Wittgenstein, Ludwig, *Tractatus Logico-Philosophicus* (1921), tr. by D.F. Pears and B.F. McGuinness, Routledge & Kegan Paul, NY, 1961.

von Wright, Georg Henrik, *The Logical Problem of Induction* (1941), Basil Blackwell, London, 1957.

_____, *A Treatise on Induction and Probability*, Harcourt, Brace & Co., Inc., London, 1951.

CHAPTER 23

ENTROPY MINIMAX INDUCTION

CHAPTER 23

ENTROPY MINIMAX INDUCTION

A.   Entropy Minimax Principles, Laws and Rules

In this chapter, we review entropy minimax as a process of induction.
We take up the matter of its justification in the next chapter.

Entropy minimax is a procedure by means of which we use observational
and background data to assign conditional outcome probabilities to
events based on the following two principles of inductive reasoning:

1.   Assign values to conditional probabilities which represent
     no more information than is contained in the observational
     and background data.

2.   Make these assignments on the basis of a definition of the
     conditions which extracts as much information as possible
     from the available observational and background data.

Mathematical formulation of these principles using information
theory leads to the following three laws of induction:

1.   Given any condition C_i, assign values to the probabilities
     $\{P(O_k \text{ given } C_i)\}$ so as to maximize the entropy

$$S(O \text{ given } C_i) = - \sum_k P(O_k \text{ given } C_i) \log P(O_k \text{ given } C_i),$$

     subject to the constraints of all available observational
     and background information. Thus, by implication, use
     values for the probabilities $\{P(C_i)\}$ which minimize
     $S(C) = -\sum_i P(C_i) \log P(C_i)$.

2.   Use a partition of the outcome space $O = \{O_k\}$ consistent
     with the objectives of the analysis. When frequency is
     the outcome variable and the purpose is to reduce uncer-
     tainty about yet-to-be-observed outcomes, partition the
     range from 0 to 1 into intervals corresponding to the
     possible effect of an outcome on the frequency.

3. Use a partition of the condition space $C = \{C_i\}$ for which the entropy

$$S(O \text{ given } C) = -\sum_i P(C_i) \sum_k P(O_k \text{ given } C_i) \log P(O_k \text{ given } C_i)$$

is a minimum for the observed data, relative to the expected value of the minimum for random data.

For practical applications, approximations to these laws can be achieved for fairly general classes of problems by using the following four rules:

1. When frequency is the outcome, use an equal interval partition. For large samples, use an infinitesimal interval approximation. For general outcomes, use a classification with as few and as highly populated classes as possible, consistent with the objectives of the analysis.

2. For the conditional outcome probabilities, use

$$P(O_k \text{ given } C_i) = \frac{x_{ki} + w_k}{n_i + w} \, ,$$

where

$$x_{ki} = \text{number of observations of outcome } O_k \text{ given condition } C_i,$$

$$w_k = \text{weight of outcome } O_k \text{ (see rule \#4),}$$

$$n_i = \sum_k x_{ki} , \text{ and}$$

$$w = \sum_k w_k \, .$$

3. Use a partition of the condition space $C = \{C_i\}$ for which the entropy $S(O \text{ given } C)$ is a minimum computed on the training sample.

4. Set the relative weights w_k/w equal to the unpartitioned training sample frequencies. Set the weight normalization w to minimize the expected value of the informational disagreement

$$\Delta S_e = \sum_k F_k \log F_k/P_k$$

between the probabilities and the observed frequencies for a random test sample. Use Monte-Carlo estimation of ΔS_e. Use an extent of freedom of condition space partition search $EOF = \{\{C_i\}\}$ which minimizes this minimum ΔS_e.

The three laws define a nonparametric approach to entropy minimax. The four rules provide a parametric approximation to them.

These rules provide a defined induction procedure which can be implemented in computer software. In actual implementation, one also considers such matters as cost in run time versus expected marginal benefits of more extensive extent of freedom.

B. Nonparametric Entropy Minimax

Fig. 132 illustrates the basic elements of the nonparametric approach to entropy minimax induction. The four starting blocks are: language, patterns, background and sample.

Language comprises the terminology we use for classifying objects, relationships, etc., together with the rules of grammar. It includes, as a subset, the classification systems developed by formal scientific methods; but includes also the informally developed written and spoken language of general conversation. For a single induction process, language may generally be regarded as constant. (Language evolution is an effect of an ensemble of many inductions, in which classification terminology and grammar may change.)

The starting point for generation of the patterns is the initial variables definition and feature selection process. The background

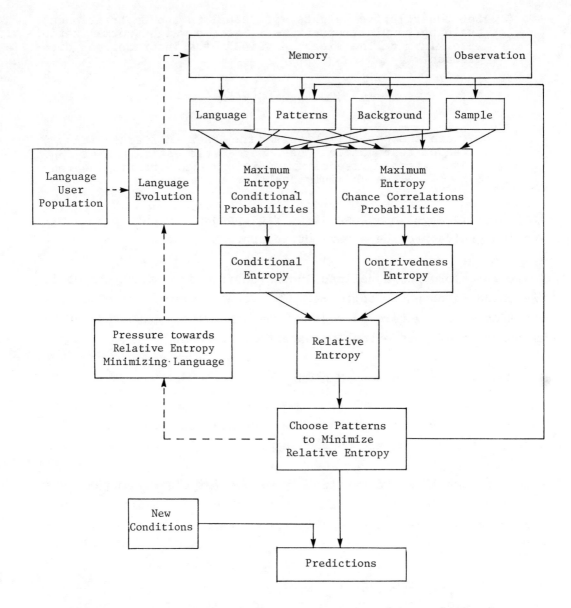

Fig. 132. Nonparametric Approach to Entropy Minimax Induction
(Dotted lines indicate ensemble of many inductions.)

weights are based on previous general experience in this feature space. The sample contains the actual observations as currently being processed.

Using these as inputs, we compute two sets of probabilities and entropies. First, we estimate the conditional probabilities based on the observational and background data. We use entropy maximization as the criterion for making these probability estimates. From these we compute the conditional entropy.

Second, we compute the maximum entropy chance correlation probabilities and the associated contrivedness entropy. This is done by computing how low a conditional entropy we could expect to obtain by training on random data, given the freedom permitted the algorithm to alter pattern boundaries.

We then calculate the difference between these two entropies. This relative entropy tells us how much better we are doing on real data than we could expect to do on random data. We then vary the pattern boundaries, within the given extent of pattern boundary freedom, and repeat the calculation. We do not need to repeat the contrivedness calculation since its results remain constant. When we have ranged over all of the permitted pattern boundaries, we choose those patterns for which the relative entropy is least.

The model building process is now complete and we are ready to make predictions for new conditions. These will use maximum entropy probabilities given minimum relative entropy pattern boundaries.

Note that the prediction process is quite straightforward once we have found the desired patterns and probabilities. We simply identify which pattern is matched by the new conditions and use its probabilities for the prediction.

C. Parametric Entropy Minimax

Fig. 133 illustrates the basic elements of the parametric approach
to entropy minimax. Here the contrivedness calculations are replaced by sample randomization, trial prediction and crossvalidation.

Advantages of the parametric approach include:

- o Substantial computer run-time savings over Monte-Carlo es-
    timation of chance correlation probabilities.

- o Valuable test case crossvalidation. (Using up one degree
    of freedom to fit the background weight has little effect
    on the crossvalidation confidence levels, considering the
    complexity of the relationships generally involved.)

The total sample is randomly split into training and test halves.
The whole procedure is more informationally efficient if this random
split is constrained so that the two halves have roughly the same
distributions on the variables. However, in the splitting process,
intervariable conditional constraints must be avoided, since the
objective is to test for the presence of relationships between the
dependent and independent variables.

The entire entropy minimizing pattern selection process is repeated
for various tentative background weights. Using the patterns corresponding to each tentative weight, the test sample probabilities
are predicted and compared to the observed frequencies. That weight
is chosen for which the probabilities and frequencies have least
disagreement. The patterns corresponding to the chosen weight are
those best suited to make predictions for new conditions.

D. Data Dependent Simplicity

It is conventional to define *simplicity* in terms of the number of variables used in a proposition and to define the meaning of *predictive*

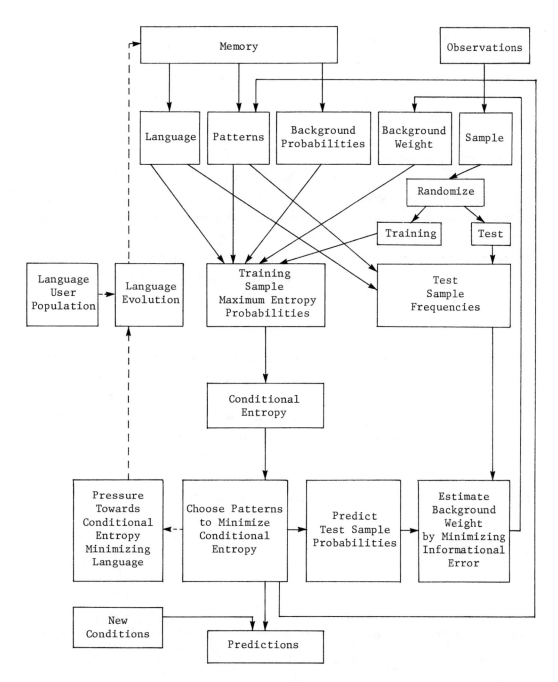

Fig. 133.   Parametric Approach to Entropy Minimax Induction
Crossvalidation is Used to Estimate a Single Parameter:
The Background Weight.   (Dotted lines indicate ensembles
of many inductions.)

success of the proposition by means of a least squared error criterion. Using such definitions, it is not possible to demonstrate,
except in special cases, that greater simplicity implies more predictive success. This is because the assertion is in general untrue
in these terms.

In entropy minimax theory, we define simplicity as the reciprocal
of the entropy, normalized by the logarithm of the number of outcome classes. The simplest proposition is the one with least entropy on our data, subject to given constraints.* We define the measure of predictive success as the entropy decrement on the data,
subject to the same constraints. These constraints define the chosen
assessment utilities (accuracy, definitiveness, risk-avoidance, etc.).
In these terms, it follows that the simplest proposition has the
highest expected value for predictive success.

It is interesting to watch this definition of simplicity in operation on the computer. For data sets with many irregularities, the
weight normalization is driven up. This results in a preference
for patterns with few conditions and high populations. For data
sets with fewer irregularities, the weight normalization is lowered.
This increases the tendency of the algorithm to accept patterns
with more numerous qualifiers, trading low population for high outcome purity. The entropic notion of simplicity is not an absolute
concept, definable entirely in terms of the logical structure and
mathematical form of the patterns. It is rather a data dependent
concept.

---

* R.A. Christensen, "Induction and the Evolution of Language," Physics Dept.,
Univ. of Calif., Berkeley, CA, 1963 (*Volume II*, p. 147); *Foundations of Induc-
tive Reasoning*, Berkeley, CA, 1964, Chapter 8: "Inductive Reasoning and the
Evolution of Language," Physics Dept., Univ. of Calif., Berkeley, CA, 1964
(*Volume II*, p. 164); "Inductive Reasoning as the Source of Human Knowledge,"
Univ. of Calif., Berkeley, CA, 1965 (*Volume II*, pp. 208-209); "A General
Approach to Pattern Discovery," Tech. Rept. No. 20, Computer Center, Univ. of
Calif., Berkeley, CA, 1967 (*Volume III*, p. 20).

QUOTES

GROSSETESTE (c. 1209)

This, therefore, is the way by which the abstracted universal is reached from singulars through the help of the senses; clearly the experimental *(experimentale)* universal is acquired by us, whose mind's eye is not purely spiritual, only through the help of the senses. For when the senses several times observe two singular occurrences, of which one is the cause of the other or is related to it in some other way, and they do not see the connexion between them, as, for example, when someone frequently notices that the eating of scammony happens to be accompanied by the discharge of red bile and does not see that it is the scammony that attracts and withdraws the red bile, then from constant observation of these two observable things he begins to form a third, unobservable thing, namely the scammony is the cause that withdraws the red bile. And from this perception repeated again and again and stored in the memory, and from the sensory knowledge from which the perception is built up, the functioning of the reasoning begins. The functioning reason therefore begins to wonder and to consider whether things really are as the sensible recollection says, and these two lead the reason to the experiment, namely, that scammony should be administered after all other causes purging red bile have been isolated and excluded. But when he has administered scammony many times with the sure exclusion of all other things that withdraw red bile, then there is formed in the reason this universal, namely, that all scammony of its nature withdraws red bile; and this is the way in which it comes from sensation to a universal experimental principle.

* * * *

That is better and more valuable which requires fewer, other circumstances being equal, just as that demonstration is better, other circumstances being equal, which necessitates the answering of a smaller number of questions for a perfect demonstration or requires a smaller number of suppositions and premisses from which the demonstration proceeds. For if one thing were demonstrated from many and another thing from fewer equally known premisses, clearly that is better which is from fewer because it makes us know more quickly, just as universal demonstration is better than particular because it produces knowledge from fewer premisses. Similarly in natural science, in moral science, and in metaphysics the best is that which needs no premisses and the better that which needs the fewer, other circumstances being equal.

> —Robert Grosseteste (c. 1168-1253)
> "Commentary on Aristotle's *Posterior Analytics*"
> (Oxford, c. 1209), *Comm. Post.* i.14, ff. 13-14;
> i.17, f.17, Venice, 1494, quoted by A.C. Crombie,
> *Robert Grosseteste and the Origins of Experimental
> Science 1100-1700*, Oxford at the Clarendon Press,
> London, 1953, pp. 73-74, 86.

DE MORGAN (1847)

By degree of probability we really mean, or ought to mean, degree of belief. It is true that we may, if we like, divide
probability into ideal and objective, and that we must do so,
in order to represent common language. It is perfectly correct
to say "It is much more likely than not, *whether you know it or
not*, that rain will soon follow the fall of the barometer." We
mean that rain does soon follow much more often than not, and
that there do exist the means of arriving at this knowledge.
The thing is so, everyone will say, and can be known. It is not
remembered, perhaps, that there is an *ideal probability*, a pure
state of the mind, involved in this assertion: namely, that the
things which have been are correct representatives of the things
which are to be. That up to this 21st of June, 1847, the above
statement has been true, ever since the barometer was used as a
weather-glass, is not denied by any who have examined it: that
the connexion of natural phenomena will, for some time to come,
be what it has been, cannot be settled by examination: we all
have strong reason to believe it, but our knowledge is *ideal*, as
distinguished from *objective*. And it will be found that, frame
what circumstances we may, we cannot invent a case of purely objective probability. I put ten white balls and ten black ones
into an urn, and lock the door of the room. I may feel well assured that, when I unlock the room again, and draw a ball, I am
justified in saying it is an even chance that it will be a white
one. If all the metaphysicians who ever wrote on probability
were to witness the trial, they would, each in his own sense and
manner, hold me right in my assertion. But how many things there
are to be taken for granted! Do my eyes still distinguish colours as before? Some persons never do, and eyes alter with age.
Has the black paint melted, and blackened the white balls? Has
any one else possessed a key of the room or got in at the window,
and changed the balls? We may be *very sure*, as those words are
commonly used, that none of these things have happened, and it
may turn out (and I have no doubt will do so, if the reader try
the circumstances) that the ten white and ten black balls will be
found, as distinguishable as ever, and unchanged. But for all
that, there is much to be assumed in reckoning upon such a result, which is not so objective (in the sense in which I have
used the word) as the knowledge of what the balls were when they
were put into the urn. We have to assume all that is requisite
to make our experience of the past the means of judging the
future.

—Augustus De Morgan (1806-1871)
Formal Logic (1847), The Open Court Co., London,
1926, pp. 198-199.

BRILLOUIN (1951)

$$\text{negentropy} \longrightarrow \text{information} \longrightarrow \text{negentropy} \qquad (1)$$

We coined the abbreviation "negentropy" to characterize entropy
with the opposite sign.

* * * *

The physicist making an observation performs the first part of
process (1); he transforms negative entropy into information.
We may now ask the question: can the scientist use the second
part of the cycle and change information into negentropy of some
sort? Without directly answering the question, we may risk a suggestion. Out of experimental facts, the scientist builds up
scientific knowledge and derives scientific laws. With these laws
he is able to design and build machines and equipment that nature
never produced before; these machines represent most unprobable
structures, and low probability means negentropy?

> —Leon Brillouin (1889-1969)
> "Maxwell's Demon Cannot Operate: Information and
> Entropy, I," *J. of Applied Physics, 22,* March 1951,
> pp. 334, 337.

JAYNES (1957)

...mathematical difficulties, however great, have no bearing on
matters of principle...the tendency of entropy to increase is not a
consequence of the laws of physics as such...An entropy increase may
occur unavoidably, due to our incomplete knowledge of the forces act-
ing on a system, or it may be an entirely voluntary act on our part.
In the latter case, an entropy increase is the means by which we simplify a prediction problem by discarding parts of the available information which are irrelevant, or nearly so, for the particular predictions desired. It is very similar to the statistician's practice
of "finding a sufficient statistic." The price we must pay for this
simplification is that the possibility of predicting other properties with the resulting equations is thereby lost.

> —Edwin T. Jaynes (1922-   )
> "Information Theory and Statistical Mechanics, II,"
> *Phys. Rev., 108,* Oct. 15, 1957, p. 178.

CHERRY (1966)

...an observer looking down a microscope, or reading instruments,
is not to be equated with a listener on a telephone receiving spoken
messages. Mother Nature does not communicate to us with signs or
language. A *communication channel* should be distinguished from a
channel of observation and, without wishing to seem too assertive,
the writer would suggest that in true communication problems the
concept of entropy need not be evoked at all.

> — Colin Cherry (1914-   )
> *On Human Communication: A Review, a Survey, and a
> Criticism* (1957), 2nd ed., The MIT Press, Camb., MA,
> 1966, p. 217.

REFERENCES

Boyle, Brian E., "Symptom Partitioning by Information Maximization," Dept. of E.E., MIT, Cambridge, MA, 1972.

Brillouin, Leon, "Maxwell's Demon Cannot Operate: Information and Entropy, I," *Journal of Applied Physics*, *22*, March 1951, pp. 334-343.

Carnap, Rudolf, "The Abstract Concept of Entropy and Its Use in Inductive Logic," *Two Essays on Entropy*, Univ. of Calif. Press, Berkeley, CA, 1977.

Cherry, C., *On Human Communication: A Review, a Survey, and a Criticism*, 2nd ed., The MIT Press, Camb., MA, 1966.

Christensen, R.A., "Induction and the Evolution of Language," Physics Dept., Univ. of Calif., Berkeley, CA, July 19, 1963. [*Chapter 7 of Volume II.*]

_____, *Foundations of Inductive Reasoning*, Berkeley, CA, 1964.

_____, "Inductive Reasoning and the Evolution of Language," Physics Dept., Univ. of Calif., Berkeley, CA, Dec. 1964. [*Chapter 8 of Volume II.*]

_____, "Inductive Reasoning as the Source of Human Knowledge," Physics Dept., Univ. of Calif., Berkeley, CA, June 1965. [*Chapter 10 of Volume II.*]

_____, "A General Approach to Pattern Discovery," Tech. Rept. No. 20, Computer Center, Univ. of Calif., Berkeley, CA, June 29, 1967 (revised Nov. 15, 1967). [*Chapter 1 of Volume III.*]

_____, "Seminar on Entropy Minimax Method of Pattern Discovery and Probability Determination," presented at Carnegie-Mellon Univ. and MIT, Feb. and March 1971, Tech. Rept. No. 40.3.75, Carnegie-Mellon Univ., Pittsburgh, PA, April 7, 1971. [*Chapter 3 of Volume III.*]

_____, "Entropy Minimax Method of Pattern Discovery and Probability Determination," Arthur D. Little, Inc., Acorn Park, Camb., MA, March 7, 1972. [*Chapter 4 of Volume III.*]

Cox, Richard T., "Of Inference and Inquiry, An Essay in Inductive Logic," *The Maximum Entropy Formalism*, ed. by R. Levine and M. Tribus, MIT Press, Camb., MA, 1979, pp. 119-167.

De Morgan, Augustus, *Formal Logic* (1847), The Open Court Co., London, 1926.

Franklin, D., "Analyses of Fission Gas Release Data by a Pattern Recognition Technique," IDG Note 2423, OECD Halden Reactor Project, Inst. for Atomenergi; Halden, Norway, Jan. 11, 1978.

Gift, D.A., William R. Schonbein and E. James Potchen, "An Introduction to Entropy Minimax Pattern Detection and In Use in the Determination of Diagnostic Test Efficacy," report prepared under NCI Grant No. CA18871-02-DHEW and DOE Grant No. EX-76-S-02-2777.A0012, Dept. of Radiology, Michigan State Univ., E. Lansing, MI, 1980.

Greeno, James G., "Evolution of Statistical Hypotheses Using Information Transmitted," *Philosophy of Science*, *37*, 279-293 (June 1970), reprinted as "Explanation and Information," in *Statistical Explanation and Statistical Reference*, ed. by Wesley C. Salmon, Univ. of Pittsburgh Press, Pittsburgh, PA, 1971, pp. 89-110.

Grosseteste, Robert, "Commentary on Aristotle's *Posterior Analytics*" (Oxford, c. 1209), *Comm. Post.* i.14, ff. 13-4; i.17, f. 17. Venice, 1494, quoted by A.C. Crombie, *Robert Grosseteste and the Origins of Experimental Science 1100-1700*, Oxford at the Clarendon Press, London, 1953.

Haken, Hermann, *Synergetics, An Introduction: Nonequilibrium Phase Transitions and Self-Organization in Physics, Chemistry and Biology*, Springer-Verlag, Berlin, 1978.

Jaynes, E.T., "Information Theory and Statistical Mechanics, II," *Phys. Rev.*, *108*, 1957, pp. 171-408.

Kemeny, John G., "The Use of Simplicity in Induction," *Phil. Rev.*, *62*, 1953, pp. 391-408.

_____, "Two Measures of Complexity," *J. of Phil.*, *52*, 1955, pp. 722-733.

Leontiades, K., "Computationally Practical Entropy Minimax Rotation, Applications to the Iris Data and Comparison to Other Methods," Bio-Medical Engr., Program, Carnegie-Mellon Univ., Pittsburgh, PA, 1976.

Lim, E., "Coolant Channel Closure Modeling Using Pattern Recognition: A Preliminary Report," FRMPM21-2, SAI-175-79-PA, Science Applications, Inc., Palo Alto, CA, June 1979.

Niiniluoto, Ilkka and Raimo Tuomela, *Theoretical Concepts and Hypothetico-Inductive Inference*, D. Reidel Pub. Co., Dordrecht, 1973.

Oldberg, S., "New Code Development Activities," *Planning Support Document for the EPRI Light Water Reactor Fuel Performance Program*, prepared by J.T.A. Roberts, F.E. Gelhaus, H. Ochen, N. Hoppe, S.T. Oldberg and D. Franklin, EPRI, NP-1024-SR, Electric Power Research Institute, Palo Alto, CA, Feb. 1979, pp. 2.32-2.39.

Rans, L., "Employment Patterns for Criminal Justice Personnel Following LEAA Supported Higher Education," *Criminal Justice Higher Education Programs in Maryland*, report to LEAA, Contract No. C-76186, Arthur D. Little, Inc., Cambridge, MA, Oct. 31, 1973.

Reichert, T.A. and A.J. Krieger, "Quantitative Certainty in Differential Diagnosis," *Proc. of the Second Int'l. Joint Conf. on Pattern Recognition*, 74-CHO-885-4C, Copenhagen, Denmark, Aug. 13-15, 1974, pp. 434-437.

Schonbein, William R., "Analysis of Decisions and Information in Patient Management," *Current Concepts in Radiology*, *2*, ed. by E. James Potchen, The C.V. Mosby Co., St. Louis, MO, 1975, pp. 31-58.

Watanabe, Satosi, *Knowing and Guessing, A Quantitative Study on Inference and Information*, J. Wiley & Sons, Inc., NY, 1969.

CHAPTER 24

THE EVOLUTION OF LANGUAGE

CHAPTER 24

THE EVOLUTION OF LANGUAGE

A. The Principle of the Evolution of Language

The central thesis of the entropy minimax justification of induction is a principle of the evolution of language:* A language e-
volves via the aggregate inductions of the many individuals using
the language. In order to appreciate this principle, it is necessary first to recognize the pervasiveness of induction which we
continually use when we generate expectations from observations.
Of course, some are more skillful than others. Consider, for example, that master of inductive reasoning, Mr. Sherlock Holmes:

> Sherlock Holmes welcomed her with the easy courtesy for
> which he was remarkable, and having closed the door, and
> bowed her into an armchair, he looked her over in the min-
> ute, and yet abstracted fashion which was peculiar to him.
>
> "Do you not find," he said, "that with your short sight it
> is a little trying to do so much typewriting?"
>
> "I did at first," she answered, "but now I know where the
> letters are without looking." Then suddenly realizing
> the full purport of his words, she gave a violent start,
> and looked up with fear and astonishment upon her broad,
> good-humored face. "You've heard about me, Mr. Holmes,"
> she cried, "else how could you know all that?"
>
> "Never mind," said Holmes, laughing. "It is my business
> to know things. Perhaps I have trained myself to see what
> others overlook. If not, why should you come to consult me?"
>
> —A. Conan Doyle, "A Case of Identity," *Strand Magazine*

Mr. Holmes' *deductions* were quite elementary, as he frequently ac-
knowledged. It was his remarkable talents of observation and *induction* that so distinguished him.

---

* R.A. Christensen, "Induction and the Evolution of Language," Physics Dept.,
Univ. of Calif., Berkeley, CA, 1963, (*Volume II*, Chapter 7); *Foundations of
Inductive Reasoning*, Berkeley, CA, 1964, Chapter 10.

Entropy minimax describes the induction behavior of reasoning beings in an approximate sense as an idealized law of engineering
describes the mechanical behavior of imprecisely characterized
materials. The inductions which people are constantly in the process of making, as millions of experiences are daily accumulated
and used to guide decision-making, involve entropy minimizing classification as well as entropy maximizing probability estimation.
As new experiences arise and become more common, classifications
are modified and words and phrases signifying them alter in meaning: new words are created, old ones pass out of use. The language evolves, enabling simpler, lower entropy, descriptions to be
given for experiences.

Consider, for example, the invention of the word "positron". Prior
to 1932 it was thought that all particles of rest mass 9.107×10^{-28}
grams had negative electric charge.* They were called "electrons".
Then C. D. Anderson observed that, in a cloud chamber, some of their
ionization tracks bent the opposite way in the presence of a magnetic field. This indicated a positive rather than a negative
charge.

At this point, two options were open to the English language. The
first was to continue to call the general category "electrons".
Then we could refer to the two now-distinguishable types by using
the intersection of "negatively charged" and "electrons" for one
class, and the intersection of "positively charged" and "electrons"
for the other.

The second option was to invent a new word. The language took the
latter course, and we now have the word "positron". Why?

From a classical logic point of view it appears that the two are
entirely equivalent. From such a viewpoint, we need new words only

---

* Though Dirac had speculated it might not be so.

when we encounter a new perceptual distinction that cannot be represented by any concatenation of existing words. (Distinctions deriving from "undescribable encounters" cannot, of course, be defined except by directly exhibiting the experience.)

However, from the viewpoint of entropy minimax theory, there is another consideration. This is the entropy of contrivedness. Anderson's experience of observing the ionization tracks in the cloud chamber is expressed with less contrivedness when we use the electron/positron terminology. Of course, the entropy savings by using the new word must be balanced, considering the frequency and domain of its use in this and other communications, against the cost of maintaining a new definition in the vocabulary inventory. In this case, the balance was struck in favor of a new word.

B. Simplicity, Elegance and the Class Structure of Theories

When there is insufficient empirical evidence to distinguish between two theories, an appeal is frequently made to some interpretation of Ockham's razor, a concept of simplicity or elegance. However, in the absence of a clear operational definition of a "simplicity" or "elegance" scale and a valid rationale for adopting it, we are faced with a situation in which researchers have a fair degree of license to dress up theories to fit the label.

One common technique is to camouflage the complexity of a theory by transforming from interfeature complexity to intrafeature complexity. Principal components analysis and analysis of variance approaches, for example, are well suited for this. The composite features often become so complex that researchers must avail themselves of one or another selection from a long menu of factor analysis techniques. These are used to constrain the compositions of the composite features to something interpretable in terms of the intended application.

The principle of entropy minimization provides an operationally unambiguous measure of simplicity. It is the conditional outcome

negentropy (a term coined by Brillouin for "negative entropy").
Thus, the simplest theory is the one which maximizes negentropy.

This concept of simplicity differs fundamentally from prior approaches. It does not key simplicity to data independent quantities, such as the number of descriptors and the number of inter-
linkings between descriptors. Rather it keys simplicity to performance of the theory relative to the data. It is a physical
rather than a metaphysical concept of simplicity.

If we fix the data and vary the theory, the minimum entropy principle says to choose the "simplest" theory. If we fix the theory
and vary the data, the simplicity of the theory will change in accord with its predictive power.

C.   Quixotic Classes

In 1955 Goodman posed the following puzzle. Define:

$$\text{Grue} \quad = \quad \begin{cases} \text{Green prior to the year 2000 A.D.} \\ \text{Blue during and after the year 2000 A.D.} \end{cases}$$

$$\text{Bleen} = \begin{cases} \text{Blue prior to the year 2000 A.D.} \\ \text{Green during and after the year 2000 A.D.} \end{cases}$$

Now consider the two statements:

    1) "Grass is green."
    2) "Grass is grue."

He pointed out that we have as much observational evidence to support the second statement as the first. In fact, there is complete
symmetry between the two, for we could as easily have defined:

$$\text{Green} = \begin{cases} \text{Grue prior to the year 2000 A.D.} \\ \text{Bleen during and after the year 2000 A.D.} \end{cases}$$

$$\text{Blue} = \begin{cases} \text{Bleen prior to the year 2000 A.D.} \\ \text{Grue during and after the year 2000 A.D.} \end{cases}$$

Actually, the situation is not so puzzling as it may at first seem, when we recognize that we are constantly rebuilding the meanings of our words, especially those that refer to physical objects. We do this with each new moment as we decide whether or not we *want* the candidate referrant to be signified by the word. Usually we quickly decide to retain the current meaning. Only when our expectations about the referrant's behavior are seriously contradicted do we experience difficulty in this decision. This is because the process of rebuilding the meanings of our words is a component of the process of inductive reasoning. Surprisal upon comparing our sense data to our expectations is the flag that signals us that either our probability estimates or our conceptual categorizations may need revision.

There is no formal logical reason for preferring "grass is green" to "grass is grue", just as there is no formal logical reason for supposing that it is impossible (or even that it is unlikely) for grass to change from green to blue on January 1, 2000. These are synthetic statements, not analytical propositions.

Further, there is no formal logical reason to state that "grass is green" has more empirical evidence than does "grass is grue." Ultimately, we have no purely logical reason for supposing we have any evidence at all about what is likely to happen, even probabilistically, in the future, other than some form of assumption that the future will be "like" the past, i.e., that we will in the future decide to refer to its aspects with the same words as we now do the aspects of the present. In this sense, Hume and his strict interpreter Wittgenstein are correct.

If we fight against the natural evolution of language, we can create imaginary worlds with these imaginary enigmas. But in the real world, our language is a physical fact, not an abstract philosophical concept.

Carnap and his followers give the term "probability" two interpretations: P_1, "the long-run frequency" probabilities applicable to infinite populations; and P_2, the "logical" probabilities expressing purely semantic relations between two sentences independent of verification against external empirical facts*. It is curious that the same people who chide measure-theorists for not giving probability a real-world interpretation feel free to wash their own hands of any possibility of being found wrong. Their logical probabilities$_2$ can convey no real-world information$_1$. More seriously, they omit a most important third interpretation of probability: P_3, the "short-run frequency" probabilities applicable to finite populations.

Infinite samples do not exist in real life, so it is certainly inappropriate to call them "empirical". Empirical probabilities necessarily pertain to the small, medium and large but finite numbers of future instances that we attempt to predict in the real-world. This may often put us in the "messy" realm of non-asymptotic behavior, medium-sized discrete samples, and other difficulties. But why should reality have to accommodate our convenience?

D. Language Evolution and Intrinsic Uncertainty

Language is elusively protean. Its evolution occurs as the combined effect of a vast network of individual inductive processes. To analyze this, we must use words. Yet the process that we study is itself shaping the meanings of these words. The very instrument we must necessarily employ to conduct our analyses, namely language, interferes intrinsically with the analysis. This does not mean that the situation is inherently unanalyzable. However, it does mean that there are limits to the degree to which uncertainty can be reduced by further analysis.

---

* The duality of the term probability has long been recognized. See I. Hacking, *The Emergence of Probability*, Camb. Univ. Press, 1975, pp. 12-13 for a review of discussions of this duality by Condorcet, Poisson, Cournot and others.

We are creatures gathering data concerning the world about us by means of observation.   Induction is a process of reasoning by means of which we record these observations in general terms.   The question whether or not we *ought* to believe such a generalization does not arise, at least at this stage.   We made the historically settled observations.   We *do* believe the generalizations.   They are merely reflections of all of, but no more than, information available to us. "Ought" implies a degree of freedom that this analysis has not referenced.   Is there any such freedom?   Perhaps.   If there is, it will be in the choice of the language in which we conduct our inductive reasoning (for, as we have seen, inductive conclusions do depend upon language).   But is there really any choice in this?

The answer is a paradox, akin to the problem of the observer unavoidably interfering with the object of his experiment in an unpredictable fashion by examining it too closely.   To the extent that we *freely* exercise choice of language, we destroy the efficacy of inductive reasoning.   To the extent that we ride along with its natural evolution, our inductive inferences are as secure as the available information will allow.

Disagreements over the meanings of words often encompass the most significant aspects of a controversy.   Quiet acceptance of another's redefinition away from a naturally evolved language is one of the subtler forms of intellectual surrender.   When language is dictated (say by manipulative lexicography, governmental decree, or controlled mass-communications), then the reality perceptions of those using the language are correspondingly altered.

The preparation and up-dating of dictionaries is an extremely important activity of scholarship, carrying a heavy burden of intellectual and social responsibility.   In dictionaries, general and specialized, we preserve a bottom-line summary of our collective experience in terms of classes of "sameness" important to our lives.   For literate societies, it rivals in importance the teaching of language to the young, the critical passing-on of enhanced predictive capacity, the ability to sense messages from the future in patterns of the past, from generation to generation in all living species.

E.   Inductive Relativity and Bootstrap Reality

So far, we have been talking as though it is always the observer who must organize his thoughts to discern patterns in a system's be-havior.   However, information transmittal, like motion, is relative. By interacting with a system sufficiently to gain information about it, we necessarily affect its state of order.   We lower its entropy a quantum by making an observation and gaining information about it (while, in accord with the second law of thermodynamics, we pay for this information with increased entropy of our measuring apparatus).

Just as the motion of a system is relative to the observer's physi-cal reference frame, so also the organization of a system is rela-tive to the observer's conceptual reference frame.   Generally, the weight of background information is sufficient to make it seem that organization has an absolute reference.   But this is only an illu-sion of large-sample thinking, like the apparent absoluteness of mo-tion is an illusion to the observer standing fixedly on the massive earth.   As space and time are relative to the observer's physical refer-ence frame, entropy is relative to the observer's conceptual frame.

Reality is the extent to which what it appears from available data that we can know about events is independent of what we happen to know.   When we have much data relative to the complexity of a sys-tem, as in cases in which we have dependable mechanistic models, the reality of the system is very intense.   When the system is com-plex and we have comparatively little data, as is frequently the case in the normal course of human events, its reality is weaker. The intensity can be turned up by gathering more data.   But no fi-nite amount of data will ever make the reality absolute.

For many systems with which we are familiar, the residual unreality can be reduced to negligible proportions by sufficient data.   In some systems, however, the intrinsic complexity is sufficiently great that there appears to be a non-negligible lower limit of un-reality.   What we can know about these systems is a strong function of our knowledge, no matter how much data we collect.

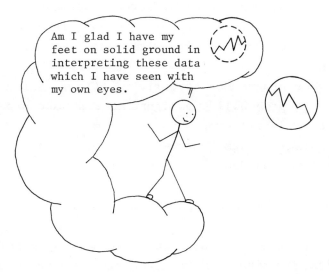

Fig. 134.   We See through the Filter of Our Mental Constructs

We are truly adrift physically and conceptually.   Just as our physi-cal measurements are relative to our physical frame of reference, our inductive judgments are relative to our conceptual frame of experience.   A major portion of this experience is submerged in the structure of our language and the meanings of the words, phrases and sentences we use.   The search for a solid "principle of induc-tion" on which to anchor our generalizations is as futile as the search for an ultimate frame of reference for Newtonian mechanics.

F.   Is There a Residual Problem of Justification?

Perhaps all our language evolution principle does is take cogni-zance of the fact that much of the information available for induction is buried in the language used to describe the situations and events. Then the justification is really another application of the minimum entropy principle.   Is there, nonetheless, a residual problem of

justifying *induction in the naturally evolving language?* Granting that we
do put faith in such inductive inferences, can we not still properly ask for a reason why we *ought to?*

Is there, however, any meaningful difference between this question
and the following: Granting that colliding billiard balls *do* re-
bound so as to conserve energy and momentum, can we not still properly
ask for a reason why they *ought to?* Or: Granting that a computerized decision-maker programmed to perform entropy minimax inductive
classifications and probability assignments *does* make a particular
decision, can we not still properly ask for a reason why it *ought to?*
Is there any meaning to "ought" when we have, by our analysis, so
constrained the situation that there is no freedom to be otherwise?
Or is there freedom but only that of a quantum logic as fundamentally
indeterminant as Copenhagenian quantum mechanics?

Of course, whether or not there is such freedom may itself be an
induction problem; in which case, we are ultimately trapped still
using words to analyze themselves. The definition of the simplest
word is akin to the deepest metaphysics.

G.   Time and Mind-Body Relationship

A basic precept of my epistemology is that only particulars exist
in the real world, generals exist solely in the mind. Now I suggest we go one step farther. The mind contains naught but generals.

The particular realities of the present are linked to the past by
a physical causality. However, the mind, which contains only gen-
eralities, lives entirely in the future. No matter how hard we
attempt to formulate a statement about a particular in the present
or past, we fail. Our statement about reality can only speak gen-
eralities about the future.

"That chair is there." This statement, in the last analysis, is a
prediction that, if people look toward a specific location, they
will agree that they see a chair.

"The defendant committed the murder." This statement, in its court-
room context, has no meaning other than as a prediction about the
future judgments of twelve persons.

Of course, when the future comes upon us and we are faced with the
predicted event (or its negation), we can only state it as another
generalization about a more distant future. "People agreed that the
chair was there." "The jury found the defendant guilty." Did they?
How do we know? Why, we predict that, if people in a new future in-
terview viewers of the putative chair or review records of the court-
room trial, they will agree about what the people saw or what the
jury found.

The mind lives not only entirely in the future, but entirely in an
ever-evolving future. With each new instant in time, we run the risk of
drifting away from reality into the entropy of contrivedness, and
must continually gather and process data to maintain the intensity
of the reality of our concepts. This is the persistent struggle of
the intellect against the entropy maximizing disorganizing of its
madness and death, the struggle to maintain the temporal separation
of mind and body.

Because of the weight normalization associated with contrivedness
introduced by physical entropy maximization, a single observation
necessarily has no reality whatsoever. Lower conceptual entropy
of our sense data, which is experienced as a more intense reality,
accompanies larger sample sizes and high outcome purity. The less
the outcome purity, the larger the sample size needed for the same
reality intensity.

The universe has two aspects. The physical aspect, consisting of
matter and energy, is a current flowing forward in time. The conceptual aspect, consisting of the patterns in which this matter
and energy is organized, is a countercurrent flowing backward.
Material bodies travel upwind, jostling into the organizing pressures of back-streaming entropic wave fronts.

The physical present is influenced by communications from the past in accord with a physical causality which is subject to the speed of light restrictions of Einsteinian separability. The carrier of these physical signals from the past is matter and energy.

The conceptual present is influenced by communications from the future, in accord with a conceptual causality of entropy minimization. We peer into a fog of chance correlations and see therein, through the image-processing capability of our language constructs, patterns of the future. The carrier of these conceptual signals from the future is form, the organization of matter and energy.

Though it may at first seem strange to conceive of messages from the future, it seems to me stranger still that we should know of something *without* exchanging information of any kind with it. To think that we do, with greater than random success, predict future events without in some way being touched by them is tantamount to having blind faith in mystical prophecies, whatever the technical or pragmatic verbiage in which we cloak our anticipations.

If we are not receiving messages from the future, how then can we know even probabilistically anything about it? If we are receiving messages from the future, then the explanation and justification of induction is evident.

QUOTES

PLATO (c. 353 B.C.)

Every existing object has three things which are the necessary
means by which knowledge of that object is acquired; and the
knowledge itself is a fourth thing; and as a fifth one must
postulate the object itself which is cognizable and true. First
of these comes the name; secondly the definition; thirdly the
image; fourthly the knowledge.

* * * *

And none of the objects, we affirm, has any fixed name, nor is
there anything to prevent forms which are now called "round"
from being called "straight," and the "straight" "round"; and
men will find the names no less firmly fixed when they have
shifted them and apply them in an opposite sense. Moreover, the
same account holds good of the Definition also, that, inasmuch
as it is compounded of names and verbs, it is in no case fixed
with sufficient firmness. And so with each of the Four, their
inaccuracy is an endless topic; but, as we mentioned a moment
ago, the main point is this, that while there are two separate
things, the real essence and the quality, and the soul seeks to
know not the quality but the essence, each of the Four proffers
to the soul either in word or in concrete form that which is
not sought; and by thus causing each object which is described
or exhibited to be always easy of refutation by the senses, it
fills practically all men with all manner of perplexity and uncertainty.

> —Plato (427-347 B.C.)
> "Epistle to Dion's Associates and Friends," *Timaeus,*
> *Critias, Cleitophon, Menexenus and Epistles*
> (c. 353 B.C.), tr. by R.G. Bury, Harvard Univ.
> Press, Cambridge, MA, 1942, pp. 533, 537.

ARISTOTLE (c. 335-323 B.C.)

Occasionally, perhaps, it is necessary to coin words, if no word
exists by which a correlation can adequately be explained...Thus
we may perhaps most easily comprehend that to which a thing is
related, when a name does not exist, if, from that which has a
name, we derive a new name, and apply it to that with which the
first is reciprocally connected...

> —Aristotle (384-322 B.C.)
> *Categoriae* (c. 335-323 B.C.), Ch. 7, tr. by
> E. Edghill, *The Basic Works of Aristotle,* ed.
> by R. McKeon, Random House, NY, 1941, p. 19.

LOCKE (1690)

How general words are made--The next thing to be considered is, how general words come to be made. For since all things that exist are only particulars, how come we by general terms, or where find we those general natures they are supposed to stand for? Words become general by being made the signs of general ideas: and ideas become general by separating from them the circumstances of time and place, and any other ideas that may determine them to this or that particular existence. By this way of abstraction they are made capable of representing more individuals than one; each of which, having in it a conformity to that abstract idea, is (as we call it) of that sort.

*   *   *   *

But yet, I think, we may say, the sorting of them under names is the workmanship of the understanding, taking occasion, from the similitude it observes amongst them, to make abstract general ideas, and set them up in the mind with names annexed to them, as patterns or forms (for in that sense the word form has a very proper signification), to which, as particular things existing are found to agree, so they come to be of that species, have that denomination, or are put into that *classis*. For when we say, this is a man, that a horse; this justice, that cruelty; what do we else but rank things under different specific names, as agreeing to those abstract ideas of which we have made those names the signs? And what are the essences of those species, set out and marked by names, but those abstract ideas in the mind; which are, as it were, the bonds between particular things that exist, and the names they are to be ranked under? And therefore the supposed real essences of substances, if different from our abstract ideas, cannot be the essences of the species we rank things into. For I demand, what are the alterations [which] may or may not be in a horse or lead, without making either of them to be of another species? In determining the species of things by our abstract ideas, this is easy to resolve. But if any one will regulate himself herein by supposed real essences, he will, I suppose, be at a loss: and he will never be able to know when anything precisely ceases to be of the species of a horse or lead.

*   *   *   *

...all the great business of *genera* and *species*, and their essences, amounts to no more but this, that men making abstract ideas, and settling them in their minds, with names annexed to them, do thereby enable themselves to consider things and discourse of them, as it were in bundles, for the easier and readier improvement and communication of their knowledge , which would advance but slowly, were their words and thoughts confined only to particulars.

—John Locke (1632-1704)
An Essay Concerning Human Understanding (England, 1690)
Oxford Univ. Press, London, 1924, pp. 228, 231-232, 237.

LEIBNIZ (1765)

§1. *Ph.* Although particular things alone exist, the largest number of words are *general terms*, because it is impossible, §2, for each particular thing to have a particular and distinct name; besides the fact that in such case a prodigious memory would be necessary, in comparison with which that of certain generals who could call by name all their soldiers would be nothing. The matter indeed becomes infinite, if every animal, every plant, and even every leaf of a plant, every grain, in short every grain of sand, which might need a name must have its name. [And how name the parts of things sensibly uniform, as water, fire?]   §3. Besides, these particular names would be useless, the principle end of language being to excite in the mind of him who listens to me an idea similar to mine. [Thus the similitude suffices, which is indicated by general terms.]   §4. And particular words alone would not serve to extend our knowledge, [nor to make us judge of the future by the past, or of one individual by another.]

> —Gottfried Wilhelm Leibniz (1646-1716)
> *New Essays On the Understanding* (Leipzig, posthumous, 1765), The Open Court Pub. Co., LaSalle, IL, 1949, p. 307.

LAPLACE (1814)

Analogy is based upon the probability, that similar things have causes of the same kind and produce the same effects. This probability increase[s] as the similitude becomes more perfect.

> —Pierre Simon de Laplace (1749-1827)
> *A Philosophical Essay on Probabilities* (Paris, 1814), tr. by F. Truscott and F. Emory, Dover Pubs., Inc., NY, 1951, p. 180.

von SCHLEGEL (1827)

In those languages which appear to be at the lowest grade of intellectual culture, we frequently observe on a closer acquaintanceship a very high and elaborate degree of art in their grammatical structure. This is especially the case with the Basque and the Lapponian, and many of the American languages.

> —Frederick von Schlegel (1722-1829)
> *Philosophy of Life* (Vienna, 1827), Bohn's Standard Library, London, p. 395, quoted by C. Staniland Wake, *Chapters on Man with the Outlines of a Science of Comparative Psychology*, Trübner and Co., London, 1868, p. 101.

MÜLLER (1870)

A much more striking analogy, therefore, than the struggle for life among separate languages, is the struggle for life among words and grammatical forms which is constantly going on in each language. Here the better, the shorter, the easier forms are constantly gaining the upper hand, and they really owe their success to their own inherent virtue.

> —Friedrich Max Müller (1823-1900)
> "The Science of Language" (a review of August Schleicher's book *Darwinism Tested by the Science of Language*), *Nature*, Jan. 6, 1870, p. 257.

DARWIN (1871)

So with languages: the most symmetrical and complex ought not to be ranked above irregular, abbreviated, and bastardised languages, which have borrowed expressive words and useful forms of construction from various conquering, conquered, or immigrant races.

> —Charles Darwin (1809-1882)
> *The Descent of Man and Selection in Relation to Sex* (1871), Modern Library, Random House, NY, p. 467.

MACH (1897)

When once the inquiring intellect has formed, through adaptation, the habit of connecting two things, A and B, in thought, it tries to retain this habit as far as possible, even where the circumstances are slightly altered. Whenever A appears, B is added in thought. The principle thus expressed, which has its root in an effort for economy, and is particularly noticeable in the work of the great investigators, may be termed the *principle of continuity*.

> —Ernst Mach (1838-1916)
> *The Analysis of Sensations* (1897), tr. by C. Williams, Open Court Pub. Co., NY, 1902, reprinted by Dover Pubs., Inc., NY, 1959, p. 57.

FREUD (1899)

...the aim of this first psychical activity was to produce a 'perceptual identity'[1] - a repetition of the perception which was linked with the satisfaction of the need.

The bitter experience of life must have changed this primitive thought activity into a more expedient secondary one. The establishment of a perceptual identity along the short path of regression within the apparatus does not have the same result elsewhere in the mind as does the cathexis of the same perception from without. Satisfaction does not follow; the need persists.

An internal cathexis could only have the same value as an external one if it were maintained unceasingly, as in fact occurs in hallucinatory psychoses and hunger phantasies, which exhaust their whole psychical activity in clinging to the object of their wish.

* * * *

Thought is after all nothing but a substitute for a hallucinatory wish; and it is self-evident that dreams must be wish-fulfilments, since nothing but a wish can set our mental apparatus at work.

> —Sigmund Freud (1856-1939)
> *The Interpretation of Dreams* (1899), ed. and tr. by J. Strachey, Avon Books, Discus Edition, NY, 1965, pp. 605, 606.

BERGSON (1907)

The affirmation of a reality implies the simultaneous affirmation of all the degrees of reality intermediate between it and nothing.

> —Henri Bergson (1860-1941)
> *Creative Evolution* (1907), tr. by A. Mitchell, Modern Library, Random House, NY, 1944, p. 351.

---

1 [I.e., something, perceptually identical with the 'experience of satisfaction'.]

WRINCH and JEFFREYS (1921)

A large fraction of known physical laws are expressible in simple mathematical forms, and we have already indicated the need for some examination of the kind of advantage that simplicity confers on a physical law...

* * * *

Is the prevalence of these simple and accurate laws due to the nature of our investigation, or to some widespread quality in the external world itself?

* * * *

...it will never be possible to attach appreciable probability to an inference if it is assumed that all laws of an infinite class, such as all relations involving only analytic functions, are equally probable *a priori*. If inference is possible, the admissable laws must not be all equally probable *a priori*. It is suggested that all admissible laws can be arranged in a well-ordered sequence, each having a finite prior probability, and such that each is more probable than any that follows it in the sequence.

* * * *

On the other hand, we cannot specify completely what the actual order must be, for if a sequence can be well ordered in one way it can be in many.

> —Dorothy Wrinch (1894-1976) and Harold Jeffreys (1891-   )
> "On Certain Fundamental Principles of Scientific Inquiry," *Phil. Mag.*, *42*, 1921, pp. 379, 380, 389, 388.

WEYL (1927)

...there is no doubt that wherever *thought* and the causative agent of *will* emerge, especially in man, that power is increasingly controlled by a purely spiritual world of images (knowledge, ideas).

> —Hermann Weyl (1885-1955)
> *Philosophy of Mathematics and Natural Science* (1927), tr. by O. Helmer, Princeton Univ. Press, Princeton, NJ, 1949, p. 300.

SAPIR (1928)

It is quite an illusion to imagine that one adjusts to reality essentially without the use of language and that language is merely an incidental means of solving specific problems of communication or reflection. The fact of the matter is that the "real world" is to a large extent unconsciously built up on the language habits of the group. No two languages are ever sufficiently similar to be considered as representing the same social reality. The worlds in which different societies live are distinct worlds, not merely the same world with different labels attached.

> —Edward Sapir (1884-1939)
> "The Status of Linguistics as a Science" (1928),
> in *Selected Writings of Edward Sapir in Language,*
> *Culture and Personality,* ed. by D.G. Mandelbaum,
> Univ. of Calif. Press, Berkeley, CA, 1949, p. 162.

HEISENBERG (1930)

It is not surprising that our language should be incapable of describing the processes occurring within the atoms, for, as has been remarked, it was invented to describe the experiences of daily life, and these consist only of processes involving exceedingly large numbers of atoms.

> —Werner Heisenberg (1901-1976)
> *The Physical Principles of the Quantum Theory*
> (Leipzig, 1930), tr. by C. Ekhart and F. Hoyt,
> Dover Pubs., Inc., NY, p. 11.

BOHR (1933)

Science is the observation of phenomena and the communication of the results to others, who must check them. Only when we have agreed on what has happened objectively, or on what happens regularly, do we have a basis for understanding...The cloud chamber is a measuring apparatus, which means that this photograph entitles us to conclude that a positively charged particle which has the properties of an electron has passed through the chamber...It is one of the basic presuppositions of science that we speak of measurements in a language that has basically the same structure as the one in which we speak of everyday experience.

> —Niels Bohr (1885-1962)
> Quoted by Werner Heisenberg, *Physics and Beyond,*
> *Encounters and Conversations,* Harper & Row, NY,
> 1971, p. 130.

SCHRÖDINGER (1935)

...David Hume pointed out that there is no intrinsic connection
between cause and effect which can be perceived and understood
by the human mind.

* * * *

...Why do we concede to what has happened in the past a controlling influence on our expectation of what is to happen in the
future?

* * * *

...the mere fact that we, human beings, have survived to raise
the question, in a certain sense indicates the required answer!

> —Erwin C. Schrödinger (1887-1961)
> *Science and the Human Temperament* (1935), reprinted
> as *Science Theory and Man*, Dover Pubs., Inc., NY,
> 1957, pp. 39, 40, 41.

BRIDGMAN (1936)

Two aspects of the question of "meaning" are involved here.
There is in the first place a general aspect; with regard to
this it seems to me that as a matter of self-analysis I am never
sure of a meaning until I have analyzed what I do, so that for
me meaning is to be found in a recognition of the activities
involved...

The more particular and important aspect of the operational significance of meaning is suggested by the fact that Einstein
recognized that in dealing with physical situations the operations which give meaning to our physical concepts should properly
be physical operations, actually carried out. For in so restrict-
ing the permissible operations, our theories reduce in the last
analysis to descriptions of operations actually carried out in
actual situations, and so cannot involve us in inconsistency or
contradiction, since these do not occur in actual physical situations.

> —Percy W. Bridgman (1882-1961)
> *The Nature of Physical Theory*, Princeton Univ.
> Press, Princeton, NJ, 1936, pp. 8-9.

WHORF (1942)

...the forms of a person's thoughts are controlled by inexorable
laws of pattern of which he is unconscious. These patterns are
the unperceived intricate systematizations of his own language
—shown readily enough by a candid comparison and contrast with
other languages, especially those of a different linguistic
family. His thinking itself is in a language - in English, in
Sanskrit, in Chinese. And every language is a vast pattern-sys-
tem, different from others, in which are culturally ordained the
forms and categories by which the personality not only communicates, but also analyzes nature, notices or neglects types of
relationship and phenomena, channels his reasoning, and builds
the house of his consciousness.

> —Benjamin Lee Whorf (1897-1941)
> "Language, Mind and Reality," *Theosophist*, Madras,
> India, January and April 1942, reprinted in
> *Language, Thought and Reality*, ed. by J. Carroll,
> The MIT Press, Cambridge, MA, 1956, p. 252.

ZIPF (1949)

For if speech consists of words that are tools which convey meanings, there is the possibility both of a more economical way, and
of a less economical way, to use word-tools for the purpose of
conveying meanings.

* * * *

We may even visualize a given stream of speech as being subject
to *two "opposing forces."* The one "force" *(the speaker's economy)*
will tend to reduce the size of the vocabulary to a single word
by unifying all meanings behind a single word...Opposed to this
...is a second "force" *(the auditor's economy)* that will tend to
increase the size of a vocabulary to a point where there will be
a distinctly different word for each different meaning.

* * * *

...the stream of speech, even in respect to its tiniest minutiae,
is organized out of deference to the primary exigencies of
economy.

> —George Kingsley Zipf (1902-1950)
> *Human Behavior and the Principle of Least Effort*,
> Addison-Wesley Press, Inc., Camb., MA, 1949,
> "On the Economy of Words," pp. 20, 21, and "Formal
> Semantic Balance and the Economy of Evolutionary
> Process," p. 57.

EINSTEIN (1949)

The system of concepts is a creation of man together with the rules of syntax, which constitute the structure of the conceptual systems. Although the conceptual systems are logically entirely arbitrary, they are bound by the aim to permit the most nearly possible certain (intuitive) and complete co-ordination with the totality of sense experiences; secondly they aim at greatest possible sparsity of their logically independent elements (basic concepts and axioms), i.e., undefined concepts and underived [postulated] propositions.

...Hume saw clearly that certain concepts, as for example that of causality, cannot be deduced from the material of experience by logical methods. Kant, thoroughly convinced of the indispensability of certain concepts, took them—just as they are selected—to be the necessary premises of every kind of thinking and differentiated them from concepts of empirical origin. I am convinced, however, that this differentiation is erroneous, i.e., that it does not do justice to the problem in a natural way. All concepts, even those which are closest to experience, are from the point of view of logic freely chosen conventions, just as is the case with the concept of causality, with which this problematic concerned itself in the first instance.

> —Albert Einstein (1879-1955)
> *Albert Einstein: Philosopher-Scientist*, ed. and tr. by P. Schilpp, The Library of Living Philosophers, Inc., Evanston, IL, 1949, p. 13.

CARNAP (1950)

(i) Probability$_1$ is the degree of confirmation of a hypothesis h with respect to an evidence statement e, e.g., an observational report. This is a logical, semantical concept. A sentence about this concept is based, not on observation of facts, but on logical analysis; if it is true, it is L-true (analytic).

(ii) Probability$_2$ is the relative frequency (in the long run) of one property of events or things with respect to another. A sentence about this concept is factual, empirical.

*   *   *   *

"How can the statement 'the probability of rain tomorrow on the evidence of the given meteorological observations is one-fifth' be verified? We shall observe either rain or not-rain tomorrow, but we shall not observe anything that can verify the value one-fifth." This objection, however, is based on a misconception concerning the nature of the probability$_1$ statement. This statement does not ascribe the probability$_1$ value 1/5 to tomorrow's rain but rather to a certain logical relation between the prediction of rain and the meteorological report. Since the relation is logical, the statement is, if true, L-true; therefore it is not in need of verification by observation of tomorrow's weather or of any other facts.

> —Rudolf Carnap (1891-1970)
> *Logical Foundations of Probability*, Univ. of Chicago Press, Chicago, IL, 1950, pp. 19, 30.

QUINE (1953)

Furthermore it becomes folly to seek a boundary between synthetic statements, which hold contingently on experience, and
analytic statements, which hold come what may. Any statement
can be held true come what may, if we make drastic enough
adjustments elsewhere in the system...Revision even of the logical law of the excluded middle has been proposed as a means of
simplifying quantum mechanics; and what difference is there in
principle between such a shift and the shift whereby Kepler
superseded Ptolemy, or Einstein Newton, or Darwin Aristotle?

                    *   *   *   *

As an empiricist I continue to think of the conceptual scheme of
science as a tool, ultimately, for predicting future experience
in the light of past experience.

                    *   *   *   *

Total science, mathematical and natural and human, is similarly
but more extremely underdetermined by experience. The edge of
the system must be kept squared with experience; the rest, with
all its elaborate myths or fictions, has as its objective the
simplicity of laws.

                    *   *   *   *

Each man is given a scientific heritage plus a continuing bar-
rage of sensory stimulation; and the considerations which guide
him in warping his scientific heritage to fit his continuing
sensory promptings are, where rational, pragmatic.

              —Willard van Orman Quine (1908-   )
              From a Logical Point of View, Harvard Univ. Press,
              Camb., MA, 1953, pp. 43, 44, 45, 46.

POLYA (1954)

Adaptation of the mind may be more or less the same thing as
adaptation of the language; at any rate, one goes hand in hand
with the other...

              —George Polya (1887-   )
              Induction and Analogy in Mathematics, Princeton
              Univ. Press, Princeton, NJ, 1954, p. 55.

BRILLOUIN (1962)

We must now consider the viewpoint repeatedly emphasized by
Bridgman: The only *physical quantities* are those which *can be
measured,* and for which a measuring operation can be stated.
Things that cannot be observed and measured have no actual
physical existence. They are creations of our imagination: they
can be useful and help our visualization, but they have no real
physical meaning.

*   *   *   *

We used to imagine that there was a real universe, outside of us,
that could persist even when we stop observing it. But (quoting
M. Planck), we must immediately add: "this real outside world is
not directly perceptible to us". We replace it with a *physical
model of the world* (ein physikalisches Weltbild) which is more
or less adapted to observations.

> —Leon Brillouin (1889-1969)
> "Observation, Information and Imagination," pre-
> sented at the symposium by the Academie Interna-
> tionale de Philosophie des Sciences, Brussels,
> 3-8 Sept. 1962, *Information and Prediction in
> Science,* ed. by S. Dockx and P. Bernays, Academic
> Press, NY, 1965, pp. 11, 13.

PEREZ and TONDL (1962)

The universe is not composed of "things" that may be absolutely
isolated, yet it is possible--by using the notion of "things" or
"entities"--to approximate the objective reality* thanks to the
relative autonomy of the above notions within the frame of a
decision problem (in the broad sense of the word).

> —Albert Perez and Ladislov Tondl (1924-   )
> "On the Role of Information Theory in Certain Sci-
> entific Procedures," presented at the symposium
> sponsored by the Academie Internationale de Phi-
> losophie des Sciences, Brussels, 3-8 Sept. 1962,
> *Information and Prediction in Science,* ed. by
> S. Dockx and P. Bernays, Academic Press, NY, 1965,
> p. 35.

---

* L. Tondl, "Cognition Role of the Abstraction," Moscow, 1960.
(In Russian.)

BLUM (1968)

We may think of the cultural pattern upon which the behavior of
a society is based as a kind of cross section of all the component individual mnemotypes plus information stored outside men's
brains in written records etc.; this I have called the *collective
mnemotype*. In the transfer and storage of information among individual members of the society and within the collective mnemotype, there is great chance for errors in copying and storage.
These errors may play a role in cultural evolution comparable to
that of mutations within genotypes in biological evolution...

But there must be a tendency to conserve cultural pattern, or
chaos rather than orderly evolution would result from such lack
of restraint. Such a tendency may stem from the similarity of
the individual mnemotypes of a given society, which are formed
within the common environment provided by the collective mnemotype, where mutual exchange of information is relatively easy.

> —Harold F. Blum (1899-   )
> *Time's Arrow and Evolution*, Princeton Univ. Press,
> Princeton, NJ, 1968, pp. 215-216.

ELSASSER (1971)

What are the epistemological problems in a statistical system
of description?...Ultimately, the pivotal questions turned out
to be two in number: first, the relationship between individual
objects and abstract classes designed to represent properties of
objects in the theory; second, the size of such classes, that is,
the number of their members.

> —Walter M. Elsasser (1904-   )
> "Philosophical Dissonances in Quantum Mechanics,"
> *Perspective in Quantum Theory*, ed. by W. Yourgrau and
> A. van der Merwe, Dover Pubs., Inc. NY, 1971, p. 216.

ROSENFELD (1971)

At the stage of concrete operations, language is simply a means
of communication of sensorimotor experience: words are incorporated into this significant combinations of sensorimotor
schemes, and constitute a symbolic representation of the latter.
Since the sensorimotor schemes are themselves symbols of the
aspects of experience retained as significant, one may say without excessive schematization that two correlated sets of symbols,
or code systems, are used for the registration and communication
of experience. At the formal stage, however, the verbal code
system pursues an autonomous development by purely abstract deviations of new concepts without immediate correspondence in the
sensorimotor field. It is this formal refinement that makes
scientific thinking possible...

> —Leon Rosenfeld (1904-   )
> "Unphilosophical Considerations on Causality in
> Physics," *Perspectives in Quantum Theory*, ed. by
> W. Yourgrau and A. van der Merwe, Dover Pubs.,
> Inc., NY, 1971, p. 234.

REFERENCES

Anderson, Carl D., "The Positive Electron," *Phys. Rev.*, *43*, 1933, p. 491.

Arbib, Michael A., *The Metaphorical Brain*, J. Wiley & Sons, Inc., NY, 1972.

Aristotle, *Categorie*, tr. by E. Edghill, *The Basic Works of Aristotle*, ed. by P. McKeon, Random House, NY, 1941.

Bergson, Henri, *Creative Evolution* (1907), tr. by A. Mitchell, Modern Library, Random House, NY, 1944.

Blum, Harold F., *Time's Arrow and Evolution* (1951), 3rd. ed., Princeton Univ. Press, Princeton, NJ, 1968.

Bohr, Niels (1933) quoted by W. Heisenberg, *Physics and Beyond, Encounters and Conversations*, Harper & Row, NY, 1971.

Bridgman, P.W., *The Nature of Physical Theory*, Princeton Univ. Press, Princeton, NJ, 1936.

Bunge, Miro, *The Myth of Simplicity*, Prentice-Hall, Inc., Englewood Cliffs, NJ, 1963.

Carnap, Rudolf, *Logical Foundations of Probability*, The Univ. of Chicago Press, Chicago, IL, 1950.

_____, *The Logical Structure of the World & Pseudoproblems in Philosophy*, tr. by R. George, Univ. of Calif. Press, Los Angeles, CA, 1967.

Christensen, R.A., "Induction and the Evolution of Language," Physics Dept., Univ. of Calif., Berkeley, CA, July 19, 1963. [*Chapter 7 of Volume II.*]

_____, *Foundations of Inductive Reasoning*, Berkeley, CA, 1964.

_____, "Inductive Reasoning and the Evolution of Language," Physics Dept., Univ. of Calif., Berkeley, CA, Dec. 1964. [*Chapter 8 of Volume II.*]

_____, "Inductive Reasoning as the Source of Human Knowledge," Univ. of Calif., Berkeley, CA, June 1965. [*Chapter 10 of Voluem II.*]

Creelan, Majorie B., *The Experimental Investigation of Meaning*, Springer Pub. Co., NY, 1966.

Darwin, Charles R., *The Descent of Man and Selection in Relation to Sex* (1871), Random House, NY.

DeWitt, Bryce S., "Quantum Mechanics and Reality," *Physics Today*, *23*, Sept. 1970.

Dockx, S. and P. Bernays (eds.), *Information and Prediction in Science*, Academic Press, NY, 1965.

Einstein, Albert, *Albert Einstein: Philosopher-Scientist*, ed. and tr. by P. Schilpp, The Library of Living Philosophers, Inc., Evanston, IL, 1949.

Elsasser, Walter M., "Philosophical Dissonances in Quantum Mechanics," *Perspectives in Quantum Theory*, ed. by W. Yourgrau and A. van der Merwe, Dover Pubs., Inc., NY, 1971.

Flew, Anthony (ed.), *Logic and Language*, Basil Blackwell, Oxford, 1953.

Fogel, L.J., A.J. Owens and M.J. Walsh, *Artificial Intelligence Through Simulated Evolution*, J. Wiley & Sons, Inc., NY, 1966.

Freud, Sigmund, *The Interpretation of Dreams* (1899), ed. and tr. by J. Strachey, Avon Books, Discus Edition, NY, 1965.

Gatlin, Lila L., *Information Theory and the Living System*, Columbia Univ. Press, New York, 1972.

Goodman, Nelson, "Axiomatic Measurement of Simplicity," *Journal of Philosophy*, *52*, 1955, pp. 709-722.

Hawkins, David, *The Language of Nature*, W.H. Freeman & Co., San Francisco, CA, 1964.

Hayakawa, S.I., *Language in Thought and Action* (1939), Harcourt, Brace & World, Inc., NY, 1949.

Heisenberg, Werner, *The Physical Principles of the Quantum Theory* (1930), tr. by C. Ekhart & F. Hoyt, Dover Pubs., Inc., NY.

Hillman, Donald J., "The Measurement of Simplicity," *Phil. of Science, 28*, 1962, p. 225.

Laplace, Pierre Simon de, *A Philosophical Essay on Probabilities* (1814), tr. by F. Truscott and F. Emory, Dover Pubs., Inc., NY, 1951.

Leibniz, Gottfried Wilhelm, *New Essays On the Understanding* (1765), The Open Court Pub. Co., La Salle, IL, 1949.

Linell, Per, *Psychological Reality in Phonology*, Cambridge Univ. Press, Cambridge, 1979.

Locke, John, *An Essay Concerning Human Understanding* (1690), Oxford Univ. Press, London, 1924.

Löfgren, Lars, "Recognition of Order and Evolutionary Systems," *Computer and Information Sciences - II*, ed. by J.T. Tou, Academic Press, NY, 1967.

Lubbock, J., *Origin of Civilization*, D. Appleton and Co., NY, 1871, p. 278.

Mach, Ernst, *The Analysis of Sensations* (1897), tr. by C. Williams, Open Court Pub. Co., NY, 1902, reprinted by Dover Pubs., Inc., NY, 1959.

Miller, George A. and Philip N. Johnson-Laird, *Language and Perception*, Harvard Univ. Press, Cambridge, 1976.

Müller, Friedrich Max, "The Science of Language," *Nature*, January 6, 1970, pp. 256-259.

Parkinson, G.H.R. (ed.), *The Theory of Meaning*, Oxford Univ. Press, London, 1968.

Plato, "Epistle to Dion's Associates and Friends," *Timaeus, Critias, Cleitophon, Menexenus and Epistles* (c. 353 B.C.), tr. by R.G. Bury, Harvard Univ. Press, Cambridge, MA, 1942.

Polya, George, *Induction and Analogy in Mathematics*, Princeton Univ. Press, Princeton, NY, 1954.

Quastler, Henry (ed.), *Essays on the Use of Information Theory in Biology*, Univ. of Ill. Press, Urbana, IL, 1953.

Quine, W.V., *From a Logical Point of View*, Harvard Univ. Press, Cambridge, MA, 1953.

Reiber, R.W. (ed.), *Psychology of Language and Thought*, Plenum Press, NY, 1980.

Reichert, T., J. Yu and R. Christensen, "Molecular Evolution as a Process of Message Refinement," *Journal of Molecular Evolution, 8,* 1976, pp. 41-54.

Rosenfeld, Leon, "Unphilosophical Considerations in Physics," *Perspectives in Quantum Theory,* ed. by W. Yourgrau and A. van der Merwe, Dover Pubs., Inc., NY, 1971.

Sapir, Edward, "The Status of Linguistics as a Science" (1928), in *Selected Writings of Edward Sapir in Language, Culture and Personality,* ed. by D.G. Mandelbaum, Univ. of Calif. Press, Berkeley, CA, 1949, pp. 160-166.

Schaff, Adam, *Language and Cognition* (1964), tr. by O. Wojtasiewicz, McGraw-Hill, Inc., NY, 1973.

von Schlegel, F., *Philosophy of Life* (Vienna, 1827), Bohn's Standard Library London, p. 395, quoted by C. Staniland Wake, *Chapters on Man with Outlines of a Science of Comparative Psychology,* Trubner and Co., London, 1868.

Schlesinger, George N., *Aspects of Time,* Hackett Pub. Co., Indianapolis, IN, 1980.

Schrödinger, Erwin C., *Science and the Human Temperament* (1935), reprinted as *Science Theory and Man,* Dover Pubs., Inc., NY, 1957.

Searle, J.R. (ed.), *The Philosophy of Language,* Oxford Univ. Press, London, 1971.

Singh, Jagjit, *Great Ideas in Information Theory, Language and Cybernetics,* Dover Pubs., Inc., NY, 1966.

Sober, Elliott, *Simplicity,* Oxford at the Clarendon Press, London, 1975.

Svenonicis, Lars, "Definability and Simplicity," *J. of Symbolic Logic, 20,* 1955, pp. 235-250.

Weyl, Hermann, *Philosophy of Mathematics and Natural Science* (1927), tr. by O. Helmer, Princeton Univ. Press, Princeton, NJ, 1949.

Whorf, Benjamin Lee, "Language, Mind and Reality," *Theosophist,* Madras, India, January and April, 1942, reprinted in *Language, Thought, and Reality,* ed. by J. Carroll, The MIT Press, Cambridge, MA, 1956.

Wrinch, Dorothy and Harold Jeffreys, "On Certain Fundamental Principles of Scientific Inquiry," *Phil. Mag., 42,* 1921, pp. 379-389.

Zipf, G.K., "The Repetition of Words, Time Perspective and Semantic Balance," *J. Gen. Psych., 32,* 1945, p. 127.

_____, *Human Behavior and the Principle of Least Effort,* Addison-Wesley, Cambridge, MA, 1949.

CHAPTER 25

ENTROPY MINIMAX AND VALUE JUDGMENT

CHAPTER 25

ENTROPY MINIMAX AND VALUE JUDGMENT

A. Uses of Information

Throughout this volume, we have attempted to restrict ourselves to a value-free concept of truth even though we have stressed its informational dependency. However, if information has domain as well as magnitude, then we must ask whether there is, in the last analysis, a necessary value judgment component to probabilistic assessments of reality.

Consider the decision theory paradigm

$$V(C_i) = \sum_k P(O_k \text{ given } C_i) \ U(O_k \text{ given } C_i),$$

where $U(O_k \text{ given } C_i)$ is the utility of outcome O_k given condition C_i. The classical decision theory principle is to select that condition C_i for which the expected value $V(C_i)$ of the utility is maximum.

The entropy minimax principles are to select the definitions of the $\{C_i\}$ so as to minimize

$$S(O \text{ given } C) = - \sum_i P(C_i) \sum_k P(O_k \text{ given } C_i) \ \log P(O_k \text{ given } C_i),$$

while selecting numerical values of the $P(C_i)$ to maximize

$$S(C) = - \sum_i P(C_i) \ \log P(C_i),$$

and numerical values of the $P(O_k \text{ given } C_i)$ to maximize

$$S(O \text{ given } C_i) = - \sum_{k} P(O_k \text{ given } C_i) \log P(O_k \text{ given } C_i),$$

subject to given information. How are the entropy minimax and decision theory principles related?

The $V = \sum PU$ formulation is not a particulary convenient representation in which to conduct this analysis. We would need to ask: If we made an error of a certain amount in $P(O_k \text{ given } C_i)$, how would this affect our choice among the $\{C_i\}$? What would be the resulting loss in max $V(C_i)$? How do we minimize this loss?

> The fact that we summarize the data as estimates of probabilities implies that the tasks the observer is asked to perform in psycho-physical experiments are tasks in which there is a probability of error...whenever one is confronted with a probability of error he is confronted with a decision theory problem.
>
> —Wilson P. Tanner, Jr.
> "Psychological Implications of Psychophysical Data," *Human Decision in Complex Systems*, *Annals of the New York Academy of Sciences, 80*, Jan. 28, 1961, pp. 751-755.

The two forms of analysis, induction and decision theory, reflect what may appear to be conceptually distinct viewpoints. The use of $S(O \text{ given } C)$ in entropy minimax induction assumes that $P(C_i)$ has meaning. The decision theory paradigm, on the other hand, allows for a free choice among the $\{C_i\}$. Thus, $P(C_i)$ has no place in decision theory.

Entropy minimax provides conditional predictive information for generalized use. Decision theory employs this information to settle on one or the other of alternative courses of action for a particular use.

To examine utility considerations in the framework of entropy minimax, we employ the assessment diagram. See Fig. 135.

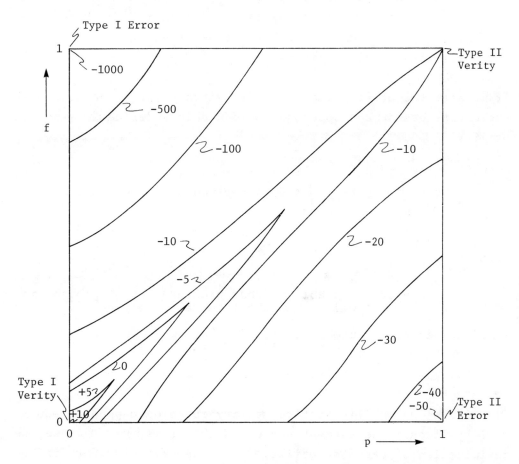

Fig. 135.   Illustrative Utility Function, U(f given p),
Defined over Assessment Space

B.   Utility Dependency of Probabilities

We saw in Chapter 16 that no single number will suffice to assess a set of predictions for all possible uses. Rather, one must, in general, consider the entire assessment diagram. This is because different types of correct and erroneous predictions have differ- ent utilities (or disutilities).

If we knew the utility function, U(f given p), for the user over the entire assessment space, we would be able to define the expected utility

$$V = \int_0^1 dp\ F(p) \int_0^1 df\ G(f\ given\ p)\ U(f\ given\ p),$$

where

$$F(p)dp = \text{probability that predicted event probability will be in the interval p to p+dp,}$$

$$G(f\ given\ p)df = \text{probability, given p, that observed event fre- quency will be in the interval f to f+df, and}$$

$$U(f\ given\ p) = \text{utility of event frequency f given p.}$$

Fig. 135 shows an illustrative utility function mapped as equi- utility lines in assessment space. Within this formalism, we use maximization of expected utility, V, as the criterion for fixing the weight normalization.

The entropy decrement minimization criterion given in Chapter 20 for determining the normalization is entirely focused on maximizing information accuracy. Expected utility maximization, on the other hand, also takes into consideration such matters as the definitive- ness of the predictions and the distinction between Type I and Type II verities and errors.

A wide variety of utility functions can be formulated. An appropriate formulation for a particular application will depend upon the circumstances and the value judgments of the user.

Consider, for example, the following function:

$$V_\beta = -\beta \Delta S_e + (1-\beta)\, \Delta S_d ;$$

where ΔS_e and ΔS_d are defined in Chapter 16. The parameter β can be adjusted to control the relative emphasis on accuracy and definitiveness. With $\beta = 1$, full emphasis is on accuracy; with $\beta = 0$ it is on definitiveness. An intermediate value of $\beta = 0.95$, for example, would put strong emphasis on accuracy but trade some accuracy for increased definitiveness.

An example of a somewhat more comprehensive model is the following function:

$$V_\gamma = -\beta \Delta S_e(\gamma_e) + (1-\beta)\, \Delta S_d(\gamma_v) ,$$

where

$$\Delta S_e(\gamma_e) = \sum_{j=1}^{L} C(\gamma_e; \hat{p}_j, \hat{f}_j) \{\hat{f}_j \ln[\hat{f}_j/\hat{p}_j] + (1-\hat{f}_j) \ln[(1-\hat{f}_j)/(1-\hat{p}_j)]\}, \text{ and}$$

$$\Delta S_d(\gamma_v) = \sum_{j=1}^{L} C(\gamma_v; \hat{p}_j, \hat{D}_j) \{(\hat{p}_j - \overline{D}_j) \log [\hat{p}_j/\overline{D}_j]$$

$$+ (\overline{D}_j - \hat{p}_j) \log [(1-\hat{p}_j)/(1-\overline{D}_j)]\}.$$

The assessment skew coefficient $C(\gamma; x, y)$ is defined as

$$C(\gamma; x, y) = \begin{cases} \gamma , & x > y \\ 0 , & x = y \\ 1-\gamma , & x < y , \end{cases}$$

where γ determines the extent of skewness toward Type II results.

The three parameters in this model can be interpreted as weights for three aspects of predictive performance assessment:

$$\beta = \begin{cases} 1 & \text{weight accuracy only} \\ 0.5 & \text{weight accuracy and definitiveness equally} \\ 0 & \text{weight definitiveness only} \end{cases}$$

$$\gamma_e = \begin{cases} 1 & \text{weight Type II errors only} \\ 0.5 & \text{weight Types I and II errors equally} \\ 0 & \text{weight Type I errors only} \end{cases}$$

$$\gamma_v = \begin{cases} 1 & \text{weight Type II verities only} \\ 0.5 & \text{weight Types I and II verities equally} \\ 0 & \text{weight Type I verities only} \end{cases}$$

Selection of these parameters is a matter of value judgment. They influence the weight normalization. This, in turn, influences the resolution of the feature space partition, which then affects the conditional probabilities. Thus, value judgment plays a role in the determination of numerical values for probabilities based on observational data.

This was recognized from the reverse perspective by Efron and Morris when they considered the problem of how far one should extend the outer boundaries of the total sample upon which to base predictions. Their conclusion: "That depends largely upon what we want to use the numbers for."

It is sometimes said that the magnitude of information does not depend upon utility. It is true that its explicit dependence is solely on probability. However, to the extent that our probability estimates are inherently a function of utility, information is implicitly dependent upon utility.

Although we have obtained this result within the context of the concept of weight normalization, it is quite general. The event classification for small samples is ultimately a matter of value judgment. Because of this, the probabilities we assign to any particular future event must, in the last analysis, contain an element of value-dependence.

The selection of test-statistics, acceptance-values and critical values, for example, has long been recognized as a matter of value judgment. Yet, there has generally been felt to be an underlying reality about which we have only limited evidence. However, if we have truly taken all available data into consideration and still have an element of value-dependence to our description, then there is no further reality in the sense that we can ever possibly know of it.

Does this violate the requirement that the probabilities reflect no more than the available information? Utility and information are, in a sense, canonical conjugates. A dependence upon utility is not a dependence upon information per se. From a purely information-theoretic viewpoint, it "only" affects how we partition the space. Within any partition we still maximize entropy, so in this sense remain unbiased. This is so despite the fact that it can significantly affect the outcome probabilities for particular events.

In the Copenhagen interpretation of quantum phenomena we have examples of observable aspects of reality which depend upon our knowledge. Here, in cases of small sample knowledge about complex systems, we have an observable aspect of reality which also depends upon the relative utility of alternative future worlds.

Is there any way around this rather bizarre conclusion?

Perhaps there is an "absolute" way of assessing predictive performance. Then we could determine the weight normalization independent of utility. However, this is contrary to the known variability of utility functions over assessment space.

Suppose two persons with different utility functions disagree about
the probability of a future outcome for a complex system based on
a small sample. See Fig. 136.

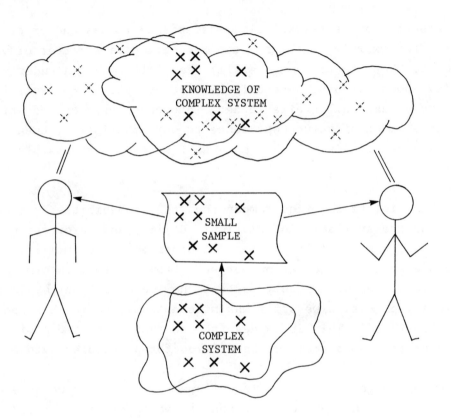

Fig. 136.   Small Sample Knowledge of Complex Systems Is,
            to Some Extent, Inherently Dependent upon
            Personal Values

Their present disagreement cannot be resolved by future gathering of more data. Suppose one looked back and said, "I was more accurate than you when the samples were still small. Therefore, my utilities are more objective than yours." The other could respond, "Accuracy is not the only consideration. Definitiveness and risk-avoidance are important also. My past decision was correct."

Perhaps we have, in the last analysis, merely defined a new sort of probability--a decision-making probability. It seems conceptually very similar to the subjective probability of Ramsey, de Finetti, and Savage. It is the probability which we act as though we believe.

We cannot distinguish between value judgment influencing our viewpoint and influencing reality itself. Each is informationally relative to the other. If we undergo a value change, we *see* reality alter. When the boundaries of sameness metamorphosize, causality reforms in past, present and future. Reality shifts are produced by innumerable value-changing events in the normal course of life: puberty, pregnancy, childbirth, aging, psycho-social environmental conditioning, advertising, religious experience, personal enlightenment, emotional shock, disease, alcohol, drugs, and joining a society speaking a different language. Some are sufficiently rapid that the shift is perceived as in-process. Others are so slow that they are only noticed when making before-after comparisons.

This leaves us wondering: What is objective probability? The only answer seems to be that it is, in some sense, the "limit" of subjective probability as it becomes increasingly independent of utility with increasingly large samples. But such limits do not, in general, exist. Often rational but different prognoses of the future can be brought into agreement by gathering more data. Many systems, however, are so complex that no humanly conceivable amount of data can make the value-dependence negligible. With complex systems and small samples, we cannot eliminate the subjective influence upon reality.

C. Selecting and Partitioning the Dependent Variable

We finally return to what it is we started out to predict. As described in Chapter 6, selecting the independent variable and its
classification is largely a matter of value judgment. Depending
upon the relative importance of different types of information for
the intended application, one will choose to analyze data relative
to one or another variable. Similarly, one will adjust the classification of this variable to correspond to the intended uses of the
information.

Just as value judgment plays a role in our analysis of the independent variables, so also information magnitude considerations play
a role in our dependent variable judgments. This was illustrated
by the burglar paradox in Chapter 2, in which the statistical significance of our findings depends upon the extent of our search.
With respect to both the independent and the dependent variables,
considerations of information magnitude and value judgment are in-
extricably intertwined.

D. Recap

In summary, probability depends upon the following:

1. The future events that we wish to predict something about.

2. The choice of the dependent variable: what it is that we
    are interested in predicting about these events.

3. The choice of a partition of the dependent variable: what
    it is about the dependent variable that we wish to predict.

4. The values of independent variables upon which the de-
    pendent variable may depend.

5. The available data concerning events more or less similar
    to the future events.

6. The utilities (or disutilities) to us of various types of
    correct and erroneous predictions.

Action at a distance is anathema to physical intuition. How can one thing affect another without *touching* it? So we invent the concept of a particle, a wave packet of energy, to represent the something which carries or mediates physical action at a distance across space-time. For each action at a distance circumstance, we conduct experiments, seeking a something mediating the touching. This something, be it called a photon, a graviton, a W-boson or a quark, is conceived as *moving*, i.e., as being the "same" thing at different times. Assignment of "sameness" to distinct entities implies classification. Here we have point classification into world-lines. But classes exist only in our mind. Thus, at the most basic level of understanding nature, we are faced with conceptual action at a distance, the problem of induction. And we can similarily ask what carries *its touch*? A packet of entropy has the correct properties.

In the past, physics has uncovered a rich variety of attributes of matter-packets (lifetime, mass, energy, velocity, momentum, electric charge, magnetic moment, baryon number, intrinsic spin, helicity, isotopic spin, strangeness, charm, color, etc.). We may similarly reveal a variety of attributes of order-packets, a variegation accompanying the diversity of observers' informational reference frames and decision-making behavior. These "orderons", which we experience macroscopically as laws of nature, represent projections of the mass structuring of physical events into the conceptual subspaces of neural networks. Their properties express which orderons interact with which events and how. A universal, zero entropy, orderon, like a zero mass particle, would have longest range, though in a different space. Our every-day expectation generation activities of mind are stimulated by finite entropy orderons of multifarious genera.

QUOTES

ARISTOTLE (c. 335-323 B.C.)

For if all propositions whether positive or negative are either true or false, then any given predicate must either belong to the subject or not, so that if one man affirms that an event of a given character will take place and another denies it, it is plain that the statement of the one will correspond with reality and that of the other will not. For the predicate cannot both belong and not belone to the subject at one and the same time with regard to the future.

* * * *

Everything must either be or not be, whether in the present or in the future, but it is not always possible to distinguish and state determinately which of these alternatives must necessarily come about.

-Aristotle (384-322 B.C.)
On Interpretation (c. 335-323 B.C.), tr. by
E. Edghill, *The Basic Works of Aristotle*, ed.
by R. McKeon, Random House, NY, 1941, pp. 46, 48.

NEWTON (1686)

We are certainly not to relinquish the evidence of experiments for the sake of dreams and vain fictions of our own devising.

-Isaac Newton (1642-1727)
*Sir Isaac Newton's Mathematical Principles of
Natural Philosophy and Its System of the World*
(1686), tr. by A. Motte (1729), ed. by F. Cajori,
Univ. of Calif. Press, Berkeley, CA, 1962, p. 398.

GAUSS (1821)

The question which concerns us here has something vague about it from its very nature, and cannot be made really precise except ,by some principle which is arbitrary to a certain degree. The determination of a magnitude by observation can be compared, with some appropriateness, to a game in which there is a loss to fear and no gain to hope for; each error committed being likened to a loss which one suffers, the relative undesirability of such a game should be expressed by the probable loss, that is to say

GAUSS (1821, cont.)

by the sum of the products of the various possible losses by
their respective probabilities. But what loss should one as-
sociate with a given error? This is something which is not
clear in itself; the evaluation depends in part on our choice.

> —Karl F. Gauss (1777-1855)
> "Theory of the Combination of Observations Which
> Leads to the Smallest Errors," (Göttingen, 15 Feb.
> 1821), *Gauss' Work (1803-1826) on the Theory of
> Least Squares* tr. by H. Trotter, Tech. Rept. No. 5,
> Dept. of Army Rept. No. 5B99-01-004, Stat. Tech.
> Res. Group, Dept. of Math., Princeton, NJ, 1957,
> pp. 7-8.

POINCARÉ (1905)

Whether we take the moral, the esthetic or the scientific point
of view, it is always the same thing. Nothing is objective
except what is identical for all; now we can only speak of such
an identity if a comparison is possible, and can be translated
into a 'money of exchange' capable of transmission from one
mind to another. Nothing, therefore, will have objective value
except what is transmissible by 'discourse,' that is, intel-
ligible.

> —Henri Poincaré (1854-1912)
> *The Value of Science* (1905), tr. by G. Halsted,
> Dover Pubs., Inc., NY, 1958, p. 137.

PLANCK (1923)

...the possibility of a completely objective scientific investigation into psychological phenomena only extends to the critical
examination of personalities other than the observer, so long as
they are independent of the observer. In so far as it is completely effaced from the mind of the investigator, it also extends to the past, but not to the present, nor to the future,
which must always be attained through the present. Thought
and research are themselves psychological phenomena in man, and
if the object of the investigation is identical with the investigator, he must change continually as his knowledge advances.

> —Max Planck (1858-1947)
> *A Survey of Physics* (1923), tr. by R. Jones and
> D. Williams, reprinted as *A Survey of Physical
> Theory*, Dover Pubs., Inc., NY, 1960, pp. 67-68.

EINSTEIN (1949)

> Physics is an attempt conceptually to grasp reality as it is
> thought independently of its being observed...In pre-quantum
> physics there was no doubt as to how this was to be understood.
> In Newton's theory reality was determined by a material point
> in space and time; in Maxwell's theory, by the field in space
> and time. In quantum mechanics it is not so easily seen...The
> probability is here to be viewed as an empirically determinable,
> and therefore certainly as a "real" quantity which I may deter-
> mine if I create the same ψ-function very often and perform a
> q-measurement each time. But what about the single measured
> value of q? Did the respective individual system have this
> q-value even before the measurement? To this question there is
> no definite answer within the framework of the [existing] theory,
> since the measurement is a process which implies a finite dis-
> turbance of the system from the outside; it would therefore be
> thinkable that the system obtains a definite numerical value for
> q (or p) the measured numerical value, only through the measure-
> ment itself.

* * * *

> One can escape from this conclusion only by either assuming that
> the measurement of S_1 (telepathically) changes the real situa-
> tion of S_2 or by denying independent real situations as such to
> things which are spatially separated from each other. Both al-
> ternatives appear to me entirely unacceptable.

> —Albert Einstein (1879-1955)
> "Autobiographical Notes," in *Albert Einstein:*
> *Philosopher-Scientist*, tr. and ed. by P. Schilpp,
> The Lib. of Living Philosophers, Evanston, IL,
> 1949, pp. 81-83, 85.

BRILLOUIN (1962)

> It is only by ignoring the human value of the information that
> we have been able to construct a scientific theory of information
> based on statistics, and this theory has already proved very
> useful. There are, however, many problems that cannot be dis-
> cussed along these lines...This new element is needed every time
> one considers information as a basis for prediction and for
> practical use.

> —Leon Brillouin (1889-1969)
> *Science and Information Theory*, Academic Press,
> NY, 1962, p. 294.

REFERENCES

Aristotle, *On Interpretation*, tr. by E. Edghill, *The Basic Works of Aristotle*, ed. by R. McKeon, Random House, NY, 1941.

Arrow, Kenneth J., *Essays in the Theory of Risk-Bearing*, North-Holland Pub. Co., Amsterdam, 1970.

Arrow, Kenneth J. and F.H. Hahn, *General Competitive Analysis*, Holden-Day, Inc., San Francisco, CA, 1971.

Brillouin, Leon, *Science and Information Theory*, Academic Press, NY, 1962.

Christensen, R.A., "Representation of Extent of Belief," excerpts from "Individual Decision-Making Under Conditions of Known Risk," Böblingen, Germany, August 1961. [*Chapter 2 of Volume II.*]

_____, "Nuclear Fuel Rod Failure Hazard Axes," *Fuel Rod Mechanical Performance Modeling, Task 3: Fuel Rod Modeling and Decision Analysis*, FRMPM33-2, RP971-2, Tenth Quarterly Progress Report, Entropy Limited, Lincoln, MA, August-October 1979, pp. 7.1-7.11.

DeGroot, Morris H., *Optimal Statistical Decisions*, McGraw-Hill Book Co., NY, 1970.

Efron, B. and C. Morris, "Stein's Paradox in Statistics," *Scientific American*, *236*, May 1977, pp. 119-127.

Einstein, Albert, "Autobiographical Notes," in *Albert Einstein: Philosopher-Scientist*, tr. and ed. by P. Schilpp, The Library of Living Philosophers, Evanston, IL, 1949.

Gauss, Karl F., "Theory of the Combination of Observations Which Lead to the Smallest Errors," (Göttingen, Feb. 15, 1821), *Gauss' Work (1803-1826) on the Theory of Least Squares*, tr. by H. Trotter, Tech. Rept. No. 5, Dept. of Army Project No. 5B99-01-004, Statistical Techniques Research Group, Dept. of Math, Princeton Univ., Princeton, NJ, 1957.

Josephson, B.D. and V.S. Ramachandran (eds.), *Consiousness and the Physical World*, Pergamon Press, Oxford, 1980.

Keeney, Ralph L. and Howard Raiffa, *Decisions With Multiple Objectives: Preferences and Value Tradeoffs*, J. Wiley & Sons, NY, 1976.

Middleton, D., "Statistical Theory of Signal Detection," *IRE Trans. on Information Theory*, PGIT-3, 1954, pp. 26-62.

von Neumann, John and Oskar Morgenstern, *Theory of Games and Economic Behavior*, Princeton Univ. Press, Princeton, NJ, 1944.

Newton, Isaac, *Sir Isaac Newton's Mathematical Principles of Natural Philosophy and Its System of the World* (1686), tr. by A. Motte (1729), ed. by F. Cajori, Univ. of Calif. Press, Berkeley, CA, 1962.

Oldberg, S., "Probabilistic Code Development," *Planning Support Document for the EPRI Light Water Reactor Fuel Performance Program*, prepared by J.T.A. Roberts, F.E. Gelhaus, H. Ocken, N. Hoppe, S.T. Oldberg, G.R. Thomas and D. Franklin, EPRI NP-737-SR, Electric Power Research Inst., Palo Alto, CA, Jan. 1978, pp. 2.52-2.60.

Peterson, W.W., T.G. Birdsall and W.C. Fox, "The Theory of Signal Detectability," *IRE Trans. on Info. Theory*, *PGIT-4*, 1954, pp. 171-212.

Planck, Max, *A Survey of Physics* (1923), tr. by R. Jones and D. Williams, reprinted as *A Survey of Physical Theory*, Dover Pubs., Inc., NY, 1960.

Poincaré, Henri, *The Value of Science* (1905), tr. by F. Halsted, Dover Pubs., Inc., NY, 1958.

Polanyi, Michael, *Personal Knowledge, Towards a Post-Critical Philosophy*, The Univ. of Chicago Press, Chicago, IL, 1958.

Potchen, E. James, "Study on the Use of Diagnostic Radiology," *Current Concepts in Radiology*, *2*, ed. by E.J. Potchen, The C.V. Mosby Co., St. Louis, MO, 1975, pp. 18-30.

Raiffa, Howard, *Decision Analysis*, Addison-Wesley Pub. Co., Reading, MA, 1970.

Tanner, Wilson P., "Psychological Implications of Psychophysical Data," *Human Decisions in Complex Systems*, conference edited by W. McCulloch, *Annals of the New York Academy of Sciences*, *80*, Jan. 28, 1961, pp. 751-765.

Tanner, W.P. and J.A. Swets, "The Human Use of Information," *IRE Trans. on Inform. Theory*, *PGIT-4*, 1954, pp. 213-221.

APPENDICES

APPENDIX 1

EXAMPLES OF CONTINUOUS
UNIVARIATE DISTRIBUTIONS

NORMAL

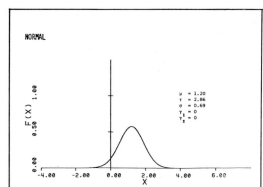

$\mu = 1.20$
$\tau = 2.86$
$\sigma = 0.69$
$\gamma_1 = 0$
$\gamma_2 = 0$

$$f(x;a,b) = \frac{1}{\sqrt{2\pi}\ b}\ e^{-\frac{1}{2}\left(\frac{x-a}{b}\right)^2}, \quad b \geq 0$$

$\mu = a$

$\sigma^2 = b^2$

$\gamma_1 = 0$

$\gamma_2 = 0$

$S_R = \frac{1}{2}(1 + \log 2\pi) + \log b$

CAUCHY

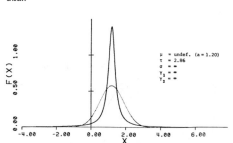

$\mu = $ undef. $(a = 1.20)$
$\tau = 2.86$
$\sigma = \infty$
$\gamma_1 = \infty$
$\gamma_2 = \infty$

$$f(x;a,b) = \frac{1}{\pi b}\ \frac{1}{1 + \left(\frac{x-a}{b}\right)^2}, \quad b > 0$$

$\mu = $ undefined (though $f(x)$ is symmetric about $x = a$)

$\sigma^2 = \infty$

$\gamma_1 = \infty$

$\gamma_2 = \infty$

$S_R = \log 4\pi + \log b$

LAPLACE

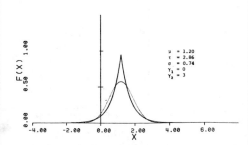

$\mu = 1.20$
$\tau = 2.86$
$\sigma = 0.74$
$\gamma_1 = 0$
$\gamma_2 = 3$

$$f(x;a,b) = \frac{1}{2b}\ e^{-\left|\frac{x-a}{b}\right|}, \quad b \geq 0$$

$\mu = a$

$\sigma^2 = 2b^2$

$\gamma_1 = 0$

$\gamma_2 = 3$

$S_R = 1 + \log 2 + \log b$

LOGISTIC

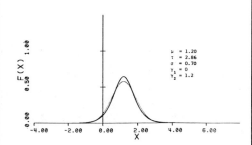

$\mu = 1.20$
$\tau = 2.86$
$\sigma = 0.70$
$\gamma_1 = 0$
$\gamma_2 = 1.2$

$$f(x;a,b) = \frac{1}{4b}\ \text{sech}^2\left\{\frac{1}{2}(x-a)/b\right\}, \quad b \geq 0$$

$\mu = a$

$\sigma^2 = \pi^2 b^2/3 \approx 3.2899b^2$

$\gamma_1 = 0$

$\gamma_2 = 6/5$

$S_R = 2 + \log b$

EXTREME VALUE

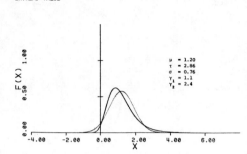

$$f(x;a,b) = \frac{1}{b} \exp\left\{-\left(\frac{x-a}{b}\right) - e^{-\left(\frac{x-a}{b}\right)}\right\} , \quad b > 0$$

$$\mu = a + \gamma b \approx a + 0.57722b$$

$$\sigma^2 = \frac{1}{6}\pi^2 b^2 \approx 1.6449b^2$$

$$\gamma_1 = -\frac{6\sqrt{6}}{\pi^3}\psi''(1) \approx 1.1395$$

$$\gamma_2 = 2.4$$

$$S_R = 1 + \gamma + \log b$$

MAXWELL-BOLTZMANN

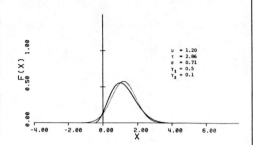

$$f(x;a,b) = \frac{4}{\sqrt{\pi}\,b}\left(\frac{x-a}{b}\right)^2 e^{-\left(\frac{x-a}{b}\right)^2}, \quad x > a,\ b \geq 0$$

$$\mu = a + 2b/\sqrt{\pi} \approx a + 1.1283b$$

$$\sigma^2 = \left(\frac{3}{2} - \frac{4}{\pi}\right)b^2 \approx 0.22676b^2$$

$$\gamma_1 = \frac{1}{\sqrt{\pi}}\left(\frac{16}{\pi} - 5\right)\left(\frac{3}{2} - \frac{4}{\pi}\right)^{-\frac{3}{2}} \approx 0.4857$$

$$\gamma_2 = \left(\frac{15}{4} + \frac{4}{\pi} - \frac{48}{\pi^2}\right)\left(\frac{3}{2} - \frac{4}{\pi}\right)^{-2} - 3 \approx 0.10816$$

$$S_R = \gamma - \frac{1}{2}(1 - \log \pi) + \log b$$

GAMMA

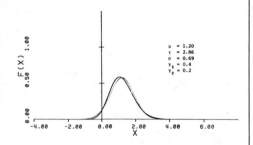

$$f(x;a,b,c) = \frac{1}{b\Gamma(c)}\left(\frac{x-a}{b}\right)^{c-1} e^{-\left(\frac{x-a}{b}\right)}, \quad x \geq a,\ b \geq 0,\ c \geq 0$$

$$\mu = a + bc$$

$$\sigma^2 = b^2 c$$

$$\gamma_1 = 2/\sqrt{c}$$

$$\gamma_2 = 6/c$$

$$S_R = c + (1-c)\psi(c) + \log\Gamma(c) + \log b$$

WEIBULL

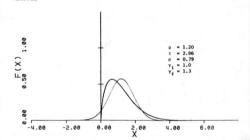

$$f(x;a,b,c) = \frac{c}{b}\left(\frac{x-a}{b}\right)^{c-1} e^{-\left(\frac{x-a}{b}\right)^c}, \quad x \geq a,\ b \geq 0,\ c \geq 0$$

$$\mu = a + b\Gamma\left(1 + \frac{1}{c}\right)$$

$$\sigma^2 = b^2\left\{\Gamma\left(1 + \frac{2}{c}\right) - \left[\Gamma\left(1 + \frac{1}{c}\right)\right]^2\right\}$$

$$\gamma_1 = \left\{\Gamma\left(1 + \frac{3}{c}\right) - 3\,\Gamma\left(1 + \frac{1}{c}\right)\Gamma\left(1 + \frac{2}{c}\right) + 2\left[\Gamma\left(1 + \frac{1}{c}\right)\right]^3\right\}$$
$$\times \left\{\Gamma\left(1 + \frac{2}{c}\right) - \left[\Gamma\left(1 + \frac{1}{c}\right)\right]^2\right\}^{-\frac{3}{2}}$$

$$\gamma_2 = \left\{\Gamma\left(1 + \frac{4}{c}\right) - 4\,\Gamma\left(1 + \frac{1}{c}\right)\Gamma\left(1 + \frac{3}{c}\right) + 6\left[\Gamma\left(1 + \frac{1}{c}\right)\right]^2\Gamma\left(1 + \frac{2}{c}\right)\right.$$
$$\left. - 3\left[\Gamma\left(1 + \frac{1}{c}\right)\right]^4\right\}\left\{\Gamma\left(1 + \frac{2}{c}\right) - \left[\Gamma\left(1 + \frac{1}{c}\right)\right]^2\right\}^{-2} - 3$$

$$S_R = 1 + \left(\frac{c-1}{c}\right)\gamma - \log c + \log b$$

LOG-NORMAL

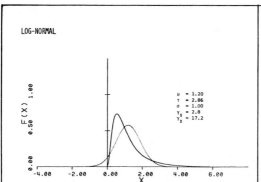

$$\mu = 1.20$$
$$\tau = 2.86$$
$$\sigma = 1.00$$
$$\gamma_1 = 2.8$$
$$\gamma_2 = 17.2$$

$$f(x;a,b,c) = \frac{1}{(x-a)c\sqrt{2\pi}} \exp\left\{-\frac{1}{2c^2}\left[\log\left(\frac{x-a}{b}\right)\right]^2\right\}, \quad x \geq a,\ b \geq 0,\ c \geq 0$$

$$\mu = a + be^{\frac{1}{2}c^2}$$

$$\sigma^2 = b^2 e^{c^2}\left(e^{c^2}-1\right)$$

$$\gamma_1 = \sqrt{\left(e^{c^2}-1\right)}\left(e^{c^2}+2\right)$$

$$\gamma_2 = \left(e^{c^2}-1\right)\left(e^{3c^2}+3e^{2c^2}+6e^{c^2}+6\right)$$

$$S_R = \tfrac{1}{2}(1 + \log 2\pi) + \log c + \log b$$

CHI-SQUARED

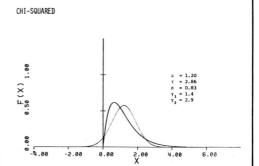

$$\mu = 1.20$$
$$\tau = 2.86$$
$$\sigma = 0.83$$
$$\gamma_1 = 1.4$$
$$\gamma_2 = 2.9$$

$$f(x;a,b) = \frac{x^{\frac{1}{2}a-1}e^{-\frac{x}{2b}}}{(2b)^{\frac{1}{2}a}\ \Gamma(\frac{1}{2}a)}, \quad x > 0,\ a > 0,\ b > 0$$

$$\mu = ab$$

$$\sigma^2 = 2ab^2$$

$$\gamma_1 = \sqrt{8/a}$$

$$\gamma_2 = 12/a$$

$$S_R = \frac{a}{2} + \log\left[2\Gamma\left(\frac{a}{2}\right)\right] + \left(1-\frac{a}{2}\right)\psi\left(\frac{a}{2}\right) + \log b$$

CHI

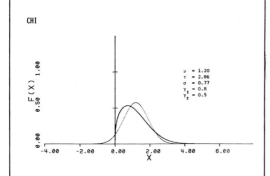

$$\mu = 1.20$$
$$\tau = 2.86$$
$$\sigma = 0.77$$
$$\gamma_1 = 0.8$$
$$\gamma_2 = 0.5$$

$$f(x;a,b) = \frac{x^{a-1}e^{-\frac{1}{2}\left(\frac{x}{b}\right)^2}}{2^{\frac{1}{2}a-1}b^a\ \Gamma\left(\frac{a}{2}\right)}, \quad x > 0,\ a > 0,\ b > 0$$

$$\mu = \sqrt{2}\ b\ \Gamma\left(\tfrac{a+1}{2}\right)/\Gamma\left(\tfrac{a}{2}\right)$$

$$\sigma^2 = ab^2 - 2b^2\left[\Gamma\left(\tfrac{a+1}{2}\right)/\Gamma\left(\tfrac{a}{2}\right)\right]^2$$

$$\gamma_1 = \frac{\sqrt{2}\,(1-2a)\,\Gamma\left(\tfrac{a+1}{2}\right)/\Gamma\left(\tfrac{a}{2}\right) + 4\sqrt{2}\left[\Gamma\left(\tfrac{a+1}{2}\right)/\Gamma\left(\tfrac{a}{2}\right)\right]^3}{\left\{a - 2\left[\Gamma\left(\tfrac{a+1}{2}\right)/\Gamma\left(\tfrac{a}{2}\right)\right]^2\right\}^{3/2}}$$

$$\gamma_2 = \frac{a(a+2) + 4(a-2)\left[\Gamma\left(\tfrac{a+1}{2}\right)/\Gamma\left(\tfrac{a}{2}\right)\right]^2 - 12\left[\Gamma\left(\tfrac{a+1}{2}\right)/\Gamma\left(\tfrac{a}{2}\right)\right]^4}{\left\{a - 2\left[\Gamma\left(\tfrac{a+1}{2}\right)/\Gamma\left(\tfrac{a}{2}\right)\right]^2\right\}^2} - 3$$

$$S_R = \frac{a}{2} + \log\left[\frac{1}{\sqrt{2}}\ \Gamma\left(\frac{a}{2}\right)\right] - \left(\frac{a-1}{2}\right)\psi\left(\frac{a}{2}\right) + \log b$$

F

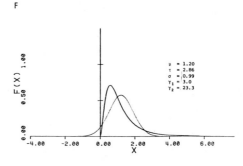

$$\mu = 1.20$$
$$\tau = 2.86$$
$$\sigma = 0.99$$
$$\gamma_1 = 3.0$$
$$\gamma_2 = 23.3$$

$$f(x;a,b) = \frac{\Gamma\left(\frac{a+b}{2}\right)}{\Gamma\left(\frac{a}{2}\right)\Gamma\left(\frac{b}{2}\right)}\ a^{a/2}\ b^{b/2}\ \frac{x^{(b-2)/2}}{(a+bx)^{(a+b)/2}}, \quad x \geq 0,\ a > 0,\ b > 0$$

$$\mu = \frac{a}{a-2}, \quad a > 2$$

$$\sigma^2 = \frac{2a^2(a+b-2)}{b(a-2)^2(a-4)}, \quad a > 4$$

$$\gamma_1 = \sqrt{\frac{8(a-4)}{(a+b-2)b}} \cdot \frac{(a+2b-2)}{(a-6)}, \quad a > 6$$

$$\gamma_2 = \frac{12[(a-2)^2(a-4)+b(a+b-2)(5a-22)]}{b(a-6)(a-8)(a+b-2)}, \quad a > 8$$

$$S_R = \log\frac{a}{b} + \log B\left(\frac{a}{2},\frac{b}{2}\right) - \left(1+\frac{a}{2}\right)\psi\left(\frac{a}{2}\right) + \left(1-\frac{b}{2}\right)\psi\left(\frac{b}{2}\right) + \left(\frac{a+b}{2}\right)\psi\left(\frac{a+b}{2}\right)$$

LOMAX

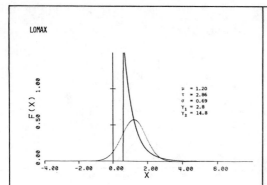

$$f(x;a,b,c) = \frac{cb^c}{(x-a)^{c+1}} \ , \ x-a \geq b > 0, \ c > 0$$

$$\mu = a + \frac{bc}{c-1} \ , \ c > 1$$

$$\sigma^2 = \frac{b^2 c}{(c-1)^2(c-2)} \ , \ c > 2$$

$$\gamma_1 = \left[\frac{c}{c-3} - \frac{3c^2}{(c-1)(c-2)} + \frac{c^3}{(c-1)^3}\right]\left[\frac{(c-1)^3(c-2)}{c}\right]\sqrt{\frac{c-2}{c}} \ , \ c > 3$$

$$\gamma_2 = \left[\frac{c}{c-4} - \frac{4c^2}{(c-1)(c-3)} + \frac{6c^3}{(c-1)^2(c-2)} - \frac{3c^4}{(c-1)^4}\right]\left[\frac{(c-1)^4(c-2)^2}{c^2}\right] - 3, \\ c > 4$$

$$S_R = 1 + \frac{1}{c} - \log c + \log b$$

RECTANGULAR

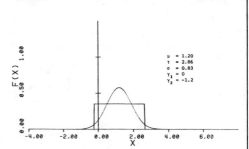

$$f(x;a,b) = \frac{1}{b} \ , \ a-\tfrac{1}{2}b \leq x \leq a+\tfrac{1}{2}b, \ b \geq 0$$

$$\mu = a$$

$$\sigma^2 = b^2/12$$

$$\gamma_1 = 0$$

$$\gamma_2 = -6/5$$

$$S_R = \log b$$

ISOSCELES TRIANGULAR

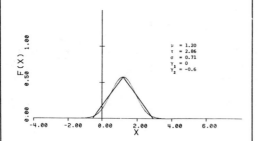

$$f(x;a,b) = \frac{1}{b} - \frac{|x-a|}{b^2} \ , \ a-b < x < a+b, \ b \geq 0$$

$$\mu = a$$

$$\sigma^2 = b^2/6$$

$$\gamma_1 = 0$$

$$\gamma_2 = -3/5$$

$$S_R = \frac{1}{2} + \log b$$

BETA

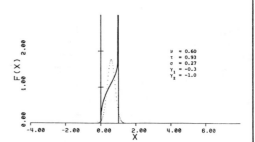

$$f(x;a,b) = \frac{x^{a-1}(1-x)^{b-1}}{B(a,b)} \ , \ 0 \leq x \leq 1, \ a \geq 0, \ b \geq 0, \ a + b > 0$$

$$\mu = a/(a+b)$$

$$\sigma^2 = ab/[(a+b)^2(a+b+1)]$$

$$\gamma_1 = \frac{2(b-a)\sqrt{a+b+1}}{(a+b+2)\sqrt{ab}}$$

$$\gamma_2 = \frac{3(a+b+1)[2(a+b)^2+ab(a+b-6)]}{ab(a+b+2)(a+b+3)} - 3$$

$$S_R = \log[B(a,b)] - (a-1)[\psi(a)-\psi(a+b)] - (b-1)[\psi(b)-\psi(a+b)]$$

APPENDIX 2

EXAMPLES OF DISCRETE
UNIVARIATE DISTRIBUTIONS

BINOMIAL

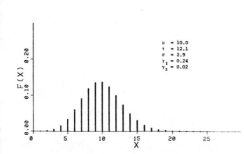

$$f(x;p,n) = \binom{n}{x} p^x (1-p)^{n-x}, \ 0 \le x \le n, \ 0 \le p \le 1, \ n > 0$$

$$\mu = np$$

$$\sigma^2 = np(1-p)$$

$$\gamma_1 = (1-2p)/\sqrt{np(1-p)}$$

$$\gamma_2 = [1-6p(1-p)]/[np(1-p)]$$

$$S = -np \log p - n(1-p)\log(1-p) - \sum_{x=1}^{n-1} \binom{n}{x} p^x (1-p)^{n-x} \log\binom{n}{x}$$

POISSON

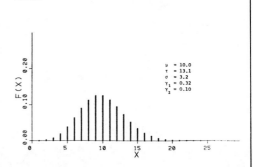

$$f(x;a) = \frac{e^{-a}a^x}{x!}, \ x \ge 0, \ a > 0$$

$$\mu = a$$

$$\sigma^2 = a$$

$$\gamma_1 = 1/\sqrt{a}$$

$$\gamma_2 = 1/a$$

$$S = a(1 - \log a) + e^{-a} \sum_{x=0}^{\infty} \frac{a^x \log x!}{x!}$$

DISCRETE RECTANGULAR

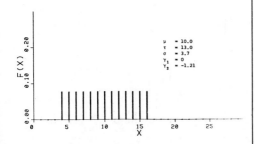

$$f(x;m,n) = \frac{1}{n}, \ x = m+1, \ m+2, \ldots, \ m+n, \ n \ge 1$$

$$\mu = m + (n+1)/2$$

$$\sigma^2 = (n^2-1)/12$$

$$\gamma_1 = 0$$

$$\gamma_2 = -\frac{6}{5} \cdot \frac{n^2+1}{n^2-1}$$

$$S = \log n$$

LOGARITHMIC SERIES

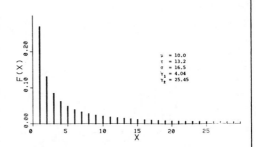

$$f(x;p) = \frac{ap^x}{x}, \ x > 0, \ 0 < p < 1, \ a = -1/\log(1-p)$$

$$\mu = \frac{ap}{1-p}$$

$$\sigma^2 = ap(1-ap)(1-p)^{-2}$$

$$\gamma_1 = \left[1-2ap+\frac{p}{1-ap}\right]\left[ap(1-ap)\right]^{-\frac{1}{2}}$$

$$\gamma_2 = \frac{1}{ap(1-ap)}\left[1+4p+\left(\frac{p^2}{1-ap}\right)\right] - 6$$

$$S = \frac{p \log p}{(1-p)\log(1-p)} - \log a + a \sum_{x=1}^{\infty} \frac{p^x \log x}{x}$$

NEGATIVE BINOMIAL (PASCAL)

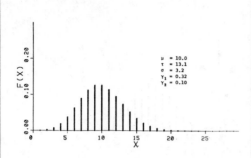

$$\mu = 10.0$$
$$\tau = 13.1$$
$$\sigma = 3.2$$
$$\gamma_1 = 0.32$$
$$\gamma_2 = 0.10$$

$$f(x;p,n) = \binom{n+x-1}{n-1} p^n (1-p)^x \ , \ x \geq 0, \ 0 < p \leq 1, \ n > 0$$

$$\mu = n(1-p)/p$$

$$\sigma^2 = n(1-p)/p^2$$

$$\gamma_1 = (2-p)/\sqrt{n(1-p)}$$

$$\gamma_2 = \frac{6-6p+p^2}{n(1-p)}$$

$$S = -\sum_x f(x;p,n) \log f(x;p,n)$$

COMPOUND POISSON (NEYMAN'S TYPE A)

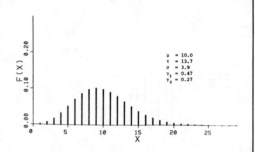

$$\mu = 10.0$$
$$\tau = 13.7$$
$$\sigma = 3.9$$
$$\gamma_1 = 0.47$$
$$\gamma_2 = 0.27$$

$$f(x;a,b) = \sum_{j=0}^{\infty} \left[\frac{(jb)^x e^{-jb}}{x!} \frac{a^j e^{-a}}{j!} \right] \ , \ x \geq 0, \ a \geq 0, \ b \geq 0$$

$$\mu = ab$$

$$\sigma^2 = ab(1+b)$$

$$\gamma_1 = (1+3b+b^2)(1+b)^{-\frac{3}{2}}(ab)^{-\frac{1}{2}}$$

$$\gamma_2 = (1+7b+6b^2+b^3)(1+b)^{-2}(ab)^{-1}$$

HYPERGEOMETRIC

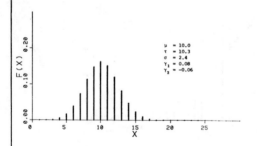

$$\mu = 10.0$$
$$\tau = 10.3$$
$$\sigma = 2.4$$
$$\gamma_1 = 0.08$$
$$\gamma_2 = -0.06$$

$$f(x;p,m,n) = \frac{\binom{pn}{x}\binom{n-pn}{m-x}}{\binom{n}{m}} \ , \ \max(0,m-n+pn) \leq x \leq \min(m,pn),$$
$$0 \leq p \leq 1, \ 1 \leq m \leq n$$

$$\mu = mp$$

$$\sigma^2 = mp(1-p)\left(\frac{n-m}{n-1}\right)$$

$$\gamma_1 = \frac{1-2p}{\sqrt{mp(1-p)}}\left(\frac{n-1}{n-m}\right)^{\frac{1}{2}}\left(\frac{n-2m}{n-2}\right)$$

$$\gamma_2 = \frac{6(5n-6)}{(n-2)(n-3)} + \frac{(n-1)n(n+1)}{(n-m)(n-2)(n-3)}\left[1 - \frac{6n}{n+1}\left(p(1-p)+\frac{m(n-m)}{n^2}\right)\right]\frac{1}{mp(1-p)}$$

NEGATIVE HYPERGEOMETRIC (BETA-BINOMIAL)

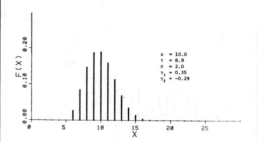

$$\mu = 10.0$$
$$\tau = 8.9$$
$$\sigma = 2.0$$
$$\gamma_1 = 0.35$$
$$\gamma_2 = -0.29$$

$$f(x;p,m,n) = \frac{\binom{pn}{m-1}\binom{n-pn}{x-m}}{\binom{n}{x-1}} \cdot \frac{pn-m+1}{n-x+1} \ , \ m \leq x \leq m+n-pn,$$
$$0 \leq p \leq 1, \ 0 < m < pn, \ n > 0$$

$$\mu = \frac{m(n+1)}{pn+1}$$

$$\sigma^2 = m(n+1)n(1-p)(pn+1-m)(pn+1)^{-2}(pn+2)^{-1}$$

$$\gamma_1 = [E_3 - 3E_2 + E_1 - 3E_1(E_2 - E_1) + 2E_1^3]\sigma^{-3}$$

$$\gamma_2 = [E_4 - 6E_3 + 7E_2 - E_1 - 4E_1(E_3 - 3E_2 + E_1) + 6E_1^2(E_2 - E_1) - 3E_1^4]\sigma^{-4} - 3$$

$$\text{where } E_i = \frac{(m+i-1)!}{(m-1)!} \frac{(n+i)!}{n!} \frac{(pn)!}{(pn+1)!}$$

APPENDIX 3

GLOSSARY AND SYMBOLS

GLOSSARY

A priori weight normalization: Total number of virtual events.

A priori weights: Numbers of virtual events in each outcome class.

Accuracy, predictive: Closeness of observed frequencies to the corresponding probability estimates. There are many different measures of predictive accuracy, of which Pearson's T, the dispersion ratio, and the entropy decrement are examples.

Amalgamation of probabilities: Formation of a single probability for an event by combining two or more probabilities for the same event.

Ambiguous: A property of the categorization of an event that the distinction between two categories does not logically pertain to the particular event; the event is simultaneously a member of both categories.

Assessment of predictions: Determination of the quality of predictions in terms of reliability, definitiveness and utility.

Asymptotically efficient (estimator property): The property of some estimators that the variance of their estimates decreases toward zero as the sample size increases without bound.

Asymptotically normal (distribution function property): The property of some distribution functions that they approach the normal distribution as the sample size grows without bound.

Asymptotically normal (estimator property): The property of some estimators that in the limit of large sample sizes their values are normally distributed.

Background information: Information relevant to the outcome probabilities for an event, other than information contained in the observational data sample.

Basis functions, entropy minimax: A specified set of M linear combinations of N observed training waveforms $\{\psi_i(t), i=1,\ldots,N\}$, given by

$$\phi_m(t) = \sum_{i=1}^{N} b_{mi}\, \psi_i(t), m=1,\ldots,M \le N,$$

where the basis coefficients $\{b_{mi}\}$ are chosen to minimize the expected conditional information loss

$$\Delta S = \int d\underline{z}\, f(\underline{z}) \sum_{k=1}^{K} p_k(\underline{z})\, \log p_k(\underline{z})/p_k'(\underline{z}),$$

where $p_k(\underline{z})$ is the probability based on the full N terms, and $p_k'(\underline{z})$ is based on a truncation to M terms.

Basis functions, Karhunen-Loève: A specified set of M linear combinations of N observed training waveforms $\{\psi_i(t), i=1,\ldots,N\}$, given by

$$\phi_m(t) = \sum_{i=1}^{N} b_{mi} \, \psi_i(t), \quad m=1,\ldots,M \leq N,$$

where the basis coefficients $\{b_{mi}\}$ are chosen to minimize the error squared upon reconstruction of all N training waveforms when only M terms are used for the reconstruction.

Bayes' (inverse probability) theorem: The theorem that if $\{O_k\}$ is a partition of the outcome space, and if the probabilities $\{P(O_j)\}$ and $\{P(C \text{ given } O_j)\}$ are fixed and based on the same data, then

$$P(O_k \text{ given } C) = \frac{P(O_k)P(C \text{ given } O_k)}{\sum_j P(O_j)P(C \text{ given } O_j)} \quad .$$

Bayes' (uniform distribution) postulate: The postulate that the range of values of a probability should be subdivided into equal length segments when using the equiprobability principle to determine the probabilities of intervals of probability.

Belief, intensity of (Peirce): The logarithm of the chance

$$J(A) = \log (C(A)),$$

where C(A) is the chance (as defined by Peirce) of A. For a single event:

$$J(A) = \log\left(\frac{P(A)}{1-P(A)}\right),$$

where P(A) is the probability of A.

Beta distribution: The continuous univariate distribution defined by the function

$$f(x) = x^{a-1}(1-x)^{b-1}/B(a,b),$$

where

$$0 \leq x \leq 1,$$

$$a,b \geq 0,$$

$$a+b > 0.$$

Beta function: The function

$$B(a,b) = \Gamma(a)\Gamma(b)/\Gamma(a+b) \quad ,$$

used in defining the beta distribution.

Biased: The property of some estimators that their expected values differ from the true values. The property of some test-statistics that the significance level of the null hypothesis is greater when the hypothesis is false than when it is true.

Binomial distribution: The discrete univariate distribution defined by the function

$$f(x) = \binom{n}{x} p^x (1-p)^{n-x},$$

where

$$0 \leq x \leq n,$$
$$0 \leq p \leq 1,$$
$$n \geq 1.$$

Bounded (distribution function property): The property of some distribution functions that they are defined over limited ranges of the random variable.

Categorization: Assignment of values for an independent variable to discrete categories.

Chance (Peirce): Single event: the ratio of the probability of being true to the probability of being false:

$$C(A) = \frac{P(A)}{1-P(A)},$$

Pair of events: the ratio of the probability of being true to the probability of being false, given truth value agreement:

$$C(A \text{ and } B) = \frac{P(A,B)}{1-P(A,B)},$$

where

$$P(A,B) = P((A \text{ and } B) \text{ given } ((A \text{ and } B) \text{ or } (\overline{A} \text{ and } \overline{B}))).$$

Chance correlations: Correlations which happen to turn up for a particular sample but which are not representative of the population as a whole.

Chi-squared distribution: The continuous univariate distribution defined by the function

$$f(x) = \frac{x^{(\nu/2-1)} e^{-x/2}}{\Gamma(\nu/2) 2^{\nu/2}},$$

where

$$x \geq 0,$$
$$\nu > 0.$$

Classification: Assignment of values for a dependent variable to discrete classes. (Sometimes also used to refer to partition of independent variable space.)

Cluster: A set of events located near to each other in feature space.

Combinatorial function: The number of combinations of n things taken x at a time:

$$\binom{n}{x} = \frac{n!}{x!\,(n-x)!},$$

used in defining the binomial distribution.

Communication: Transmission of information from a source O to a receiver C.

    S(O) = average transmitted information
    S(C) = average received information
    S(C given O) = noise
    S(O given C) = equivocation
    ΔS(O mutual C) = average mutual information exchange

Complex: The property of a description being expressed in terms of a large amount of information.

Conditional entropy: The difference between the joint entropy of a specified outcome and condition, and the entropy of the condition alone.

Conditional information: The difference between the joint information of a specified outcome and condition, and the information of the condition alone.

Conditional probability: The probability of a specific outcome assuming a given condition to be true.

Confidence, predictive: The extent of statistical assurance that the observed accuracy of a set of predictions was not achieved as a result of chance. There are many different test-statistics which can be used to define predictive confidence, of which Pearson's T, the log-likelihood ratio, and the entropy decrement are examples.

Consistent (estimator property): The property of some estimators that their estimates converge (in probability) to the true value as the sample size increases without bound.

Consistent (test-statistic property): The property of some test-statistics that their discriminatory power increases with sample size.

Constant predictor line, unbiased: A predictive performance line of a constant probability estimate at the sample average.

Contrivedness: Excessive complexity of boundaries of patterns, increasing their fit to training data but reducing their reliability on test data. Among-features contrivedness: Excessive complexity in the way in which features are used to form patterns. Within-features contrivedness: Excessive complexity in the definition of individual features.

Convenient (estimator property): The property of some estimators that they are well-understood by users and do not require excessive human and computer resources to use.

Convenient (test-statistic property): The property of some test-statistics that their critical values are either easily calculable or conveniently available in tabular form, and that it is easy to determine whether or not they are appropriate for a particular application.

Critical value: A test-statistic value with respect to which the significance level is defined, i.e., the distribution of the statistic is integrated from the critical value to the extremum to determine the significance level.

Crossvalidation: Assessment of predictions on independent test data for which the outcome was not known at the time the predictions were made.

Cumulative function: The integral of the distribution function f(x) up to a given value:

$$F(x) = \int_{-\infty}^{x} f(x')dx'.$$

Data: Observed values of dependent and independent variables for sample events.

Definitiveness, predictive: The degree by which a prediction differs from overall population averages and approaches an extremum (probability of 0 or 1). There are many different measures of definitiveness, of which the skill ratio and the definitiveness entropy decrement are examples.

Definitiveness entropy decrement: A measure of predictive definitiveness as the average information disagreement between the performance of the actual predictor and that of a worst predictor.

Dependent variable: The attribute which is to be predicted of the event.

Discriminant analysis: A method of separating events into clusters in feature space according to outcome distribution by maximizing the ratio of the among-groups score to the within-groups score, where the scores represent sum of squared distance measures of separation.

Discriminatory power: (See: Power, discriminatory.)

Dispersion ratio test-statistic: The test-statistic defined by

$$R^2 = \sum_{i=1}^{L} \left(\frac{p_i - f_i}{u_i} \right)^2 ,$$

a generalization of the dispersion as defined by Lexis.

Distribution-free (test-statistic property): The property of some test-statistics that they are calculable without assuming any specific functional form for the distribution.

Distribution function: A mathematical function f(x) of a random variable x with the intended interpretation:

f(x)Δx = probability that the value of the random variable is in the interval from x to x+Δx.

There are many forms of distribution function, of which the uniform, normal, equiprobability and binomial are common examples.

Disutility, expected error: Expected value of the disutility due to erroneous prediction.

Disutility minimizing (estimator property): A property of some estimators that they minimize the expected value of the disutility resulting from error.

Efficient (estimator property): The property of some estimators that the estimates tend to have small variances, so large samples are not needed for reasonable confidence of being near the true value.

Entropy: The expected value of information:

$$S(C) = \sum_{i=1}^{L} P(C_i) I(C_i).$$

Joint entropy: $S(O \text{ and } C) = \sum_{i=1}^{L} \sum_{k=1}^{K} P(O_k \text{ and } C_i) I(O_k \text{ and } C_i).$

Conditional entropy: $S(O \text{ given } C) = S(O \text{ and } C) - S(C).$

Mutual entropy exchange: $S(O \text{ mutual } C) = S(O) - S(O \text{ given } C).$

Renormalized distributional entropy: $S_R = -q'(1) = -\int f(x) \log f(x) \, dx.$

Entropy decrement test-statistic: the test-statistic defined by

$$\Delta S = \sum_{k=1}^{K} f_k \log (f_k/p_k),$$

where

$$f_k = \text{frequency of } k^{th} \text{ outcome,}$$
$$p_k = \text{probability of } k^{th} \text{ outcome.}$$

ΔS is the mutual entropy exchange between the set of observations $\{f_k\}$ and the condition of having been observed.

Entropy minimax estimation: A parametric estimation method in which probability estimates are taken at values for which the entropy is a maximum, in an event classification for which the conditional entropy is a minimum, where the weight normalization parameter is taken at a value for which the entropy decrement is a minimum. A nonparametric version of the method involves maximization of reality.

Entropy of chance correlations: The expected value of the minimax entropy for a random permutation of the dependent variable values:

$$S^c(\nu) = \sum_{i} S_{min}(i) \text{ Prob}(i^{th} \text{ permutation}).$$

Entropy of specification: The expected value of the information used in defining the patterns.

Equal moments estimation: A parametric estimation method in which the parameters are estimated at values for which the leading moments of the data equal the corresponding moments of the distribution (using a number of moments equal to the number of parameters being estimated).

Equiprobability distribution: The discrete univariate distribution defined by the function

$$f(x) = \frac{1}{K},$$

where x has K discrete rational values

$$0 \leq x \leq 1,$$
$$K \geq 1.$$

Equiprobability estimation: A nonparametric estimation method in which all probability estimates are set equal.

Equivocation: The doubtfulness that a specified outcome will occur given that specified data are observed.

Estimator: An estimated value for a random variable. Parametric estimator: an estimator based on an assignment of values to one or more parameters in an assumed distribution function. Nonparametric estimator: an estimator not based on any specifically assumed distribution function.

Evidence (Fisher): The amount of support which a set of data supplies for a particular value of a parameter θ in an assumed distribution function, measured as the curvature of the negative support function with respect to a specific parameter at a specified value $\hat{\theta}$ of the parameter:

$$E(\hat{\theta}) = - \left. \frac{\partial^2 U(\theta)}{\partial \theta^2} \right|_{\theta = \hat{\theta}}.$$

Excluded middle, law of the: The law that a proposition is either true or false; it cannot be both true and false; nor can it be neither true nor false.

Expected value: The probability weighted average value of a random variable.

Feature: An independent variable or a function of independent variables.

Feature space: A multi-dimensional space, the axes of which are the independent variables or functions of the independent variables.

Filter: A many-to-one transformation of a variable. Filtering, in general, reduces the detailed information content of the variable.

Fourier analysis: A decomposition of a waveform into sine-wave basis functions of various frequencies and amplitudes. Discrete:

$$F(t) = \sum_{k=-\infty}^{\infty} a_k e^{ik\pi t/T},$$

$$a_k = \frac{1}{2T} \int_{-T}^{T} F(t) e^{-ik\pi t/T} dt.$$

Continuous:

$$F(t) = \frac{1}{\sqrt{2\pi}} \int_{-\infty}^{\infty} D(\omega) e^{i\omega t} d\omega,$$

$$D(\omega) = \frac{1}{\sqrt{2\pi}} \int_{-\infty}^{\infty} F(t) e^{-i\omega t} dt.$$

Freedom, degrees of: The equivalent number of linearly independent parameters.

Freedom, extent of: The set of candidate patterns inventoried in a pattern search.

Frequency: The ratio of the number of occurrences of a particular outcome to the total number of events:

$$F = \frac{x}{n}.$$

Frequency limit hypothesis: The hypothesis that the frequency of an outcome approaches its probability as the number of observed events increases without bound:

$$\lim_{n \to \infty} \frac{x}{n} = p.$$

Gamma function: A generalization of the factorial to a continuous domain:

$$\Gamma(x) = \int_0^\infty e^{-t} t^{x-1} dt \text{ for } x \neq -1, -2, -3, \ldots,$$

$$\Gamma(n) = (n-1)! \text{ for } n = \text{integer} > 0.$$

Generalization: A proposition which makes an assertion (categorical or probabilistic) about at least one particular beyond those particulars actually observed.

GENPDP: The acronym for GENeral Pattern Discovery Program.

Hazard: A function $h(x)$ of a variable x such that $h(x)\Delta x$ is the probability of a specimen which is unfailed at x failing between x and $x+\Delta x$.

Histogram: A plot of counts or frequencies versus a variable.

Hume's dilemma: The dilemma that, if the rationale for a generalization is theoretical, it fails as a mere tautology lacking empirical substance; while if it is empirical, it fails because of an empirically unjustified assumption of applicability to unobserved cases.

Hypothesis testing: The process of determining the significance of the difference between a test hypothesis and the null hypothesis along a selected test-statistic for a given test sample. When a null hypothesis cannot be conveniently formed, a chosen measure of test hypothesis significance level is used.

Independent events: Events which are such that the outcome probabilities of one do not depend upon the outcome of the other.

Independent variables: Attributes of events which are used as information to help form estimates of the dependent variable.

Induced probability: A probability estimate based on a principle of induction. Examples include the flat distribution estimate:

$$P = \frac{x+t}{n+t+f},$$

and the curved distribution estimate:

$$P = \frac{x+t+a}{n+t+f+a+b}.$$

Induced proposition: A generalization from observational data.

Information: The extent to which one would be surprised if an outcome were to occur, measured as the logarithm of the reciprocal of the probability:

$$I(O) = -\log P(O).$$

Joint information: $I(O \text{ and } C) = -\log P(O \text{ and } C)$.

Conditional information: $I(O \text{ given } C) = I(O \text{ and } C) - I(C)$.

Mutual information exchange: $\Delta I(O \text{ mutual } C) = I(O) - I(O \text{ given } C)$.

Information, received: Information in the available observational data.

Information, transmitted: Information in the signal sent from the source.

Information channel: Medium through which information in transmitted.

Information generating function (distribution function property): The function $q(u) = \int [f(x)]^u dx$.

Informationally maximal (estimator property): The property of some parametric estimators that they contain no less information about the parameter being estimated than any other parametric estimator.

Irrelevant: A property of the categorization of an event that the categories are defined by specifications which do not logically pertain to the particular event; the event is a member of no category.

Justification of induction: A rationale for basing beliefs on a principle of induction.

K-nearest neighbors analysis: A method of estimating the outcomes for a test event by polling the K-nearest neighbors in independent variable space, where nearness is defined according to a selected metric.

Knowledge: The difference between the expected value of the information content of observed data and the prior uncertainty that this data would be observed given that a specified outcome will occur. Equivalent alternative definition: The difference between the expected value of the information content of a specified outcome and the doubtfulness that this outcome will occur given the observed data.

Kurtosis (distribution function property): A measure of the peakedness of a distribution relative to that of a normal distribution.

Law of large numbers: The law that, for a fixed outcome probability p, the frequency $\frac{x}{n}$ converges in probability to p as n increases without bound:

$$\lim_{n \to \infty} P(\left|\frac{x}{n} - p\right| < \varepsilon) = 1 \text{ for any } \varepsilon > 0.$$

where

$\quad x$ = number of observations of the outcome,
$\quad n$ = number of observations of the event.

Likelihood function: The product of the values of the frequency distribution evaluated at each data point in the sample.

Likelihood ratio: The ratio of the likelihood L when a parameter (upon which the frequency distribution depends) is replaced by its estimate, to the likelihood L_o when the parameter is given its true value.

Linear regression: A method of estimating the dependent variable as a linear combination of the independent variables by selecting coefficients giving least squared error on the training sample.

Linear shrinkage estimation (Stein-James-Lindley estimation): A nonparametric estimation method in which probability estimates are taken at a linear combination of sample averages for the specific event type and grand averages, the shrinkage factors depending upon the number of types and the standard deviations.

Log-likelihood ratio test-statistic: The test-statistic defined by

$$G^2 = -2 \log \lambda,$$

where

$$\lambda = \text{likelihood ratio.}$$

Logical combination: A combination of feature conditions used in defining a pattern boundary.

Intersection: Both conditions are satisfied.

Union: Either condition (or both) is satisfied.

Complement of union: Both conditions fail to be satisfied.

Complement of intersection: Either condition (or both) fails to be satisfied.

Maximum entropy estimation: A nonparametric estimation method in which probability estimates are taken at values for which the entropy is a maximum subject to given constraints.

Maximum likelihood estimation: A parametric estimation method in which the parameters are estimated at values for which the likelihood is a maximum.

Mean (distribution function property): The first central moment of the distribution indicating the value about which the bulk of the distribution is centered.

Minimum entropy decrement estimation: A parametric estimation method in which the parameters are estimated at values for which the entropy decrement is a minimum.

Minimum T estimation: A parametric estimation method in which the parameters are estimated at values for which Pearson's T is a minimum. (Also referred to as minimum chi-squared estimation because the distribution of T is approximately chi-squared for large sample sizes.)

Moments (distribution function properties): The distribution weighted integrals of the powers of the random variable, i.e., the kth moment of $f(x)$ is

$$\mu_k' = \int f(x) \, x^k \, dx.$$

The kth moment about the mean (also called k^{th} central moment) is

$$\mu_k = \int f(x) \, (x-\mu_1')^k \, dx.$$

Monotonic in mismatch (test-statistic property):   The property of some test-statistics that they deviate further from their perfect match values as the fit becomes progressively worse (in the sense defined by the test-statistic).

Monte-Carlo analysis: A method of determining expected values for a complex system by means of a sequence of simulations on a computer driven by a random number generator for the appropriate distribution.

Multinomial distribution: The discrete multivariate distribution defined by the function

$$f(x_1, x_2, \ldots, x_K) = n! \prod_{k=1}^{K} \frac{p_k^{x_k}}{x_k!} \, ,$$

where

$$0 \le x_k \le n,$$
$$0 \le p_k \le 1,$$
$$\sum_{k=1}^{K} p_k = 1, \quad \sum_{k=1}^{K} x_k = n,$$
$$K, n \ge 1.$$

Mutual entropy exchange: The difference between the entropy of a specified outcome and the conditional entropy of the outcome given a specified condition.

Mutual information exchange: The difference between the unconditional information of a specified outcome, and the conditional information of this outcome given a specified condition.

Noise: The prior uncertainty that specified data would be observed given that a specified outcome will occur.

Normal distribution: The continuous univariate distribution defined by the function

$$f(x) = \frac{1}{\sigma\sqrt{2\pi}} \, e^{-\frac{1}{2}\left(\frac{x-\mu}{\sigma}\right)^2} \, ,$$

where

$$\sigma \ge 0.$$

Often also called the "Gaussian" distribution, although used by researchers such as De Moivre before Gauss.

Normalized units of entropy: In normalized units,

$$S = \frac{1}{\log L} \sum_{i=1}^{L} P_i I_i ,$$

the range of entropy is $0 \le S \le 1$.   For $L = 2,3,4,5,6,7,8,9,10$, etc., the units are bits, trits, quadits, quinits, sextits, septits, octits, nonits, decits, etc., respectively.

Null hypothesis: The hypothesis which one attempts to separate statistically from the hypothesis being tested.   For example, when testing the hypothesis that two variables are dependent upon each other, the null hypothesis is that they are independent; when testing the hypothesis that two variables are different, the null hypothesis is that they are equal.

Ockham's razor: The principle that no more principles than necessary should be used to make explanations.

Outcome: The value of the dependent variable for a particular event.

Partition of feature space: A set of mutually disjoint patterns which completely covers feature space.

Pattern: A subset of feature space.

PDPLZR: The acronym for Pattern Discovery Program anaLyZeR.

PDPREP: The acronym for Pattern Discovery PREparation Program.

Perfect predictor band: A scatter band about a predictive performance line defining the points where observed frequency equals probability estimate.

Planck's constant: The fundamental quantum of physical action; fixes the lower limit to the product of uncertainties in pairs of cononically conjugate variables.

Poisson distribution: The discrete univariate distribution defined by the function

$$f(x) = e^{-\mu} \frac{\mu^x}{x!},$$

where

$$x, \mu \geq 0.$$

Possibility: A value of a dependent variable.

Potential function: A function $\gamma(z, z_i)$ which expresses the "intensity", at an arbitrary point z in feature space, of a data event at point z_i.

Power, discriminatory: The ability of a test-statistic to discriminate between the test hypothesis and the null hypothesis, measured as the probability that the null hypothesis will be rejected at the chosen significance level given that the test hypothesis is true.

Powerful (test-statistic property): The property of some test-statistics that they have high discriminatory power against the null hypothesis.

Prediction: An estimate, based on available data, of the frequency that a statement about the physical world will be found to be (or to have been) the case.

Predictive performance diagram: A plot of observed frequency versus probability estimate.

Principal components analysis: A form of linear regression in which the variables are defined to be uncorrelated, and are ordered in terms of fraction of variance in the given data which is modeled.

Principle of entropy maximization: The principle that probabilities should be given values which maximize entropy subject to given constraints.

Principle of entropy minimax: The principle that events should be classified in a way that minimizes conditional entropy while assigning values to probabilities which maximize entropy subject to given constraints.

Principle of entropy minimization: The principle that events should be classified in a way that minimizes conditional entropy.

Principle of evolution of language: The principle that a language will evolve so as to enable simpler discriptions of the experiences of the society using the language.

Principle of frequency correspondence: The principle that induced probabilities approach observational frequencies as the number of observed events increases without bound.

Principle of induction: A prescription for making generalizations from observational data. There are many different principles of induction, of which sample average, linear shrinkage, maximum likelihood, and entropy minimax are examples.

Principle of insufficient reason: The principle that probabilities should be taken as equal when there is no information to the contrary.

Prior probability: Ratio of the a priori weight to the weight normalization.

Probability: An estimator of the classification frequency of a yet to be observed outcome for an indefinite repetition of the event.

Joint probability: $P(O$ and $C) =$ probability of both O and C.

Implicative probability: $P(C$ implies $O) = 1-P(C)+P(O$ and $C)$.

Conditional probability: $P(O$ given $C) = \dfrac{P(O \text{ and } C)}{P(C)}$, undefined if $P(C) = 0$.

Induced probability: A probability estimate based on a principle of induction.

Probability domain: The set of events to which a particular probability applies.

Psi function: The first derivative of the logarithm of the gamma function:
$$\psi(a) = \frac{d}{da} \log \Gamma(a).$$

Robust (estimator property): The property of some estimators that they are relatively insensitive to the correctness of the choice of the underlying distribution.

Random event: An event, the outcome of which is not known with certainty.

Random predictor band: A scatter band about a predictive performance line of constant observed frequency at the sample average.

Random variable: A variable, the value of which is not known with certainty.

Reality: The extent of independence of what one apparently can know about events from what one does happen to know. A measure is the difference between the entropy of chance correlations and the entropy of the observed data for the number of degrees of freedom used:
$$R_j(\nu) = S^c(\nu)-S_j.$$

Reliability: Assessment of expected predictive performance in terms of expected accuracy and confidence level.

Rotation: Generation of a new axis for feature space by performing a rotation in the space of independent variables.

Rule of succession: The rule that the probability of an outcome which has occurred on x of n occurrences of an event is estimated at
$$p = \frac{x+1}{n+2}.$$

Sample: A collection of events upon which data are available.

Sample average estimation: A nonparametric estimation method in which sample averages are used as probability estimates.

Self-information: (See: Information.)

Significance level: The fraction of the area under the distribution function for the test-statistic from the critical value to the extremum.

Similarity: The degree to which two events are alike in terms of the independent variables in ways that are relevant to their being alike in terms of the dependent variable.

Simplicity: The property of a description being expressed in terms of little information.

Skewness (distribution function property): A measure of the degree of asymmetry of the distribution about its mean.

Skill ratio: A measure of predictive definitiveness as the ratio of the amount by which the actual number of correct predictions exceeds the chance number of correct predictions, to the chance number of incorrect predictions.

Spectral analysis, maximum entropy: A procedure for obtaining an estimate $\hat{D}(\omega)$ of a spectral density $D(\omega)$, by choosing values of the unknown autocorrelations for which the entropy of $D(\omega)$ is a maximum.

Standard deviation (distribution function property): Square root of the variance.

Standard width (distribution function property): A measure of the width of (range of values spanned by) a distribution as the exponential of its entropy: $\tau = e^S$.

Step-wise approximation: An algorithm for approximating the minimum of a total of many terms by sequentially seeking the minimum for each term one at a time.

Stitch-wise approximation: An algorithm for approximating the minimum of a total of many terms by sequentially seeking subtotal minima for pairs of terms.

Student's-t distribution: The continuous univariate distribution defined by the function

$$f(x) = \frac{\Gamma\left(\frac{\nu+1}{2}\right)}{\sqrt{\pi\nu}\,\Gamma\left(\frac{\nu}{2}\right)}\,(1+x^2/\nu)^{-(\nu+1)/2},$$

where
$$\nu > 0.$$

Support function: The logarithm of the likelihood:

$$U(\theta) = \log L(\theta).$$

SWAPDP: The acronym for Step-Wise Approximation Pattern Discovery Program.

T test-statistic: The test-statistic defined by

$$T = n\sum_{i=1}^{L}\frac{(p_i-f_i)^2}{p_i}$$

Also referred to as Pearson's T, since Pearson first used it as a test-statistic, based on its mathematical properties worked out earlier by Helmert.

Test sample: Data for which the dependent variable values are unknown when the patterns are generated, and which are used for pattern crossvalidation.

Test-statistic: A function of predicted and observed values which is zero when and only when the observations all exactly match the predictions and which is in some sense monotonic in extent of mismatch. There are many different forms of test-statistics, of which Pearson's T, the log-likelihood ratio, and the entropy decrement are examples.

Threshold: A value of an independent variable which can be used to separate a sample into two groups with different distributions for the dependent variable.

Training sample: Data with known values for independent and dependent variables, used as the basis for forming predictive patterns and making associated probability estimates.

Typicality: The degree to which the independent variables for an event have values which are representative of events with the same value for the dependent variable.

Unbiased (estimator property): The property of some estimators that the expected value of the estimate is the true value.

Unbiased (test-statistic property): The property of some test-statistics that the significance level of the null hypothesis is greater if it is true than if it is false.

Unbounded (distribution function property): The property of some distribution functions that they are defined over an unlimited range of values for the random variable.

Uncertain: A property of the categorization of some events that there is insufficient information to determine definitively the categories of the events.

Uncertainty in probability: The uncertainty associated with a probability,

$$P_{-\ell}^{+u},$$

specifies a scatter band about P for the data upon which the estimate is based.

Uniform distribution: The continuous univariate distribution defined by the function

$$f(x) = 1,$$

where

$$0 \leq x \leq 1.$$

Unique at perfect match (test-statistic property): The property of some test-statistics that they have a uniquely defined value corresponding to a perfect match.

Variance (distribution function property): The second moment about the mean, indicating how widely the distribution is dispersed about its mean.

Virtual events: Events defined to represent background information.

Weight normalization: (See: A priori weight normalization.)

Worst predictor line: A predictive performance line extending from the Type I error point, through the sample average, to the Type II error point.

SYMBOLS

(Definitions which apply unless otherwise noted.)

$B(a,b)$	beta function of a and b
C	condition
$C(O)$	chance (Peirce) of O
$D(\omega)$	spectral density of frequency ω
f	number of ways in which a specific outcome can fail to occur
F_k	frequency of outcome class k
$f(x)$	distribution function of x
$F(x)$	cumulative function of x
$h(x)$	hazard function of x
H_o	null hypothesis
H_1	test hypothesis #1
$I(O)$	information of O
$I(O \text{ and } C)$	joint information of O and C
$I(O \text{ given } C)$	conditional information of O given C
$J(O)$	intensity of belief (Peirce) in O
K	number of outcome classes
L	number of conditions
$L(\underset{\sim}{\theta})$	likelihood function of parameter vector $\underset{\sim}{\theta}$
M	number of independent variables
M_k^d	k^{th} moment of distribution (also denoted μ_k')
M_k^s	k^{th} moment of sample
N	number of observed events
n	number of observed events
n_k	number of observed events with k^{th} outcome
O	outcome
O_k	k^{th} way of outcome O occurring
$\bar{O}_k$	k^{th} way of outcome O failing to occur
P	probability
P_k	probability of kth outcome
$P(O \text{ and } C)$	joint probability of O and C
$P(O \text{ given } C)$	conditional probability of O given C
$q(u)$	information generating function (u-1st frequency moment)
R^2	dispersion ratio test-statistic
$R_j(\nu)$	reality using j^{th} partition for ν degrees of freedom
S	entropy
S_R	renormalized entropy
$S^c(\nu)$	entropy of chance correlations for ν degrees of freedom
$S(O \text{ and } C)$	joint entropy of O and C
$S(O \text{ given } C)$	conditional entropy of O given C
t	number of ways in which a specific outcome can occur
T	Pearson's T test-statistic
$U(\underset{\sim}{\theta})$	support function of parameter set $\underset{\sim}{\theta}$
U_j	uncertainty associated with probability P_j

w	total number of virtual events
w_k	number of virtual events with k^{th} outcome
x	random variable
$x:n$	x out of n observations
$\binom{n}{x}$	number of combinations of n things taken x at a time
α_0	significance level of null hypothesis H_0
α_1	significance level of test hypothesis H_1
β_k	k^{th} shape factor: $\beta_1 = \mu_3^2/\mu_2^3$, $\beta_2 = \mu_4/\mu_2^2$
ΔI(O mutual C)	mutual information exchange between O and C
ΔS(O mutual C)	mutual entropy exchange between O and C (shortened to simply ΔS when used as entropy decrement test-statistic)
ΔS_d	entropy decrement measure of predictive definitiveness
ΔS_e	entropy decrement measure of predictive error
κ_k	k^{th} cumulant: $\kappa_1 = \mu$, $\kappa_2 = \sigma^2$, $\kappa_3 = \mu_3$, $\kappa_4 = \mu_4 - 3\mu_2^2$
$\Gamma(a)$	gamma function of a
γ	Euler's constant: $\gamma = 0.5772156649...$
γ_1	skewness: $\gamma_1 = \mu_3/\sigma^3$
γ_2	kurtosis: $\gamma_2 = \mu_4/\sigma^4 - 3$
$\gamma(\underline{z}, \underline{z}_i)$	potential function of $\underline{z}$ for data point at $\underline{z}_i$
$\phi_m(t)$	m^{th} basis function of t
λ	likelihood ratio
μ	mean: $\mu = \mu_1'$
μ_k	k^{th} moment about the mean
μ_k'	k^{th} moment about zero
ν	number of degrees of freedom
$\psi(a)$	psi function of a
$\psi_i(t)$	i^{th} sample waveform (function of t)
$\rho(x)$	density function of x
σ	standard deviation
σ^2	variance: $\sigma^2 = \mu_2$
τ	standard width
$\underline{\theta}$	parameter vector
ω	frequency
K	skill ratio

INDEX OF NAMES

INDEX OF NAMES

INDEX OF NAMES

INDEX OF NAMES

INDEX OF NAMES

INDEX OF NAMES

INDEX OF NAMES

INDEX OF SUBJECTS

INDEX OF SUBJECTS

INDEX OF SUBJECTS

INDEX OF SUBJECTS

INDEX OF SUBJECTS

INDEX OF SUBJECTS

INDEX OF SUBJECTS

INDEX OF SUBJECTS

INDEX OF SUBJECTS

INDEX OF SUBJECTS

INDEX OF SUBJECTS

INDEX OF SUBJECTS

INDEX OF SUBJECTS

INDEX OF SUBJECTS

INDEX OF SUBJECTS